Adenovirus Cell and Immune Interactions

Adenovirus Cell and Immune Interactions

Guest Editor

Glen R. Nemerow

MDPI

Basel • Beijing • Wuhan • Barcelona • Belgrade • Novi Sad • Cluj • Manchester

Guest Editor
Glen R. Nemerow
Department of Immunology
and Microbial Science
Scripps Research Institute
La Jolla
USA

Editorial Office
MDPI AG
Grosspeteranlage 5
4052 Basel, Switzerland

This is a reprint of the Special Issue, published open access by the journal *Viruses* (ISSN 1999-4915), freely accessible at: https://www.mdpi.com/journal/viruses/special_issues/Adenovirus_Immune.

For citation purposes, cite each article independently as indicated on the article page online and as indicated below:

Lastname, A.A.; Lastname, B.B. Article Title. *Journal Name* **Year**, *Volume Number*, Page Range.

ISBN 978-3-7258-6604-5 (Hbk)
ISBN 978-3-7258-6605-2 (PDF)
https://doi.org/10.3390/books978-3-7258-6605-2

Contents

About the Editor

Glen R. Nemerow

Glen R. Nemerow is Professor Emeritus at the Scripps Research Institute, La Jolla, California. His research area is on the mechanisms involved in virus host cell and immune interactions. Collaborations with structural biologists and cell biologists hold a major emphasis toward obtaining a more complete understanding of these processes.

Preface

Ever since their discovery in the early 1950s, human adenoviruses (HAdVs), the cause of respiratory diseases of young children, have been the subject of intense basic research as well as efforts to engineer HAdV vectors to ameliorate human diseases. Although much information has been gleaned over the past 67 years from laboratory and clinical investigations, we still have an incomplete picture of the critical associations of HAdV with host cells and specific components of the immune system. An ability to uncover additional information will likely influence the future success (or failure) for exploiting HAdV vectors for gene and vaccine therapies.

Each contribution in this series will highlight the efforts and progress made to acquire basic knowledge of adenovirus–host interactions and will indicate particular areas that remain to be uncovered. Topics in this new series include key steps in HAdV–host cell interactions as well as structure-based investigations of virus and virus capsid protein associations with cell receptors and innate immune molecules. These topics will culminate with discussions on examples of HAdV vector design and how these approaches are influenced by the accumulation of knowledge of host–vector interactions.

Glen R. Nemerow
Guest Editor

viruses

MDPI

Article

Refined Capsid Structure of Human Adenovirus D26 at 3.4 Å Resolution

Vijay S. Reddy [1,*], Xiaodi Yu [1,†] and Michael A. Barry [2,3,4,5]

1 Department of Integrative Structural and Computational Biology, The Scripps Research Institute, La Jolla, CA 92037, USA; xyu6@its.jnj.com
2 Department of Internal Medicine, Mayo Clinic, Rochester, MN 55902, USA; barry.michael@mayo.edu
3 Division of Infectious Diseases, Mayo Clinic, Rochester, MN 55905, USA
4 Department of Molecular Medicine, Mayo Clinic, Rochester, MN 55905, USA
5 Department of Immunology, Mayo Clinic, Rochester, MN 55905, USA
* Correspondence: reddyv@scripps.edu
† Current address: SMPS, Janssen Research and Development, Spring House, PA 19044, USA.

Abstract: Various adenoviruses are being used as viral vectors for the generation of vaccines against chronic and emerging diseases (e.g., AIDS, COVID-19). Here, we report the improved capsid structure for one of these vectors, human adenovirus D26 (HAdV-D26), at 3.4 Å resolution, by reprocessing the previous cryo-electron microscopy dataset and obtaining a refined model. In addition to overall improvements in the model, the highlights of the structure include (1) locating a segment of the processed peptide of VIII that was previously believed to be released from the mature virions, (2) reorientation of the helical appendage domain (APD) of IIIa situated underneath the vertex region relative to its counterpart observed in the cleavage defective (*ts1*) mutant of HAdV-C5 that resulted in the loss of interactions between the APD and hexon bases, and (3) the revised conformation of the cleaved N-terminal segments of pre-protein VI (pVIn), located in the hexon cavities, is highly conserved, with notable stacking interactions between the conserved His13 and Phe18 residues. Taken together, the improved model of HAdV-D26 capsid provides a better understanding of protein–protein interactions in HAdV capsids and facilitates the efforts to modify and/or design adenoviral vectors with altered properties. Last but not least, we provide some insights into clotting factors (e.g., FX and PF4) binding to AdV vectors.

Keywords: adenovirus structure; protein network; protein–protein interactions; cement proteins; clotting factors

1. Introduction

Human adenovirus D26 (HAdV-D26, or Ad26 for short) causes acute conjunctivitis and a prolonged enteric infection among humans [1]. In addition to differences in the hypervariable regions (HVRs), Ad26 displays a short-shafted fiber comprising eight β-spiral repeats, compared with long-shafted fibers found in archetypal HAdV-C5. Significantly, Ad26 is being used as a vaccine vector against the diseases such as SARS-CoV-2 and AIDS [2–7]. A few years ago, we determined the cryo-EM structure of replication-defective HAdV-D26 (RD-HAdV-D26) at 3.7 Å resolution, which showed the conservation in the structural organization of minor/cement proteins among human adenoviruses [7]. With the advent of improvements in data processing and image reconstruction programs, we revisited the dataset that was collected previously and obtained an improved cryo-EM structure at 3.38 Å resolution. Using this higher resolution map, we obtained a refined model for Ad26 capsid by performing real-space refinement using Phenix [8] and model building in Coot [9,10], providing new insights into the structures and organization of minor proteins, IIIa, VI, and VIII, which were not previously considered. The summary of the changes made to the models of individual capsid proteins (CPs) is listed in Table 1.

Figure 1 shows the overview of the Ad26 virion structure and organization of various proteins in the AdV capsid.

Table 1. Specific changes made to the CPs in the refined model of Ad26 virion.

Protein	Residue Changes	Remarks
Hexon	None	N/A
PB	None	N/A
Fiber	None	N/A
IX	None	N/A
IIIa	283–301; 314–396 (APD)	Directionality of a.a. 283–301 is reversed and is now designated as unassigned. The APD domain is adjusted and remodeled.
VIII	140–149	A segment of processed VIII peptide (a.a. 140–149) is identified for the first time.
VI	2–31 (pVIn)	Directionality of pVIn chains is reversed and remodeled.

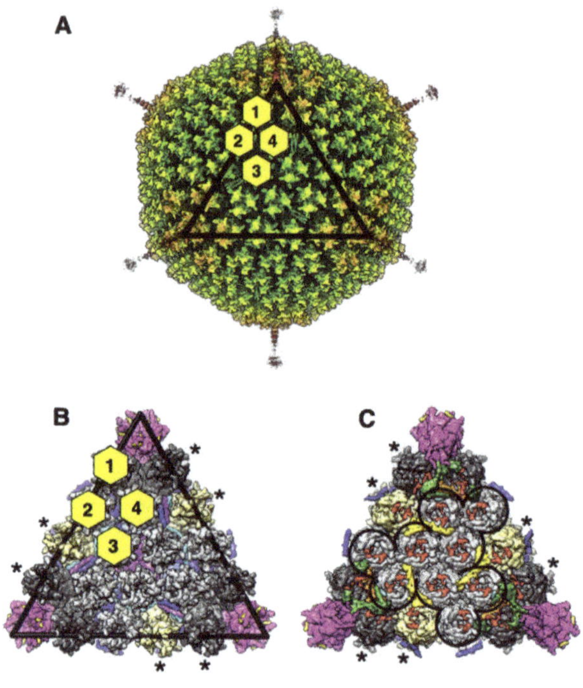

Figure 1. Overall structure and organization of HAdV-D26 (Ad26) capsid: (**A**) illustration of the radially color-coded surface of the cryo-EM reconstruction of Ad26 virion, shown as a view down the icosahedral threefold axis. A rainbow color gradient from blue to red was used to represent the regions of the map between the radii 300 Å and 500 Å, respectively. The icosahedral facet is identified by the black triangle and the four structurally distinct hexon positions are distinguished by the yellow hexagons labeled 1–4; (**B**) a zoomed-in view of the icosahedral facet seen from the outside of the virion. Hexons are depicted in surface representation and those that belong to a group of nine hexons (GONs) are shown in light gray, while the peripentonal hexons (PPHs) are colored in dark gray. The hexons that belong to neighboring facets are identified by asterisks and/or shown

in khaki color. The pentamers of penton base (PB), located at the icosahedral vertices are shown magenta. The short fiber N-terminal (FNT) peptides (a.a. 4–20), bound at the interfaces of PB subunits, are shown as gold-colored surfaces. The triskelion and 4-HLXB structures formed by the 12 protein-IXs-3 copies of 4 structurally distinct IXs, identified as P, Q, R, and S, present in an icosahedral facet, are shown in different colors blue, light blue, cyan, and purple, respectively; (**C**) a vertically flipped view of panel B, showing the inside of the facet. The outline of the GON structure is represented by the black line. The surface representation of ordered regions of IIIa, VIII-U, and VIII-V and VI are shown in dark green, orange, yellow, and red colors, respectively. The model fragments corresponding to islands of unassigned densities are shown as turquoise surfaces. The C-terminal helix of the IX(P) can be seen in blue.

2. Results and Discussion

2.1. All Hypervariable Regions (HVRs) of the Hexon Are Ordered in Ad26 Virion

As it has been reported, the overall structure of Ad26 hexon is very similar to that of HAdV-C5 (Ad5) with an RMSD (root-mean-square deviation) of 0.80 Å for 811 pairs of aligned Cα-atoms [7]. Not surprisingly, the significant structural differences between the hexons from Ad26 and Ad5 occur in the HVR loops. It is noteworthy that all HVRs, including HVR1, are ordered in the Ad26 hexon (Figure 2A). Notably, the HVR1 in Ad26 hexon is shorter by 12 a.a., compared with its counterpart in Ad5 hexon [7] (Figure S1). As previously noted, the structural differences that occur in the highly exposed HVRs 5 and 7 appear to correlate well with the observations that Ad26 does not bind coagulation factor X (FX) [11–13]. Apart from overall improvements in the quality of hexon models, no specific revisions were made to the hexons in the refined structure of Ad26.

2.2. Penton Base and Fiber

The structure of Ad26-PB is also highly conserved, compared with other adenoviruses, with an RMSD of 0.77 Å for 425 pairs of aligned Cα-atoms, with respect to its counterpart in the Ad5 structure (Figure 2B). Even though shorter by 40 a.a., compared with Ad5-PB, the Arg-Gly-Asp (RGD)-containing loop that is known to bind to cell surface integrin molecules is disordered in Ad26-PB. Other differences between the PBs of Ad26 and Ad5 structures occur in the loop (a.a. 139–153) that connects β3 and β4 strands, where there is a six-residue insertion found in Ad26 (Figure S2). In addition, a 12-residue deletion is noted at the N-terminus of Ad26 PB, relative to species-C viruses [7].

We found a 17-residue peptide comprising a.a. 4–20 at the N-terminus (NT) of fiber protein bound tightly at the interface of a pair of PB subunits in the pentameric penton (Figure 2C). The quality of the density for this fiber N-terminal (FNT) peptide is comparable to that of neighboring PBs. Moreover, the FNT-peptide adopts "elbow" shaped conformation that appears to hook the helix at the base of the disordered RGD containing loop, which perhaps is critical for fiber latching onto PB. Notably, even though we observed five copies of FNT-peptide in the Ad26 cryo-EM map, this is due to an artifact of imposing icosahedral fivefold symmetry on the threefold symmetric fiber molecule. As previously inferred, only three out of five copies should correspond to the trimeric fiber molecule [7,14–16]. Apart from overall improvements in the quality of PB and FNT models, no specific revisions were made to these models in the refined structure of Ad26.

Ad26 has been reported to utilize sialic acid, CD46, and integrins as receptors. The short-shafted Ad26 fiber is rigid, as reflected by the ability to observe it in cryo-EM studies. This rigidity likely also affects its ability to interact with receptors. In most cases, these interactions are low affinity in the range of μM and are orders of magnitude weaker than Ad5 fiber's interactions with CAR. While sialic acid and CD46 interactions were expected to be mediated by the Ad26 fiber [17], recent research actually demonstrated that CD46 interactions are surprisingly mediated by the hexon protein [18]. While interactions with sialic acid and CD46 have been somewhat controversial, interactions with integrin appear to be mediated as expected by the Ad26 penton base (PB).

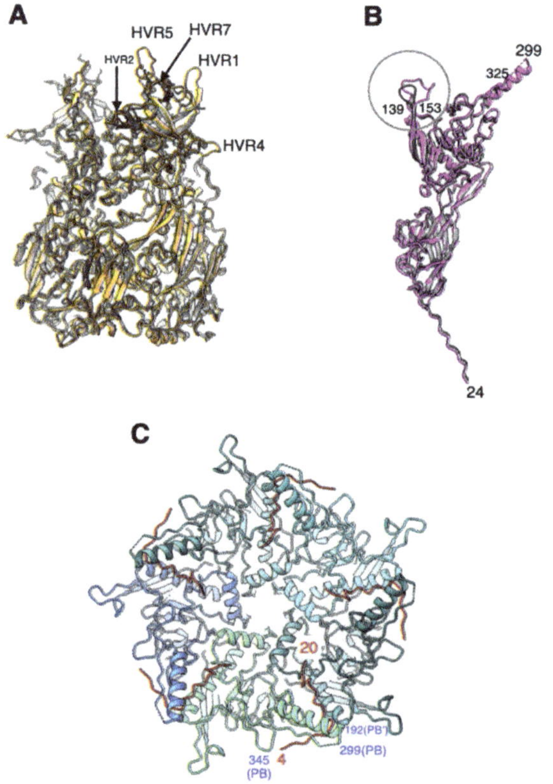

Figure 2. Structural similarities and differences between the major capsid proteins of Ad26 and Ad5 viruses: (**A**) superposition of peripentonal hexons from Ad26 (orange) and Ad5 (gray). Few selected HVR loops ordered in the Ad26 structure are identified; (**B**) superposition of penton base (PB) subunits from Ad26 (magenta) and Ad5 (gray) structures. Major structural differences occur in the loop containing residues 139–153, identified by a circle, where there exists a six-residue insertion in Ad26-PB. The RGD containing loop between the residues 299–325 is disordered; (**C**) An illustration showing the "elbow" shaped fiber N-terminal (FNT) tails (in red) interacting with the PB subunits, which are shown as ribbons in different shades of blue. Few selected PB residues are labeled in blue, while the residues of the FNT peptides are indicated in red.

2.3. Interactions Involving Protein-IX Are Conserved with Respect to Species-C HAdVs

The Ad26 structure provided the first report of a well-ordered full-length protein-IX (IX) [7] (Figure 3A) that revealed the connection (linker region) between the triskelion and coiled-coil forming regions in HAdVs. Even though the linker region is fully ordered in one (R) of the four structurally distinct IXs-P, Q, R, and S, it clearly demonstrated that the N-terminal triskelion forming regions (a.a. 1–61) are structurally conserved, while the middle linker regions (a.a. 62–93) and C-terminal coiled coils (a.a. 94–134) adopt varied conformations "molded" by the hexameric bases of the hexons [19–21] (Figure 3B). In total, 240 copies of the malleable IXs, by displaying 4 distinct conformations, facilitate the formation of the continuous hexagonal lattice comprising triskelions and four-helical bundles (4-HLXB) that represent the nodes (hubs) of the network while the linker regions act as connections between the nodes (Figure 3C). As it was observed previously, the 4-HXLB (coiled coils) is formed by four helices belonging to four structurally unique IXs (P, Q, R, and S) that remarkably come from four different triskelions: three from the same facet and the fourth one from the neighboring facet [19,20]. It is noteworthy that the antiparallel helix, contributed by the IX (P) that comes from the neighboring facet, plays the role of "handshaking" interactions with three other helices in the 4-HLXB and is essential for forming the continuous hexagonal lattice. Moreover, the hexagonal network of IXs interlaces the hexons that form a group of nine (GON) hexons (2–4) but not the peripentonal hexons, throughout the AdV capsid. Notably, a different arrangement of IXs, distinct from the above hexagonal network, has been observed in animal and reptilian adenoviruses [22–24], as well as in species-F HAdVs [25,26]. Apart from overall improvements in the quality of IX models, no specific revisions were made to these models in the refined structure of Ad26.

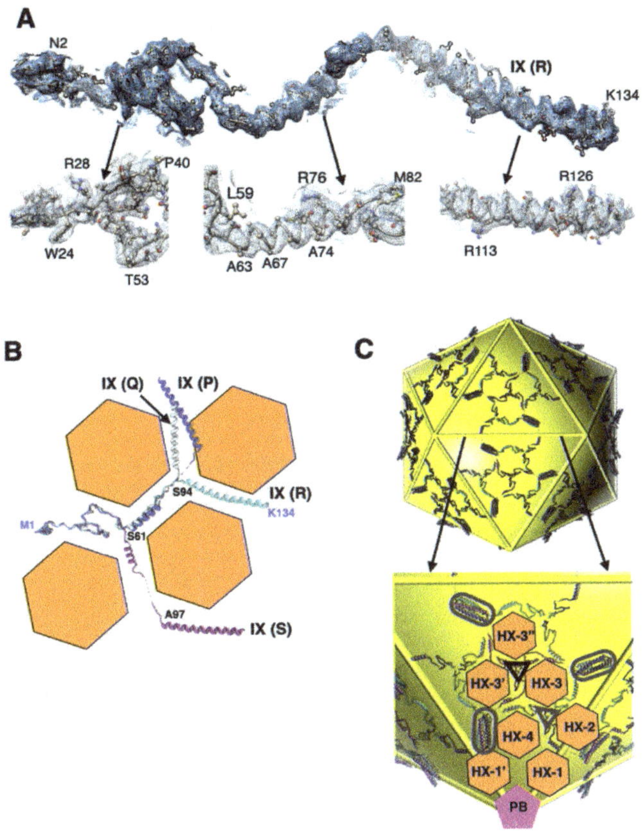

Figure 3. Structure and organization of IX: (**A**) fit of the density for the fully ordered protein-IX (R). Shown below are the zoomed-in views of different regions indicated by the arrows. Few representative residues are labeled; (**B**) superposition of four structurally distinct IXs—P, Q, R, and S—which are shown in colors blue, light blue, cyan, and purple, respectively. The triskelion forming region consists of a.a. 1–61 and the coiled-coil region is formed by residues 94–134, while the in-between residues (62–93) form the linker region. The orange hexagons represent the shape of the bases of major capsid protein, hexon, that appear to mold the conformation of IXs by bending them ~120° at the locations, identified by residues labeled in black, where they contact the vertices of the hexagons. However, the antiparallel helix contributing IX (P) (in blue) appears to deviate from the above angular constraints imposed by the hexon bases; (**C**) the extended protein network shows the overall organization of 240 IXs on the surface of an icosahedron shown in yellow (top). Shown at the

bottom is a zoomed-in view of the IX network within an icosahedral facet. The structurally distinct IXs are color coded as in panel B. The location of triskelions and coiled coils (4-HLXB) are identified by triangles and elongated ellipses, respectively. The gray and black colored triangles identify the triskelions located at the quasi (local) threefold and strict threefold axes of symmetry, respectively. The locations of selected hexons and PB are identified by orange hexagons and magenta-colored pentagons, respectively.

2.4. The Appendage Domain of IIIa Is Oriented Differently from Its Precursor Observed in ts1 Mutant of HAdV5

The structure of the majority of the visible Ad26-IIIa comprising N-terminal domain (NTD) (a.a. 4–132) and middle domain (MDLD) (a.a. 133–282) is conserved with respect to its counterpart in mature Ad5-IIIa [27]. However, there is an extra helical domain (a.a. 314–396), termed appendage domain (APD), which is ordered—albeit weakly—in the Ad26 structure but appears to be completely disordered in the mature Ad5-IIIa (Figure 4A). However, the APD has also been reported recently in the structure of a maturation deficient and thermostable Ad5 virion (Ad5-*ts1*) [28]. Although the APD in Ad26 is not well ordered as in Ad5-*ts1*, it is oriented differently relative to its counterpart in the *ts1* virion (Figure 4B). It appears that the APD domain in Ad26-IIIa is rotated by ~120°, compared with its counterpart in the Ad5-ts1 structure, which results in a longer distance (by ~25 Å) between the visible ends of MDLD and APD domains (Figure 4B). It is noteworthy that this reorganization of the APD domain in the Ad26 structure results in the loss of interactions with the contacting hexon subunits. The calculated buried surface area (BSA) between the APD and hexon subunits in the Ad26 structure (838 Å^2) is reduced by half, compared with a similar calculation of BSA (1554 Å^2) in the ts1 structure (PDB-ID: 7S78), suggesting the loss of interactions between the APD and hexon bases in the mature Ad26 virions. The BSA values were calculated using ChimeraX [29]. In other words, the association of APD with the hexon subunits is stronger in the *ts1* structure, which agrees with the greater thermostability attributed to the *ts1* virions [30,31]. We surmise that the differences in the organization of APD domains are most likely due to conformational changes associated with the maturation of IIIa, as a.a. sequence of the APD domains are highly conserved between the Ad26 and Ad5 viruses. Based on these observations, we speculate that the mature Ad26 particles are likely to be more stable than mature Ad5 virions, as the APD domain is completely disordered in the latter. However, it is unclear why the APD domain is completely disordered in the structures of mature Ad5 and Ad41 virions [25–27]. However, it is noteworthy that IIIa in Ad26 is the smallest in size, composed of 560 a.a. residues, compared with 585 and 579 in Ad5 and Ad41 viruses, respectively. Moreover, the previously assigned residues 283–301, as the continuation of the polypeptide chain of residues 251–282, are now designated as unassigned because of the revised directionality of the helix formed by some of these residues (283–301) and, therefore, are not compatible with this assignment. The remaining a.a. residues 397–560 of Ad26-IIIa, as well as the residues 283–313 connecting the MDLD and APD domains, are disordered.

2.5. A Segment of the Processed Peptide (a.a. 140–149) of VIII Is Seen Mediating Interactions between VIII and the N-Terminus of a VI (pVIn) in Hexons 1 and 4

The sequence and structure of VIII are highly conserved among different HAdVs (Table S1). However, the cleaved peptide comprising a.a. residues 112–157 was not seen before, believed to be released upon proteolytic processing by the AVP [7,32]. In the improved Ad26 structure, we identified 10 residues of the processed peptide (a.a. 140–149) wedged between the a.a.165–172 of VIII and the N-terminus (a.a. 2–8) of the VI (pVIn) (Figure 5). It is noteworthy that in the original Ad26, part of this peptide was assigned to pVIn [7]. It is notable that this short peptide is seen at both the independent locations of VIII on the Ad26 capsid interior. Such an arrangement results in the formation of a three-stranded antiparallel β-sheet involving residues 87–91, 161–171, and 144–146 of VIII, which interacts with the N-terminus (a.a. 2–8) of the closest pVIn (Figure 5A), the remainder of pVIn meanders into the hexon cavity (not shown). Together with the nearby pVIn, the above polypeptides of VIII appear to form a four-stranded β-sheet. In addition, a.a. 140–142 of VIII-U interact with the residues 26–30 in the B subunit of hexon-1, while the same residues in VIII-V interact with the equivalent residues (26–30) in the K subunit of hexon-4. Notably, except for the ordered short segment of the processed peptide (140–149), the rest of the VIII structure is very similar in Ad26 and Ad5 virions (Figure 5B).

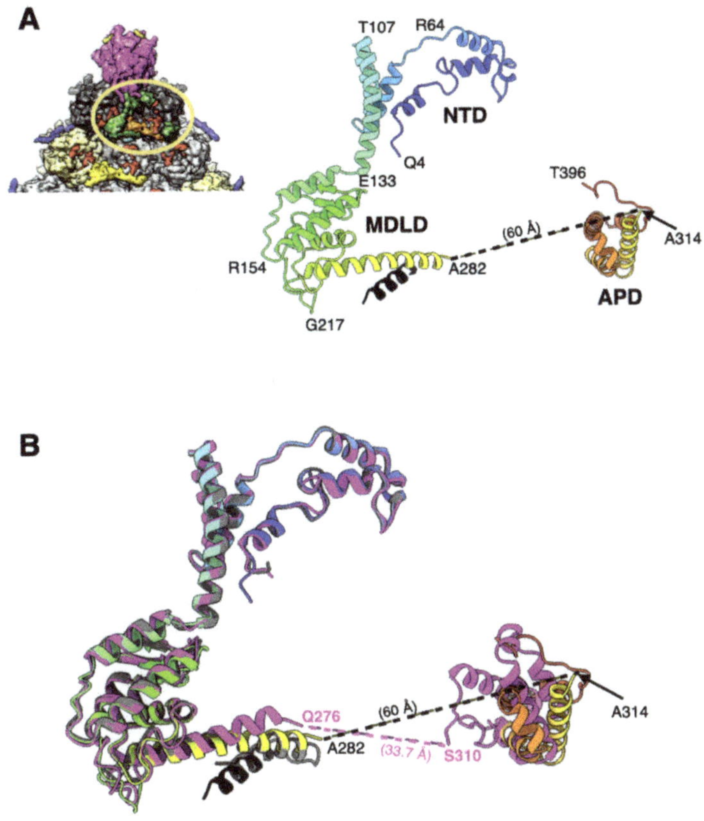

Figure 4. Structure and organization of IIIa: (**A**) shown on the top left is a portion of the icosahedral facet seen from the inside of the Ad26 virion (see Figure 1C for details), indicating the interior location of IIIa represented by green colored surface, and encircled by the orange-colored oval. Illustrated on right is a ribbon diagram of IIIa depicted in rainbow color gradient showing the structure of ordered regions. Different domains of IIIa, along with a few representative residues along the polypeptide chain, are identified. The helix shown in black was built in an unassigned island of density adjacent to the helix formed by a.a. 251–282; (**B**) superposition of ordered regions of IIIa from Ad26 (rainbow gradient), from Ad5-ts1(magenta; PDB-ID: 7S78) and from mature Ad5 (gray; PDB-ID: 6B1T). Even though the structures of NTD and MDLD domains are virtually superimposable in the Ad26 and Ad5-ts1 structure, the orientation of the APD domains seems to be different. The distances between the visible ends of MDLD and APD domains in Ad26 and Ad5-ts1 structures are indicated. Notably, the APD domain is completely disordered in the mature Ad5 structure and the unidentified helix (shown in black), which is closely associated with Ad26-IIIa overlapping with the helix formed a.a. 291–301 of mature Ad5-IIIa. However, the directionality of these helices appears to be the opposite.

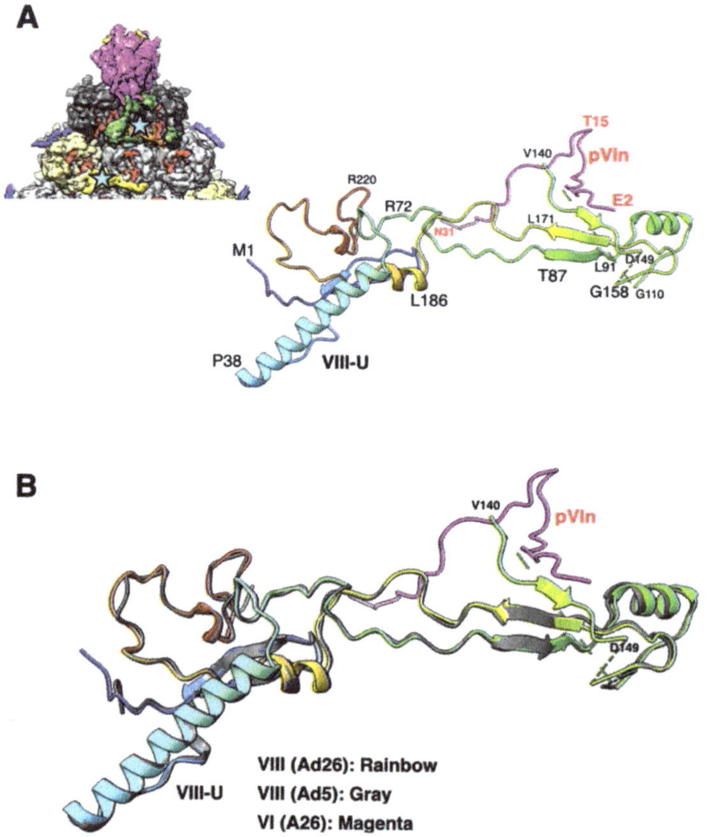

Figure 5. Structure and association of VIII with pVIn: (**A**) shown on the top left is a portion of the icosahedral facet seen from the inside of the virion (see Figure 1C for details), identifying the interior locations of two structurally independent VIII molecules, which are shown in orange (VIII-U) and yellow (VIII-V) colored surfaces and identified by cyan-colored stars. Shown on the right is a ribbon diagram of the VIII-U molecule, depicted in rainbow colors, interacting with pVIn, whose polypeptide backbone is shown in purple. Few selected residues of VIII and VI are labeled in black and red colors, respectively; (**B**) superposition of VIII (U) molecules from Ad26 (rainbow colors) and Ad5 (gray; PDB-ID: 6B1T) structures showing the overall structural similarity with an RMSD of 0.61 Å. However, the residues 140–149 are disordered in the Ad5 structure. The structure of pVIn is shown in purple.

2.6. The Structures of pVIn Are Highly Conserved

It is now clearly established that VI molecules bind and are released from the hexon cavities [27,28,33–35]. In the mature virions of HAdVs, multiple copies of the cleaved N-terminal fragments of VI (pVIn), composed of a.a. 1–33, are seen bound in the hexon cavities [7,27]. In the refined Ad26 structure, the directionality of the polypeptide chains of pVIn in the previous structure was reversed in agreement with the recent Ad5-wt and ts1 virion structures [27,28]. Furthermore, we identified two, two, three, and two copies of pVIn bound in the hexons 1–4, respectively. It is noteworthy that these VI segments were modeled at lower contour levels in varied qualities of densities. The structures of all these polypeptide segments are highly conserved with RMSDs < 1.0 Å for various comparisons between different pairs, most likely influenced by their interactions with the hexon subunits (Figure 6A). Notably, we see conserved stacking interactions between

His13 and Phe18 residues in all the segments of pVIn. Strong interactions between pVIn and hexon subunits occur at the entrance of hexon cavity, between residues 6–27 of VI and 26–48 of one hexon subunit (e.g., HX1-B); 16–19; 47–55 of adjacent hexon subunit (e.g., HX1-C) (Figure 6B). As already noted, the residues 2–8 of pVIn interact with a.a. 140–147 of VIII at two locations. This interaction results in the β-sheet complementation by the pVIn (a.a. 2–8) residues that seem to extend the three-stranded β-sheet of VIII into a four-stranded β-sheet (Figure 5). Moreover, it appears that on average 2 copies of VI are bound to each hexon, which amounts to 480 copies of VI packaged in each virion. It is noteworthy that this number is considerably higher, compared with the estimates (359 ± 24) from mass spectrometry-based proteomics analysis of Ad5 virion [36]. It is possible that the above differences in the copy numbers of VI could be due to a different virus (Ad5) was used in the proteomics analysis.

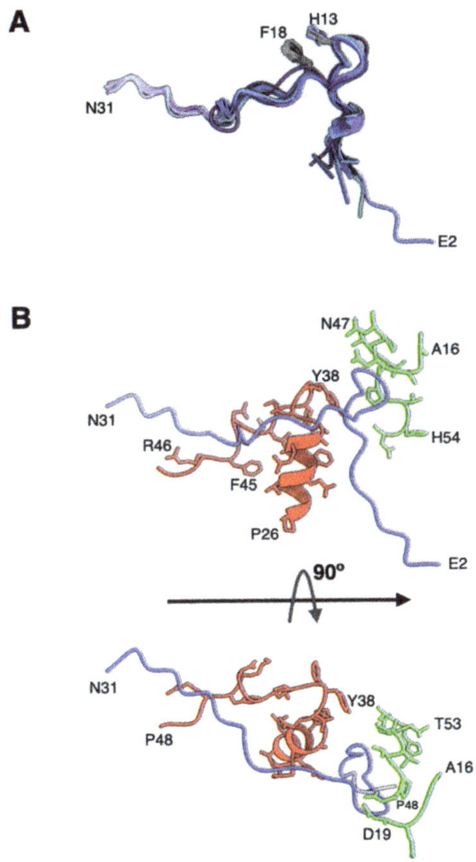

Figure 6. Structure and association of pVIn with hexon: (**A**) superposition of different N-terminal fragments (a.a. 2–31) of VI (pVIn) that are observed in the hexon cavities of Ad26 structure. The structure of pVIn in Ad26 virion appears to be well conserved, likely influenced by its interaction with hexon subunits. The His13 and Phe18 residues that are involved in stacking interactions are identified; (**B**) interaction of pVIn with the selected regions from two hexon subunits, depicted in red and green colors, at the entrance of hexon cavity. Shown at the bottom is a 90° rotated view of the top.

3. Conclusions

The overall structure and capsid protein organization of the HAdV-D26 capsid is very similar to that of HAdV-C5. However, the main phenotypic differences between Ad26 and Ad5 virions are likely to occur due to differences in the hexon HVRs as well as in the differences in the receptor binding fiber molecules. The new structural insights from the refined Ad26 structure include (1) identification of a segment of processed VIII peptide (a.a. 140–149), which was previously believed to be released from the mature virions, was found to be interacting with the N-terminus (a.a. 2–8) of VI at both the independent locations (Figure 5); (2) the rearrangement of the APD domain of IIIa in Ad26 structure relative to the corresponding domain in the Ad5-*ts1* structure (Figure 4) [28]; (3) revision of the directionality of polypeptide segments of pVIn (a.a. 1–33) sequestered inside the hexon cavities and their structural conservation that includes the stacking interactions between the His13 and Phe18 residues (Figure 6). Taken together, the new structural information on HAdV-D26 capsid provides opportunities to better understand the aspects of biology that are specific to Ad26 vectors (see below) and overall advance the efforts in designing the improved adenovirus vectors.

One clinically significant difference between Ad26 and species C adenoviruses such as Ad5 is their ability to bind vitamin K-dependent blood clotting factors [11,37]. The hexons of Ad5 and other species C adenoviruses bind coagulation factor X (FX), which appears to partially shield these viruses and protect them from destruction by human liver Kupffer cell after intravenous injection [11,37]. In contrast, Ad26 does not bind FX [11], and it is, therefore, more severely targeted by natural antibodies and complement and destruction by Kupffer cells [38]. The effect of this interaction is profound, as evinced by our recent observation that Ad5 mediates 1000-fold higher transduction of the liver than Ad26 after intravenous injections in CD46 transgenic mice [39].

This differential FX binding may also have a considerable effect on vaccine and vector safety and side effects [40–43]. Thrombosis and thrombocytopenia have been reported after the COVID-19 vaccination the Ad26.COV2.S COVID-19 vaccine (Johnson & Johnson/Janssen) [40,44–55]. While this COVID-19 vaccine has been injected intramuscularly, some fraction of injected vaccine will leak into the blood, thereby potentially inducing effects on the clotting system. Similar to most interactions of Ad26 with receptors and proteins, the binding of blood clotting factor, platelet factor 4 (PF4), also has a low affinity with binding constants for chimpanzee adenovirus vector (ChAdOx1), Ad5, and Ad26 of 0.661, 0.789, and 0.301 μM, respectively [56]. This binding has a 1000-fold lower affinity than the 0.000229 μM binding of FX to Ad5 [37]. The computational modeling studies reported in [56] identified PF4 binding amino acid residues in hexon, including E440 and D442. These residues are notably in close proximity to or overlap FX binding residues in Ad5 (e.g., E424 and T425) [13] (Figure 7).

How FX and PF4 might interact with different adenoviruses may also be strongly driven by their native concentrations in the blood or tissue. The median concentration of PF4 in human blood is 7 ng/mL [57]. The concentration of FX is approximately 8 μg/mL [11]. Therefore, the 1000-fold higher concentration and 1000-fold higher affinity of FX over PF4 may drive FX to coat species C Ads to limit the binding of PF4. In contrast, adenoviruses that do not bind FX, e.g., Ad26, would not be shielded by FX and could be more easily bound by PF4 and may increase the likelihood of the production of anti-PF4 antibodies. Whether there is competition between FX and PF4 and whether FX binding mitigates the risks of thrombotic thrombocytopenia remains to be determined. There are of course many more proteins and cells that could bind different adenoviruses beyond FX and PF4. These interactions are as diverse as the genomes of adenoviruses and likely contribute to AdV biology, efficacy, and safety as therapeutics and vaccines.

If PF4 binding by Ad26 is indeed involved in the thrombotic thrombocytopenia side effects observed during COVID-19 vaccinations, there is perhaps a simple solution to the problem. Do not inject the vaccine into the muscle. Instead, apply the vaccine by the intranasal or another mucosal route where Ad26 is much less likely to encounter the blood and PF4. Of course, other routes have their own unique side effects. However, given that the route of entry by SARS-CoV-2 is intranasal and respiratory, injecting a vaccine to counter this pathogen into the muscle may not be the best strategy.

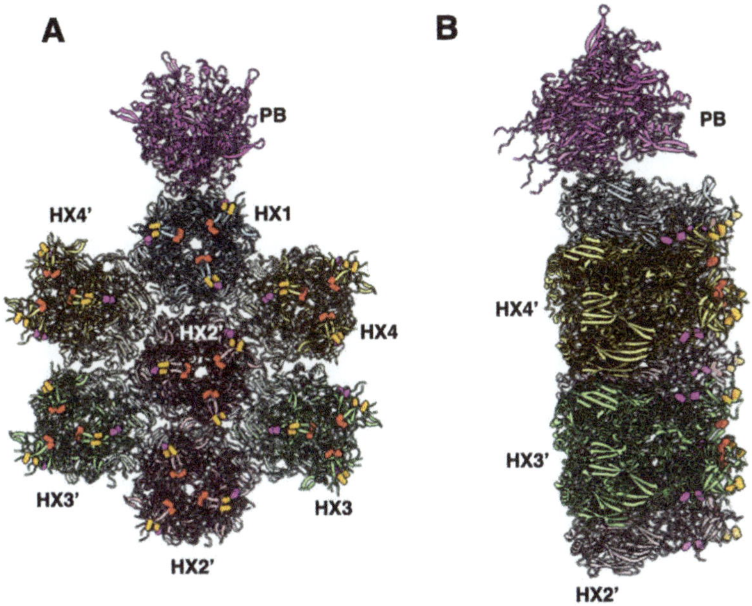

Figure 7. The location of FX and PF4 binding residues in the Ad5 structure (PDB-ID: 6B1T): (**A**) a top view showing a group of contiguous hexons identifying the FX and PF4 binding residues. The residues E424 and T425 that are suggested to bind to the GLA domain of FX [13] are shown in the red spheres, while a few of the predicted PF4 binding residues [56]-D290, D292 shown in magenta and E440 and D442 are shown in orange; (**B**) a side view of panel A showing the relative radial disposition of putative FX and PF4 binding residues on hexons.

4. Methods

The details of the production of RD-HAdV-D26 virions, cryo-EM sample preparation, and data collection were previously described in detail [7].

4.1. Improved Cryo-EM Data Processing Structure Determination

A total of 2017 micrographs were obtained after performing whole-frame alignment and correcting for beam-induced motion, using MotionCorr program [58] were loaded into the cisTEM program [59]. After performing the CTF correction using the CTFfind program [60] within cisTEM, we automatically picked 37,026 virus particles using the maximum and characteristic radius of 400 Å and a threshold peak height of 2. These particles were subjected to reference-free 2D classification, followed by 3D refinement as a single class that included all the picked particles, using the map from the polyalanine model of previously deposited Ad26 structure (PDB-ID:5TX1) as the reference volume. This step resulted in a map of 3.5 Å resolution. Subsequently, we performed 3D classification of the dataset as two classes and selected the best (one) of the classes that included 99.1% (36,706) particles. This was followed by another round of 3D classification as 2 classes. The class chosen for the final round of 3D refinement contained 30,834 particles that resulted in a 3.38 Å resolution map, which after sharpening with a B factor of −90 $Å^2$, was used for model building in Coot. The details of cryo-EM data statistics are shown in Table S2. The FSC plot of Ad26 cryo-EM reconstruction is shown in Figure S3A.

4.2. Model Building and Refinement

The existing model of the Ad26 structure (PDB-ID: 5TX1) was positioned into the icosahedral asymmetric unit of the newly reconstructed 3D map using chimera [61]. Then, manually adjusted the model wherever the fit to the density was not perfect using the "density fit analysis" tool in Coot [10]. Significantly, however, we corrected the directionality (polarity) of the N-terminal fragment of pre-protein VI (pVIn), in agreement with the recent reports [27,28]. Furthermore, we adjusted the model of the APD domain (314–396) of IIIa by the structural superposition of the well-ordered APD domain in the ts1 structure [28]. Additionally, the previously assigned residues 283–301 are now designated as unassigned because the revised directionality of the helix, formed by some of these residues (283–301), is not consistent with the continuation of the polypeptide of residues 251–282. Moreover, significantly, we built a short segment of residues (140–149) of the protease cleaved fragment (112–157) of VIII into the density seen at both the structurally independent locations. Lastly, we performed real-space refinement of the above model into the density using Phenix [8]. Various structural analyses and generation of figures were carried out using Chimera and ChimeraX [29,61]. The details of model composition and quality statistics are also shown in Table S2. A bar plot of the map-to-model correlation coefficient (CC) values for individual chains in the Ad26 model is shown in Figure S3B.

Supplementary Materials: The following supporting information can be downloaded at: https://www.mdpi.com/1999-4915/14/2/414/s1, Figure S1: Amino acid sequence alignments of hexon subunits from Ad26 and Ad5 viruses; Figure S2: Amino acid sequence alignments of penton base (PB) subunits from Ad26 and Ad5 viruses; Figure S3. (A) FSC plot showing the resolution estimation of the cryo-EM reconstruction of HAdV-D26(Ad26) virion. The partial FSC values used in the graph as a function of resolution were calculated in cisTEM [59]. (B) A bar plot of map-to-model correlation coefficient (CC) values for individual chains reported in the structure of Ad26 virion. The corresponding capsid proteins of the individual chains are identified at the top the bars. The average map-to-model CC values for all the main chain and side chain atoms are 0.83 and 0.81, respectively. The map-to-model CCs were calculated in Phenix [8]; Table S1: Sequence identity and structural similarity (RMSD) values for VIII pairs in selected HAdVs; Table S2: Cryo-EM data and model quality statistics of HAdV-D26 virion.

Author Contributions: V.S.R. and M.A.B. designed the research; M.A.B. provided the purified Ad26 virions; X.Y. carried out the initial data processing and obtained the reconstructions; V.S.R. reprocessed the data, improved the icosahedral reconstructions, and built and refined the 3D models; V.S.R. and M.A.B. wrote the manuscript. All authors have read and agreed to the published version of the manuscript.

Funding: This research was funded by the NIH grant R21 AI146644 and resources from the Mayo Clinic COVID-19 Task Force.

Institutional Review Board Statement: Not applicable.

Informed Consent Statement: Not applicable.

Data Availability Statement: The cryo-EM map and coordinates of the refined Ad26 virion have been deposited in EMDB and PDB with the accession codes EMD-25786 and 7TAU, respectively.

Acknowledgments: We are deeply grateful for the support and guidance provided by Bridget Carragher and Clint Potter, the directors of the National Resource for Automated Molecular Microscopy (NRAMM) in obtaining cryo-EM data and structure determination of Ad26. We sincerely acknowledge the help of David Veesler, Melody Campbell, and Jeffrey Speir with cryo-EM sample preparation and data collection. We also like to acknowledge the support of J.C. Ducom and Lisa Dong for installing and maintaining cryo-EM software on the HPC cluster at TSRI and for the computational support in general. This research was supported by the NIH grant R21 AI146644 to V.S.R and the funds from "Mayo Clinic COVID-19 Task Force" to M.A.B.

Conflicts of Interest: The authors declare no conflict of interest.

References

1. Kasel, J.A.; Evans, H.E.; Spickard, A.; Knight, V. Conjunctivitis and enteric infection with adenovirus types 26 and 27: Responses to primary, secondary and reciprocal cross-challenges. *Am. J. Hyg.* **1963**, *77*, 265–282. [CrossRef]
2. Baden, L.R.; Walsh, S.R.; Seaman, M.S.; Tucker, R.P.; Krause, K.H.; Patel, A.; Johnson, J.A.; Kleinjan, J.; Yanosick, K.E.; Perry, J.; et al. First-in-human evaluation of the safety and immunogenicity of a recombinant adenovirus serotype 26 HIV-1 Env vaccine (IPCAVD 001). *J. Infect. Dis.* **2013**, *207*, 240–247. [CrossRef]
3. Weaver, E.A.; Barry, M.A. Low seroprevalent species D adenovirus vectors as influenza vaccines. *PLoS ONE* **2013**, *8*, e73313. [CrossRef]
4. Senac, J.S.; Doronin, K.; Russell, S.J.; Jelinek, D.F.; Greipp, P.R.; Barry, M.A. Infection and killing of multiple myeloma by adenoviruses. *Hum. Gene Ther.* **2010**, *21*, 179–190. [CrossRef]
5. Bos, R.; Rutten, L.; van der Lubbe, J.E.M.; Bakkers, M.J.G.; Hardenberg, G.; Wegmann, F.; Zuijdgeest, D.; de Wilde, A.H.; Koornneef, A.; Verwilligen, A.; et al. Ad26 vector-based COVID-19 vaccine encoding a prefusion-stabilized SARS-CoV-2 Spike immunogen induces potent humoral and cellular immune responses. *NPJ Vaccines* **2020**, *5*, 91. [CrossRef]
6. Mercado, N.B.; Zahn, R.; Wegmann, F.; Loos, C.; Chandrashekar, A.; Yu, J.; Liu, J.; Peter, L.; McMahan, K.; Tostanoski, L.H.; et al. Single-shot Ad26 vaccine protects against SARS-CoV-2 in rhesus macaques. *Nature* **2020**, *586*, 583–588; Erratum in *Nature* **2021**, *590*, E25. [CrossRef]
7. Yu, X.; Veesler, D.; Campbell, M.G.; Barry, M.E.; Asturias, F.J.; Barry, M.A.; Reddy, V.S. Cryo-EM structure of human adenovirus D26 reveals the conservation of structural organization among human adenoviruses. *Sci. Adv.* **2017**, *3*, e1602670. [CrossRef]
8. Afonine, P.V.; Poon, B.K.; Read, R.J.; Sobolev, O.V.; Terwilliger, T.C.; Urzhumtsev, A.; Adams, P.D. Real-space refinement in PHENIX for cryo-EM and crystallography. *Acta Cryst. D Struct. Biol.* **2018**, *74 Pt 6*, 531–544. [CrossRef]
9. Emsley, P.; Cowtan, K. Coot: Model-building tools for molecular graphics. *Acta Cryst. D Biol. Cryst.* **2004**, *60 Pt 1*, 2126–2132. [CrossRef]
10. Emsley, P.; Lohkamp, B.; Scott, W.G.; Cowtan, K. Features and development of Coot. *Acta Cryst. D Biol. Cryst.* **2010**, *66 Pt 4*, 486–501. [CrossRef]
11. Waddington, S.N.; McVey, J.H.; Bhella, D.; Parker, A.L.; Barker, K.; Atoda, H.; Pink, R.; Buckley, S.M.; Greig, J.A.; Denby, L.; et al. Adenovirus serotype 5 hexon mediates liver gene transfer. *Cell* **2008**, *132*, 397–409. [CrossRef]
12. Alba, R.; Bradshaw, A.C.; Parker, A.L.; Bhella, D.; Waddington, S.N.; Nicklin, S.A.; van Rooijen, N.; Custers, J.; Goudsmit, J.; Barouch, D.H.; et al. Identification of coagulation factor (F)X binding sites on the adenovirus serotype 5 hexon: Effect of mutagenesis on FX interactions and gene transfer. *Blood* **2009**, *114*, 965–971. [CrossRef]
13. Doronin, K.; Flatt, J.W.; Di Paolo, N.C.; Khare, R.; Kalyuzhniy, O.; Acchione, M.; Sumida, J.P.; Ohto, U.; Shimizu, T.; Akashi-Takamura, S.; et al. Coagulation factor X activates innate immunity to human species C adenovirus. *Science* **2012**, *338*, 795–798. [CrossRef]
14. Zubieta, C.; Schoehn, G.; Chroboczek, J.; Cusack, S. The structure of the human adenovirus 2 penton. *Mol. Cell* **2005**, *17*, 121–135. [CrossRef]
15. Liu, H.; Wu, L.; Zhou, Z.H. Model of the trimeric fiber and its interactions with the pentameric penton base of human adenovirus by cryo-electron microscopy. *J. Mol. Biol.* **2011**, *406*, 764–774. [CrossRef]
16. Cao, C.; Dong, X.; Wu, X.; Wen, B.; Ji, G.; Cheng, L.; Liu, H. Conserved fiber-penton base interaction revealed by nearly atomic resolution cryo-electron microscopy of the structure of adenovirus provides insight into receptor interaction. *J. Virol.* **2012**, *86*, 12322–12329. [CrossRef]
17. Baker, A.T.; Mundy, R.M.; Davies, J.A.; Rizkallah, P.J.; Parker, A.L. Human adenovirus type 26 uses sialic acid-bearing glycans as a primary cell entry receptor. *Sci. Adv.* **2019**, *5*, eaax3567. [CrossRef]
18. Persson, B.D.; John, L.; Rafie, K.; Strebl, M.; Frangsmyr, L.; Ballmann, M.Z.; Mindler, K.; Havenga, M.; Lemckert, A.; Stehle, T.; et al. Human species D adenovirus hexon capsid protein mediates cell entry through a direct interaction with CD46. *Proc. Natl. Acad. Sci. USA* **2021**, *118*, e2020732118. [CrossRef]
19. Reddy, V.S. The Role of Hexon Protein as a Molecular Mold in Patterning the Protein IX Organization in Human Adenoviruses. *J. Mol. Biol.* **2017**, *429*, 2747–2751. [CrossRef]
20. Matteson, N.L.; Barry, M.A.; Reddy, V.S. Structure-based assessment of protein-protein interactions and accessibility of protein IX in adenoviruses with implications for antigen display. *Virology* **2018**, *516*, 102–107. [CrossRef] [PubMed]
21. Flint, S.J. Viral Moulds and Cement: How Interactions among Human Adenovirus Hexons and Their Protein IX Cement May Buttress Human Adenovirus Particles. *J. Mol. Biol.* **2017**, *429*, 2752–2754. [CrossRef] [PubMed]
22. Schoehn, G.; El Bakkouri, M.; Fabry, C.M.; Billet, O.; Estrozi, L.F.; Le, L.; Curiel, D.T.; Kajava, A.V.; Ruigrok, R.W.; Kremer, E.J. Three-dimensional structure of canine adenovirus serotype 2 capsid. *J. Virol.* **2008**, *82*, 3192–3203. [CrossRef]
23. Cheng, L.; Huang, X.; Li, X.; Xiong, W.; Sun, W.; Yang, C.; Zhang, K.; Wang, Y.; Liu, H.; Huang, X.; et al. Cryo-EM structures of two bovine adenovirus type 3 intermediates. *Virology* **2014**, *450–451*, 174–181. [CrossRef] [PubMed]
24. Menendez-Conejero, R.; Nguyen, T.H.; Singh, A.K.; Condezo, G.N.; Marschang, R.E.; van Raaij, M.J.; San Martin, C. Structure of a Reptilian Adenovirus Reveals a Phage Tailspike Fold Stabilizing a Vertebrate Virus Capsid. *Structure* **2017**, *25*, 1562–1573.e5. [CrossRef]
25. Rafie, K.; Lenman, A.; Fuchs, J.; Rajan, A.; Arnberg, N.; Carlson, L.A. The structure of enteric human adenovirus 41-A leading cause of diarrhea in children. *Sci. Adv.* **2021**, *7*, eabe0974. [CrossRef]

26. Perez-Illana, M.; Martinez, M.; Condezo, G.N.; Hernando-Perez, M.; Mangroo, C.; Brown, M.; Marabini, R.; San Martin, C. Cryo-EM structure of enteric adenovirus HAdV-F41 highlights structural variations among human adenoviruses. *Sci. Adv.* **2021**, *7*, eabd9421. [CrossRef]
27. Dai, X.; Wu, L.; Sun, R.; Zhou, Z.H. Atomic Structures of Minor Proteins VI and VII in Human Adenovirus. *J. Virol.* **2017**, *91*, e00850-17. [CrossRef]
28. Yu, X.; Mullen, T.M.; Abrishami, V.; Huiskonen, J.T.; Nemerow, G.R.; Reddy, V.S. Structure of a Cell Entry Defective Human Adenovirus Provides Insights into Precursor Proteins and Capsid Maturation. *J. Mol. Biol.* **2021**, *434*, 167350. [CrossRef]
29. Goddard, T.D.; Huang, C.C.; Meng, E.C.; Pettersen, E.F.; Couch, G.S.; Morris, J.H.; Ferrin, T.E. UCSF ChimeraX: Meeting modern challenges in visualization and analysis. *Protein Sci.* **2018**, *27*, 14–25. [CrossRef]
30. Perez-Berna, A.J.; Marabini, R.; Scheres, S.H.; Menendez-Conejero, R.; Dmitriev, I.P.; Curiel, D.T.; Mangel, W.F.; Flint, S.J.; San Martin, C. Structure and uncoating of immature adenovirus. *J. Mol. Biol.* **2009**, *392*, 547–557. [CrossRef] [PubMed]
31. Perez-Berna, A.J.; Ortega-Esteban, A.; Menendez-Conejero, R.; Winkler, D.C.; Menendez, M.; Steven, A.C.; Flint, S.J.; de Pablo, P.J.; San Martin, C. The role of capsid maturation on adenovirus priming for sequential uncoating. *J. Biol. Chem.* **2012**, *287*, 31582–31595. [CrossRef] [PubMed]
32. Liu, H.; Jin, L.; Koh, S.B.; Atanasov, I.; Schein, S.; Wu, L.; Zhou, Z.H. Atomic structure of human adenovirus by cryo-EM reveals interactions among protein networks. *Science* **2010**, *329*, 1038–1043. [CrossRef]
33. Reddy, V.S.; Nemerow, G.R. Structures and organization of adenovirus cement proteins provide insights into the role of capsid maturation in virus entry and infection. *Proc. Natl. Acad. Sci. USA* **2014**, *111*, 11715–11720. [CrossRef] [PubMed]
34. Snijder, J.; Benevento, M.; Moyer, C.L.; Reddy, V.; Nemerow, G.R.; Heck, A.J. The Cleaved N-Terminus of pVI Binds Peripentonal Hexons in Mature Adenovirus. *J. Mol. Biol.* **2014**, *426*, 1971–1979. [CrossRef]
35. Hernando-Perez, M.; Martin-Gonzalez, N.; Perez-Illana, M.; Suomalainen, M.; Condezo, G.N.; Ostapchuk, P.; Gallardo, J.; Menendez, M.; Greber, U.F.; Hearing, P.; et al. Dynamic competition for hexon binding between core protein VII and lytic protein VI promotes adenovirus maturation and entry. *Proc. Natl. Acad. Sci. USA* **2020**, *117*, 13699–13707. [CrossRef]
36. Benevento, M.; Di Palma, S.; Snijder, J.; Moyer, C.L.; Reddy, V.S.; Nemerow, G.R.; Heck, A.J. Adenovirus Composition, Proteolysis and Disassembly Studied by in-depth Qualitative and Quantitative Proteomics. *J. Biol. Chem.* **2014**, *289*, 11421–11430. [CrossRef]
37. Kalyuzhniy, O.; Di Paolo, N.C.; Silvestry, M.; Hofherr, S.E.; Barry, M.A.; Stewart, P.L.; Shayakhmetov, D.M. Adenovirus serotype 5 hexon is critical for virus infection of hepatocytes in vivo. *Proc. Natl. Acad. Sci. USA* **2008**, *105*, 5483–5488. [CrossRef]
38. Xu, Z.; Qiu, Q.; Tian, J.; Smith, J.S.; Conenello, G.M.; Morita, T.; Byrnes, A.P. Coagulation factor X shields adenovirus type 5 from attack by natural antibodies and complement. *Nat. Med.* **2013**, *19*, 452–457. [CrossRef]
39. Hemsath, J.R.; Liaci, A.M.; Rubin, J.D.; Parrett, B.J.; Lu, S.C.; Nguyen, T.V.; Turner, M.A.; Chen, C.Y.; Cupelli, K.; Reddy, V.S.; et al. Ex Vivo and In Vivo CD46 Receptor Utilization by Species D Human Adenovirus Serotype 26 (HAdV26). *J. Virol.* **2021**, *96*, 0082621. [CrossRef]
40. Muir, K.L.; Kallam, A.; Koepsell, S.A.; Gundabolu, K. Thrombotic Thrombocytopenia after Ad26.COV2.S Vaccination. *N. Engl. J. Med.* **2021**, *384*, 1964–1965. [CrossRef] [PubMed]
41. Greinacher, A.; Selleng, K.; Palankar, R.; Wesche, J.; Handtke, S.; Wolff, M.; Aurich, K.; Lalk, M.; Methling, K.; Volker, U.; et al. Insights in ChAdOx1 nCoV-19 vaccine-induced immune thrombotic thrombocytopenia. *Blood* **2021**, *138*, 2256–2268. [CrossRef]
42. McCrae, K.R. Thrombotic thrombocytopenia due to SARS-CoV-2 vaccination. *Cleve. Clin. J. Med.* **2021**. [CrossRef] [PubMed]
43. Bhuyan, P.; Medin, J.; da Silva, H.G.; Yadavalli, M.; Shankar, N.K.; Mullerova, H.; Arnold, M.; Nord, M. Very rare thrombosis with thrombocytopenia after second AZD1222 dose: A global safety database analysis. *Lancet* **2021**, *398*, 577–578. [CrossRef]
44. Sadoff, J.; Davis, K.; Douoguih, M. Thrombotic Thrombocytopenia after Ad26.COV2.S Vaccination—Response from the Manufacturer. *N. Engl. J. Med.* **2021**, *384*, 1965–1966. [CrossRef] [PubMed]
45. See, I.; Su, J.R.; Lale, A.; Woo, E.J.; Guh, A.Y.; Shimabukuro, T.T.; Streiff, M.B.; Rao, A.K.; Wheeler, A.P.; Beavers, S.F.; et al. US Case Reports of Cerebral Venous Sinus Thrombosis With Thrombocytopenia After Ad26.COV2.S Vaccination, March 2 to April 21, 2021. *JAMA* **2021**, *325*, 2448–2456. [CrossRef]
46. George, G.; Friedman, K.D.; Curtis, B.R.; Lind, S.E. Successful treatment of thrombotic thrombocytopenia with cerebral sinus venous thrombosis following Ad26.COV2.S vaccination. *Am. J. Hematol.* **2021**, *96*, E301–E303. [CrossRef]
47. Dhoot, R.; Kansal, A.; Handran, C.; Haykal, T.; Ronald, J.; Kappus, M.; Arepally, G.M.; Graham, M.; Strouse, J.J. Thrombocytopenia and splanchnic thrombosis after Ad26.COV2.S vaccination successfully treated with transjugular intrahepatic portosystemic shunting and thrombectomy. *Am. J. Hematol.* **2021**, *96*, 1180–1182. [CrossRef]
48. Abou-Ismail, M.Y.; Moser, K.A.; Smock, K.J.; Lim, M.Y. Vaccine-induced thrombotic thrombocytopenia following Ad26.COV2.S vaccine in a man presenting as acute venous thromboembolism. *Am. J. Hematol.* **2021**, *96*, E346–E349. [CrossRef]
49. Clark, R.T.; Johnson, L.; Billotti, J.; Foulds, G.; Ketels, T.; Heard, K.; Calvello Hynes, E. Early Outcomes of Bivalirudin Therapy for Thrombotic Thrombocytopenia and Cerebral Venous Sinus Thrombosis After Ad26.COV2.S Vaccination. *Ann. Emerg. Med.* **2021**, *78*, 511–514. [CrossRef]
50. Banerjee, S.; Sandhu, M.; Tonzi, E.; Tambe, A.; Gambhir, H.S. Immune-Mediated Thrombocytopenia Associated With Ad26.COV2.S (Janssen; Johnson & Johnson) Vaccine. *Am. J. Ther.* **2021**, *28*, e604–e606.
51. Al-Samkari, H.; Leaf, R.K.; Goodarzi, K. Transient Thrombocytopenia With Glycoprotein-Specific Platelet Autoantibodies After Ad26.COV2.S Vaccination: A Case Report. *Ann. Intern. Med.* **2021**, *174*, 1632–1633. [CrossRef] [PubMed]

52. Charidimou, A.; Samudrala, S.; Cervantes-Arslanian, A.M.; Sloan, J.M.; Dasenbrock, H.H.; Daneshmand, A. Vaccine-Induced Immune Thrombotic Thrombocytopenia with Concurrent Arterial and Venous Thrombi Following Ad26.COV2.S Vaccination. *J. Stroke Cerebrovasc. Dis.* **2021**, *30*, 106113. [CrossRef] [PubMed]
53. Kulkarni, P.A.; Prasad, V. Understanding risk of thrombosis with thrombocytopenia syndrome after Ad26.COV2.S vaccination. *Front. Med.* **2021**, *15*, 938–941. [CrossRef] [PubMed]
54. Rodriguez, E.V.C.; Bouazza, F.Z.; Dauby, N.; Mullier, F.; d'Otreppe, S.; Jissendi Tchofo, P.; Bartiaux, M.; Sirjacques, C.; Roman, A.; Hermans, C.; et al. Fatal vaccine-induced immune thrombotic thrombocytopenia (VITT) post Ad26.COV2.S: First documented case outside US. *Infection* **2021**, 1–6. [CrossRef] [PubMed]
55. Rodriguez-Pardo, J.; Gilo-Arrojo, F.A.; Ruiz-Ares, G.; Sanchez-Manso, J.C.; Valiente-Gordillo, E.; de Celis, E.; Fuentes, B.; Ximenez-Carrillo, A.; Alonso de Lecinana, M.; Rigual-Bobillo, R.; et al. Thrombosis And Thrombocytopenia Syndrome Causing Isolated Symptomatic Carotid Occlusion After Ad26.COV2.S (Janssen) COVID-19 Vaccine. *Thromb. Haemost.* **2022**, *122*, 300–303. [PubMed]
56. Baker, A.T.; Boyd, R.J.; Sarkar, D.; Teijeira-Crespo, A.; Chan, C.K.; Bates, E.; Waraich, K.; Vant, J.; Wilson, E.; Truong, C.D.; et al. ChAdOx1 interacts with CAR and PF4 with implications for thrombosis with thrombocytopenia syndrome. *Sci. Adv.* **2021**, *7*, eabl8213. [CrossRef]
57. Chesterman, C.N.; McGready, J.R.; Doyle, D.J.; Morgan, F.J. Plasma levels of platelet factor 4 measured by radioimmunoassay. *Br. J. Haematol.* **1978**, *40*, 489–500. [CrossRef]
58. Li, X.; Mooney, P.; Zheng, S.; Booth, C.R.; Braunfeld, M.B.; Gubbens, S.; Agard, D.A.; Cheng, Y. Electron counting and beam-induced motion correction enable near-atomic-resolution single-particle cryo-EM. *Nat. Methods* **2013**, *10*, 584–590. [CrossRef]
59. Grant, T.; Rohou, A.; Grigorieff, N. cisTEM, user-friendly software for single-particle image processing. *eLife* **2018**, *7*, e35383. [CrossRef]
60. Mindell, J.A.; Grigorieff, N. Accurate determination of local defocus and specimen tilt in electron microscopy. *J. Struct. Biol.* **2003**, *142*, 334–347. [CrossRef]
61. Pettersen, E.F.; Goddard, T.D.; Huang, C.C.; Couch, G.S.; Greenblatt, D.M.; Meng, E.C.; Ferrin, T.E. UCSF Chimera—A visualization system for exploratory research and analysis. *J. Comput. Chem.* **2004**, *25*, 1605–1612. [CrossRef] [PubMed]

viruses MDPI

Article

IgG-Complexed Adenoviruses Induce Human Plasmacytoid Dendritic Cell Activation and Apoptosis

Thi Thu Phuong Tran [1,2], Tuan Hiep Tran [1,3,4] and Eric J. Kremer [1,*]

1. Institut de Génétique Moléculaire de Montpellier, Université de Montpellier, CNRS, 34090 Montpellier, France; tranphuongdkh@gmail.com (T.T.P.T.); tuanhiepdkh@gmail.com (T.H.T.)
2. Department of Life Sciences, University of Science and Technology of Hanoi Vietnam Academy of Science and Technology, Hanoi 100000, Vietnam
3. Faculty of Pharmacy, PHENIKAA University, Hanoi 12116, Vietnam
4. PHENIKAA Research and Technology Institute (PRATI), A&A Green Phoenix Group JSC, Hanoi 11313, Vietnam

* Correspondence: eric.kremer@igmm.cnrs.fr

Abstract: Following repeat exposure to many human adenoviruses (HAdVs), most adults harbour long-lived B- and T-cell responses. Combined, this response typically protects us for years from re-infection by the same HAdV type. In spite of these immune responses, some HAdV types are associated with persistent infections that constitute a life-threatening risk when an individual's T-cell response is compromised. By contrast, patients with B-cell deficiencies do not appear to be at a greater risk of HAdV disease. This dichotomy begs the question of the secondary role of anti-HAdV antibodies during host defence. In this study, we explored IgG-complexed (IC)-HAdV5 and primary human plasmacytoid dendritic cell (pDC) interactions. We found that IC-HAdV5 are efficiently internalized in pDCs, stimulate their activation through TLR9 signalling, and cause apoptosis. These data may help reconcile the enigma of robust immune response to HAdVs, while concurrently allowing persistence.

Keywords: adenovirus; antibodies; immune complexes; plasmacytoid dendritic cells; innate immunity

1. Introduction

Plasmacytoid dendritic cells (pDCs) are a subset of phagocytes continuously produced in the bone marrow and, via the peripheral blood, are recruited to breaches in tissue homeostasis [1]. During a response to viral infections, the hallmark of pDCs is their ability to secrete relatively high levels of type 1 interferons (IFN-1). Combined with the innate immune responses from infected cells and other recruited immune cells, the cytokine profile of pDCs biases the adaptive immune response toward an antiviral Th1 response [2].

Members of the family *Adenoviridae* have nonenveloped icosahedral particles of approximately 90 nm, containing a double-stranded linear DNA genome of 36,000 ± 7000 bp [3]. They are found on every continent and now include more than 300 types isolated from mammals, fish, birds, and reptiles. Human adenoviruses (HAdVs) belong to the genus *Mastadenoviridae* and are grouped into seven species (A–G), which include approximately 100 types that are grouped by serology and/or sequence phylogeny [4]. Epidemiological data generated during the last six decades suggest that greater than 90% of us have been infected by a handful of HAdV types by the time we are a few years old [5]. Most HAdV infections cause type-specific, asymptomatic to self-limiting disease, in the respiratory, ocular, or gastrointestinal tracts. In spite of the robust humoral and cellular immune responses found in most adults, HAdVs induce latent/persistent infections that last for decades [6]. Persistent infections are also coherent with data demonstrating that in patients undergoing pharmaceutical or disease-induced immune suppression, subsequent adenoviremia can be traced to the seroprevalence of the same type before immune suppression [6].

It is in this context that we address the impact of pre-existing humoral immunity against HAdVs on the innate immune response by primary human phagocytes. We argue that understanding the divergent roles of pDCs in antiviral immunity vs. tolerance will help resolve the discord between the robust immune response to HAdV and the ability to induce latent infections [7,8]. Our data demonstrate that pDCs take up HAdV-5 more efficiently when complexed with IgGs, the pDCs become activated via a TLR9-associated pathway, which leads to an increase in autophagy and apoptosis.

2. Materials and Methods

2.1. Ethics Statement

Blood samples from >120 anonymous donors from the local blood bank (Etablissement Français du Sang (EFS); Montpellier, France) were used during this study. All donors provided written informed consent and an internal ethics committee at IGMM and the EFS approved the study.

2.2. Cells and Culture Conditions

pDCs were negatively selected from PBMCs using Plasmacytoid Dendritic Cell Isolation Kit II, human (MACS Miltenyi Biotec, Auburn, CA, USA) and cultured in RPMI 1640 medium (Cellgro, Swedesboro, NJ, USA) supplemented with 10% FBS, 10 mM HEPES, 100 units/mL penicillin, non-essential amino acids, and 1 mM sodium pyruvate at 37 °C/5% CO_2.

2.3. HAdV Vectors & Hexon Peptides

Adβgal is a ΔE1/E3 HAdV5 harbouring a *lacZ* expression cassette [9]. Ad^{L40Q} is an HAdV5-based vector with a leucine to glutamine mutation of an amino acid in protein VI that decreases its membrane lytic activity [10]. Alexa555- and Alexa488-HAdV5 were generated from Adβgal using an Alexa555 or Alexa488 Protein Labeling Kit (Life Technologies, Villebon-sur-Yvette, France), as previously described [11]. Ad2ts1 harbours a mutation in protease and results in several unprocessed capsid proteins and a hyper-stable capsid [12]. All HAdV viruses/vectors were produced in 293 or 911 cells and purified by double banding on CsCl density gradients, as previously described [13]. Vector purity typically reaches >99%. HAdV concentrations (physical particles/mL) were determined, as previously described [14].

2.4. Antibodies

Anti-human CD83-FITC (#556910), anti-human HLA-ABC-PE (#555553), anti-human HLA-DR-PE (#555812), anti-human CD80-FITC (#557226), anti-human CD86-APC (#555660), anti-human CD123-FIT C (#555432), anti-human CD303-PE (#561697), and anti-human CD40-APC (#313008) were from BioLegend. (San Diego, CA, USA).

2.5. Pharmacological Assays

Imiquimod is a TLR7 agonist and can inducing caspase-mediated apoptosis, and inhibits adenosine signalling in some cells. While its mechanism of action is only partially understood, in pDCs, imiquimod induces TNF secretion but not a type 1 IFN production. Chloroquine is a diprotic weak base that can sequester protons and prevent endosomal acidification. ODN 2216 is a short oligonucleotide that contains unmethylated CpG dinucleotides that can be found in pathogen genomes. The short oligonucleotides prevent TLR9-mediated signalling by inhibiting TLR9 clustering. Spautin-1, a quinazolin compound, is an inhibitor of autophagy and promotes Vps34 PI 3 (phosphoinositide 3-kinases complex) degradation by blocking the activity of USP 10. Wortmannin is a non-specific covalent inhibitor of PI 3-kinases.

2.6. Immune Complex Formation and DC Stimulations

pDCs (1×10^5 in 200 µL of complete medium) were incubated with HAdV5 or IC-HAdV5 (or IC) (2×10^4 or 4×10^4 physical particles (pp)/cell, unless otherwise indicated) for the indicated times. IC-HAdV5s were generated by mixing the virus (8×10^9 physical particles) with 2.5 µL of IVIg (human IgG pooled from 5000 to 10,000 donors/batch) (Baxter SAS; Guyancourt, France) for 15 min at room temperature. IVIg is used in patients with primary or acquired immunodeficiency as well as autoimmune diseases. Chloroquine (1–10 nM) or ODN 2216 (tlrl-2216) (InvivoGen, San Diego, California, USA) was added 1 or 2 h before stimulation.

2.7. Quantification of mRNA

Expression levels of cytokine and chemokine genes were evaluated using RT-qPCR assays. Total RNA was isolated from cells using the high pure RNA isolation Kit (Roche; Berlin, Germany) with a DNase I treatment during the purification and subsequent elution in 50 µL of RNase-free water (Qiagen, Germantown, MD, USA). Reverse transcription was performed with the superscript first-strand synthesis system (Invitrogen) using 10 µL of total RNA and random hexamers. The cDNA samples were diluted 1:20 in water and analysed in triplicate using a LightCycler 480 (Roche; Meylan, France). SYBR green PCR conditions were as follows: 95 °C for 5 min and 45 cycles of 95 °C for 15 s, 65 °C or 70 °C for 15 s, and 72 °C for 15 s, using *GAPDH* as a standard. Relative gene expression levels of each respective gene were calculated using the threshold cycle ($2^{-\Delta\Delta CT}$) method and normalized to *GAPDH* mRNA levels.

2.8. Co-Stimulatory Protein Levels

Surface levels of CD83, MHCII, CD80, CD40, and CD86 were assessed by flow cytometry. Cell membrane integrity was assessed by collecting cells via centrifugation at 800× *g*; the cell pellets were then resuspended in PBS containing 10% FBS, propidium iodide (PI) (Sigma-Aldrich, St. Louis MO, USA), or 7-aminoactinomycin D (7AAD) (Becton-Dickinson; Franklin Lakes, NJ, USA). The cell suspension was incubated for the indicated times and analysed using a FacsCalibur flow cytometer (Becton-Dickinson, Franklin Lakes, NJ, USA) and FlowJo software (https://www.flowjo.com/, accessed on 14 July 2021).

2.9. Annexin V Staining

The proportion of apoptotic thymocytes was measured by flow cytometry by staining cells for DNA using propidium iodide (PI) and surface expression of phosphatidylserine using Annexin V-FITC (Invitrogen, Waltham, MA, USA).

2.10. Cytokine Secretion: ELISA

Supernatant from the cells were collected and cytokine secretion was measured by ELISA and Luminex assays. The secretion of TNF and IFN-α was quantified by ELISA using an OptEIA human TNF ELISA Kit (Becton Dickinson) and human IFN-α ELISA (R&D Systems; Lille, France) following the manufacturer's instructions.

2.11. Statistical Analyses

All experiments were performed at least in duplicate a minimum of three independent times, and the results are expressed as the mean ± SD unless otherwise stated. The statistical analyses were performed using the Student's *t*-test unless otherwise stated. A p value < 0.05 is denoted as significant. Statistical analyses of the global cytokine profiles (pie chart) were performed by partial permutation tests using the SPICE software.

2.12. Data Availability

All data generated or analysed during this study are included in this published article.

3. Results

3.1. pDCs Take up IgG-Complexed HAdV5

pDCs were isolated from buffy coats from random blood bank donors. The purity (>95%) and phenotype were controlled by flow cytometry using anti-CD303 and anti-CD123 antibodies (Figure 1A). Loré et al. [15] previously reported that HAdV5 vectors poorly infect pDCs. Our results are consistent with these findings: at 20,000 physical particles (pp)/cell only ~4% of the pDCs took up HAdV5 (Figure 1B). We therefore asked if the neutralizing antibodies (NAbs) in IVIg modified HAdV5 uptake. In addition to an ~3-fold increase in the number of pDCs that took up HAdV5 (Figure 1B), a striking result was the >10^4-fold increase in fluorescence (Figure 1C).

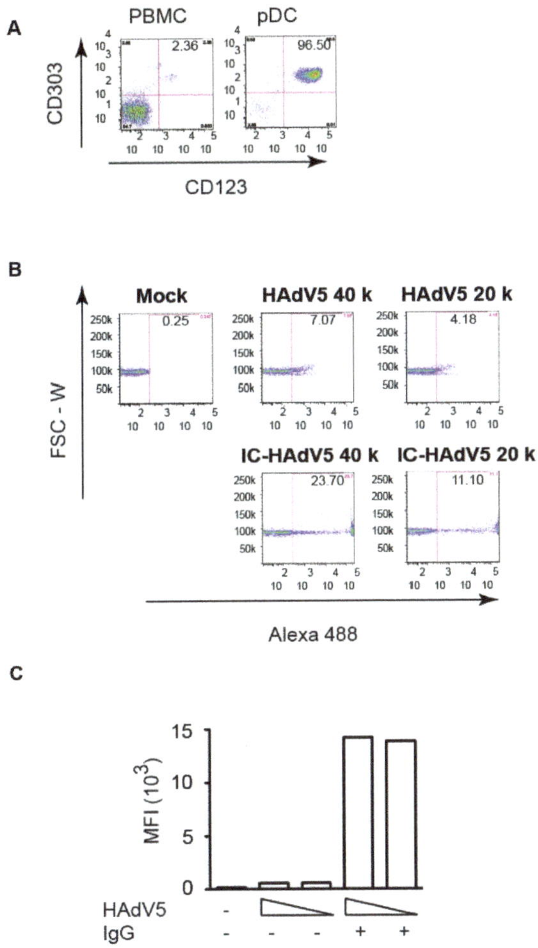

Figure 1. pDC purity and infection by HAdV5 and IC-HAdV5. (**A**) pDCs were isolated from PBMC by negative selection kit. pDCs and PBMCs were stained CD123 and CD303 to check pDC purity. (**B**,**C**) pDCs were challenged with IgG, HAdV5-Alexa488, or IC-HAdV5-Alexa488 at 20,000 and 40,000 pp/cell) (an example from a single donor). (**C**) Cumulative data from 3 donors. Median fluorescent intensity (MFI).

3.2. pDC Activation by IgG-Complexed HAdV5

Increased uptake of IC-HAdV5 by primary human monocyte-derived DCs induces phenotypic and functional maturation [8,13,16–18]. The changes in phenotype include dendritic morphology. We therefore qualitatively examined pDC morphology by scanning electron microscopy (SEM). Compared to mock- and HAdV5-treated pDCs, the majority of cells in the IC-HAdV5-challenged wells had larger and more complex processes (Figure 2A). We then focused on markers of activation. A prerequisite for increased cytokine secretion is the transcriptional upregulation of the respective genes. We therefore quantified *IFNα* and *IFNβ* mRNA levels in IC-HAdV5-challenged pDCs and found >100-fold increases compared to controls (Figure 2B). The increase in mRNA levels was also mirrored by an IC-HAdV5-induced increase ($p < 0.05$) in IFN-α and TNF secretion at 18 and 48 h post-challenge (Figure 2C).

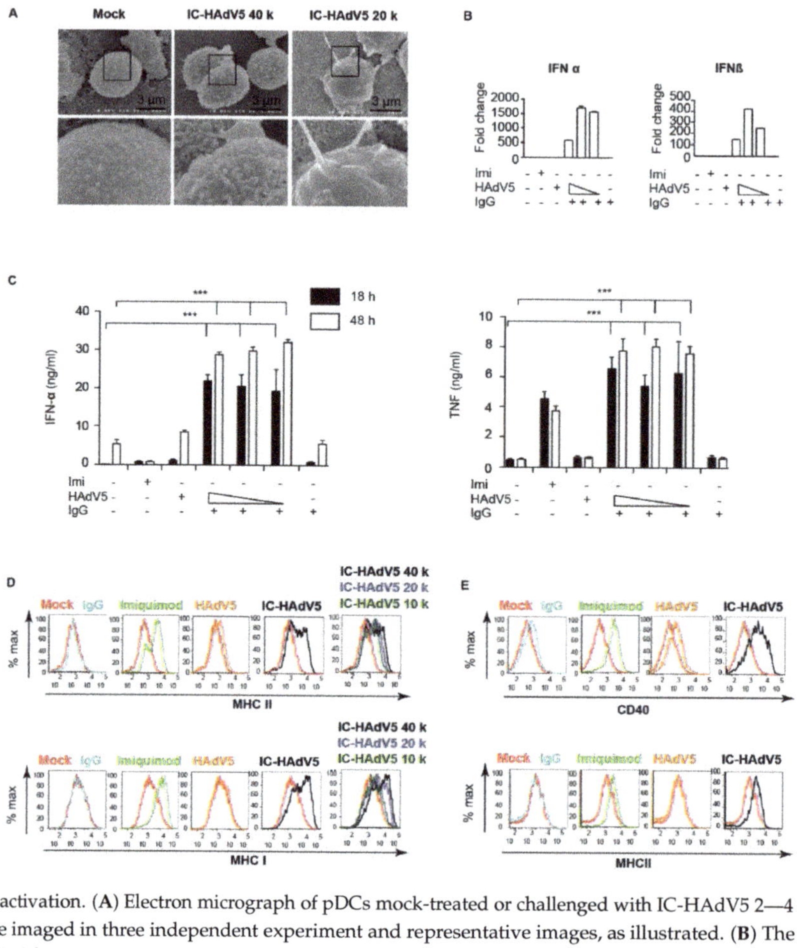

Figure 2. pDC activation. (**A**) Electron micrograph of pDCs mock-treated or challenged with IC-HAdV5 2—4 × 10^5 pp/cell. Sixty cells were imaged in three independent experiment and representative images, as illustrated. (**B**) The mRNA levels of *IFNα* and *IFNβ* from pDCs exposed to the milieu from IgG, HAdV5-, and IC-HAdV-challenged at 4 h post-activation by qPCR. The cytokine sequences were designed in-house. (**C**) The IFN-α and TNF levels in the supernatant of pDCs challenged with IgG, HAdV5-, and IC-HAdV. Challenged at 18 h and 48 h post-activation. $n \geq 3$ donors. p values were derived from a one-way ANOVA with Dunnett's post-test. *** corresponds to $p < 0.0001$ and denotes significant difference. Imiquimod was used as a positive control. (**D**,**E**) The cell surface levels of MHC I and II (**D**) and CD40 (**E**). Quantified 18 h post-stimulation with Imiquimod, HAdV5, or increasing doses of IC-HAdV5. $n = 3$ donors with similar results.

As pDCs mature, they also increase cell surface levels of activation markers. pDCs are capable of presenting both soluble and particulate exogenous antigens on both major MHC class I and II molecules [19]. We therefore used flow cytometry to determine whether the levels of CD40, and MHC I and II change when challenged by IC-HAdV5. We found that, compared to controls (IgGs, imiquimod, and HAdV5), IC-HAdV5-challenged pDCs increase the cell surface levels of all the above at 18 h post-stimulation (Figure 2D,E). Together, these data demonstrate that IC-HAdV5 induce pDC activation.

3.3. pH-Dependent TLR9 Engagement of IC-HAdV5

When bound by antibodies, viruses are taken up by the Fcγ receptors expressed by numerous cells, including phagocytes [20]. It is likely that FcγR-mediated IC-HAdV5 internalization delivers HAdV components to distinct endosomal compartments for degradation, followed by presentation by MHC II and cross-presentation by MHC I molecules [21]. In contrast to prototype antigens, HAdV5 influences intracellular trafficking. As endocytic vesicles start to acidify, the metastable HAdV5 particle begins to dissociate and release protein VI [13,22]. The endosomolytic activity of protein VI allows the release of the capsid into the cytoplasm. In the context of pDCs, this process is typically slower than epithelial cells to allow efficient antigen presentation (and cross-presentation) [23]. If this were the case for IC-HAdV5, then preventing endosomal acidification should impact pDC activation. We therefore inhibited endosomal acidification with chloroquine and assayed MHC II levels, and TNF and IFN-α secretion (Figure 3A,B). Our data demonstrate that endosomal acidification was linked to pDC activation by IC-HAdV5.

As the genomes of *Adenoviridae* are double-stranded DNA, a plausible pattern recognition receptor in pDCs to probe was TLR9 [24]. TLR9 can be inhibited using unmethylated CpG oligonucleotides (ODN 2216) sequences to prevent its oligomerization. We found that TLR9 inhibition modestly decreased MHC II levels but significantly decreased TNF and IFN-α secretion (Figure 3C,D). These data are coherent with the mechanism that the HAdV5 capsid that escapes the endocytic vesicles is partially degraded, and double-stranded DNA is available for detection by TLR9 [18,25,26].

To address this mechanism using another approach, we used AdL40Q, a HAdV5 capsid containing a mutation (L40Q) in protein VI that attenuates its membrane lytic activity [10,27,28]. Consistent with the above data, IC-AdL40Q also had a lower capacity to induce TNF, an IFN-α secretion, compared to IC-HAdV5 (Figure 3E). Together, these data demonstrate that IC-HAdV5 are taken up by pDCs in endocytic vesicles that acidify, releasing an internal HAdV5 capsid protein that then allows partially degraded capsids to be detected by TLR9, and induce pDC activation.

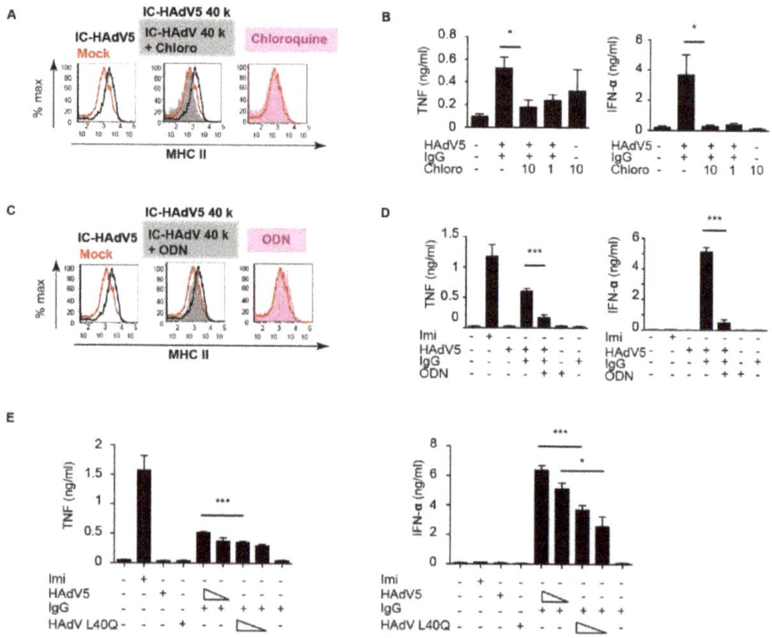

Figure 3. pDCs deliver IC-HAdV5 to endosomal vesicles. (**A**) pDCs were pre-treating with chloroquine (10 or 1 nM) for 2 h, then challenged with IgG, HAdV5, or IC-HAdV5. MHC II cell surface levels were quantified by flow cytometry ($n = 3$, with similar results between donors). (**B**) pDCs were pre-treating with chloroquine for 2 h, then challenged with IgG, HAdV5, or IC-HAdV5. TNF and IFN-α levels in the supernatant were quantified by ELISA ($n = 3$, p values were derived from t-test, * corresponds to $p < 0.01$). (**C**) pDC activation was assessed by pre-treating cells with 50 nM of ODN 2216 for 2 h, then stimulated with IgG, HAdV5, or IC-HAdV5. MHC II cell surface levels were quantified by flow cytometry ($n = 3$, with similar results between donors). (**D**) pDC activation was assessed by pre-treating cells with 50 nM of ODN 2216 for 2 h then stimulated with IgG, HAdV5, or IC-HAdV5. TNF and IFN-α levels in the supernatant were quantified by ELISA ($n = 3$, p values were derived from t-test, *** corresponds to $p < 0.0001$). (**E**) pDCs were challenged with IgG, HAdV5, HAdV L40Q, or IC-HAdV5; IC-HAdV L40Q in dose dependent and pDCs in each condition were collected, and detected IFN α and TNF secretion. Experiments were carried out in three independent experiments. p values were derived from t-tests. * and *** correspond to $p < 0.01$ and $p < 0.0001$, respectively.

3.4. IC-HAdV5 Induced Loss of Cell Membrane Integrity

One of the downstream effects of IC-HAdV uptake is whether pDC survival was affected. To determine if IC-HAdVs induced the loss of plasma membrane integrity, we quantified 7AAD + entry into cells by flow cytometry. Using time-dependent assays, we found that IC-HAdV5 induce a notable increase in the number of $7AAD^+$ cells at 18 h post challenge that continued to increase (Figure 4A). To determine whether the loss of plasma membrane integrity is a direct or indirect effect of IC-HAdV5, we used IC-HAdV-555, an immune complex made using Alexa555-labeled HAdV-C5 capsid. We found that $7AAD^+$ pDCs are associated with IC-HAdV-555 (Figure 4B), suggesting that a direct interaction was responsible for the loss of plasma membrane integrity.

We then addressed a potential synergy of autophagy and apoptosis. We used spautin-1 to increase the degradation of PI3K complexes, and wortmannin to directly inhibit PI3K activity: both inhibitors reduced pDC activation as quantified by CD40 cell surface levels (Figure 4C), and IFN-α and TNF secretion ((Figure 4D,E). To address apoptosis, we incubated IC-HAdV5-challenged pDCs with 7AAD and quantified cell surface levels of Annexin V, a prototype marker of apoptosis [29]. We found that IC-HAdV5-challenged pDCs undergo apoptosis, as the majority of $7AAD^{+}$ cells were also Annexin V^{+} (Figure 5A,B). We did not find a notable change in Annexin V^{+} cells at the doses used. Together, these data demonstrate that IC-HAdV5 induce autophagy and apoptosis in human pDCs.

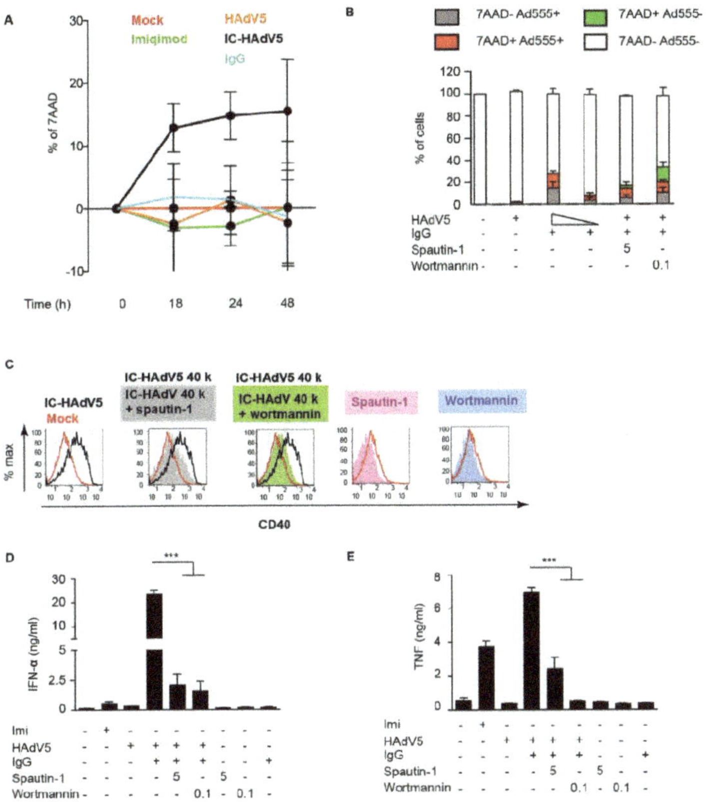

Figure 4. IC-HAdV5 induced loss of cell membrane integrity. (**A**) pDCs were incubated with IgG, HAdV5, and IC-HAdV5 and cell membrane integrity was assessed by intercalation of 7AAD into cellular DNA from 0 to 48 h. (**B**) pDCs were pre-treated with spautin-1 (5 μM) or wortmannin (0.1 μM) for 2 h, then were incubated with HAdV-C5-Alexa555-, IC-HAdV-555 (40,000 or 20,000 pp/cell), or IgG and assayed by flow cytometry at 18 h. The quadrants were set for 7AAD- Ad555- (white), 7AAD- Ad555 + (grey), 7AAD + Alexa555 + (red), and 7AAD + Ad555- (green). The percentage of each subpopulation at each condition is depicted in the histogram. These assays were performed in three donors with similar results. (**C**) pDCs were pre-treated with spautin-1 or wortmannin for 2 h then were incubated with IgG, HAdV5, or IC-HAdV5 in 18 h. CD40 surface expression was detected by flow cytometry. These assays were performed in three donors, with similar results. (**D**,**E**) pDCs were pre-treated with spautin-1 or wortmannin and then were incubated with IgG, HAdV5, or IC-HAdV5 in 18 h. IFN-α (**D**) and TNF secretion (**E**) were quantified by ELISA. Experiments were carried out in three donor experiments. *p* values were derived from *t*-tests. *** corresponds to $p < 0.0001$, respectively.

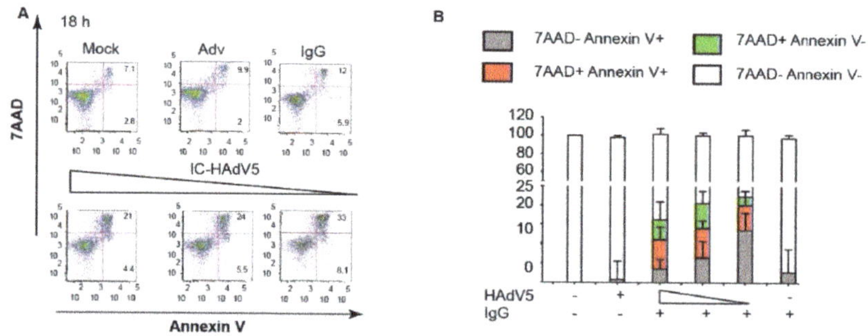

Figure 5. IC-HAdV5 caused autophagy and apoptosis in pDCs. (**A**) Representative flow cytometry profile of pDCs challenged with IgG, HAdV5, and IC-HAdV5 (40,000 to 10,000 pp/cell), then incubated with anti-Annexin V antibodies and assayed by flow cytometry at 18 h. (**B**) pDCs were challenged with IgG, HAdV5, and IC-HAdV5 (40,000 or 20,000 pp/cell), then incubated with Annexin V and assayed by flow cytometry at 18 h. The quadrants were set for 7AAD-/Annexin V- (white), 7AAD-/Annexin V+ (grey), 7AAD+/Annexin V+ (red), and 7AAD+/Annexin V- (green). The percentage of each subpopulation at each condition is depicted in the histogram. These assays were performed in three donors, with similar results.

4. Discussion

In this study, we explored the interaction of a human adenovirus and pre-existing humoral immunity, with primary human plasmacytoid DCs. We found that a relatively high MOI was needed to infect pDCs with HAdV5. However, pDCs more readily took up IgG-complexed HAdV5, and the susceptible cells took up more particles/cell (based on the level of transgene expression). The increased susceptibility of pDCs to IC-HAdV5 was mirrored by an increase in the morphological and functional activation. IC-HAdV5 also induced autophagy and apoptosis in pDCs.

It would be naïve to assume that pDC purification by negative selection from human blood will generate a distinct and homogeneous population. The state-of-the-art approach to purifying pDCs from human blood may co-select a small population of pre-DCs, and therefore interpretation of our results must take this into account. pDC heterogeneity is likely linked to functional heterogeneity [30]. Typically, pDCs are resistant to most viral infections unless high doses are used. Reizis suggested that this allows them to resist virus-induced perturbations [31]. Low infection is also the case for HAdV5. The above two facts beg the question of whether the HAdV5- or IC-HAdV5-infected pDCs are a unique subset. Or, are there different subsets that are infected by HAdV5 versus those that take up IC-HAdV5? One way to address this would be intracellular staining for transcription factors (e.g., TCF4) [32] and/or IFN-1. It is not surprising that HAdV types that use CD46 as a receptor (e.g., HAdV35) more efficiently (60 fold) infect pDCs [15]. The physiological consequence of HAdV35-infected pDCs has not been examined thoroughly, and it is interesting to note that seroprevalence to HAdVs that use CD46 as a receptor (and would therefore more efficiently infect pDCs) is notably lower than HAdVs that preferentially use the coxsackievirus adenovirus receptor [33,34].

One question that remains to be addressed is how pDC activation is regulated and how, in the event of dysregulation, does this affect host homeostasis. pDCs play a unique role in linking innate and adaptive immune responses. Their ability to influence an adaptive immune response to viruses, via the secretion of type 1 IFN, clearly impacts the duration of infections and, paradoxically, the potential for latency. The fact that antibodies that neutralize HAdV5 infection of epithelial cells increase the uptake of pDCs is noteworthy. However, increased uptake is not unique to pDCs: conventional DCs are more susceptible to IgG-complexed HAdV infections via FcγR-mediated uptake [8, 13,18,35]. While pDCs contribute to an antiviral response in a naïve host, their role in controlling or promoting HAdV persistence needs attention [36,37].

In our study, we focused on the initial phase of infection and did not take into account the possible impact that HAdV replication could have in pDCs secreting IFN-1. Hearing and colleagues showed that while IFNs fail to inhibit HAdV replication in cancer cell lines, replication in "normal" human cells could be inhibited by IFN-α and IFN-γ via repression of transcription of the E1A immediate early gene product. Both IFN-α and IFN-γ impede the association of E4TF-1 with the E1A enhancer region during the early phase of HAdV infection [37]. This study demonstrated that IFN signalling suppresses lytic virus replication and promotes persistent infections. Therefore, it is possible that, in some environments, IgG-complexed HAdVs are taken up by pDCs and cause the production and secretion of IFNs, which then silences the expression of HAdV genomes in bystander cells. Once transcriptionally silenced for an extended period (days?), the mechanism by which a HAdV would need to set up a "replication-friendly" environment is unknown because uptake and intracellular trafficking prime the cell for a permissive replication environment.

Finally, in our setting, it is unknown if pDCs take up HAdV particles or genomes from other pDCs undergoing apoptosis. Viral transfer from infected cells to pDCs does occur for hepatitis C, [38], dengue, West Nile [39], and Epstein Barr viruses [40]. pDC death by the relatively quiet process of apoptosis (versus a more inflammatory death like pyroptosis [41]) could favour pathogen persistence.

Author Contributions: Conceptualization, E.J.K.; methodology, T.T.P.T.; validation, T.T.P.T. and T.H.T.; formal analysis, T.T.P.T., T.H.T. and E.J.K..; investigation, T.T.P.T. and T.H.T.; resources, E.J.K.; data curation, T.T.P.T. and T.H.T.; writing first draft— T.T.P.T.; writing—review and editing, T.T.P.T. and E.J.K.; supervision, E.J.K.; project administration, E.J.K.; funding acquisition, E.J.K. All authors have read and agreed to the published version of the manuscript.

Funding: This study was funded by the Université de Montpellier, Vietnamese Minister of Education, and l'Institut de Génétique Moléculaire de Montpellier.

Institutional Review Board Statement: The study was conducted according to the guidelines of the Declaration of Helsinki and approved by the Institutional Ethics Committee of IGMM.

Informed Consent Statement: Informed consent was obtained from all subjects involved in the study.

Data Availability Statement: All data are included in the manuscript.

Acknowledgments: We thank E.K.L. members for technical help and constructive comments. We thank the staff of the MRI for help with microscopy studies and flow cytometry. We thank T. Gostan for help with statistical analyses and Sylvie Grandemange and Claire Daien for reagents and advice. E.J.K. is an Inserm fellow.

Conflicts of Interest: The authors declare no conflict of interest. The funders had no role in the design of the study, in the collection, analyses, or interpretation of data, in the writing of the manuscript, or in the decision to publish the results.

References

1. Reizis, B.; Bunin, A.; Ghosh, H.S.; Lewis, K.L.; Sisirak, V. Plasmacytoid dendritic cells: Recent progress and open questions. *Annu. Rev. Immunol.* **2011**, *29*, 163–183. [CrossRef]
2. Geginat, J.; Nizzoli, G.; Paroni, M.; Maglie, S.; Larghi, P.; Pascolo, S.; Abrignani, S. Immunity to Pathogens Taught by Specialized Human Dendritic Cell Subsets. *Front. Immunol.* **2015**, *6*. [CrossRef] [PubMed]
3. Greber, U.F.; Arnberg, N.; Wadell, G.; Benko, M.; Kremer, E.J. Adenoviruses—From pathogens to therapeutics: A report on the 10th International Adenovirus Meeting. *Cell. Microbiol.* **2013**, *15*, 16–23. [CrossRef]
4. Harrach, B.; Benkő, M. Adenoviridae. In *Encyclopedia of Virology; Reference Module in Biomedical Sciences. Human and Animal Viruses;* Elsevier: Amsterdam, The Netherlands, 2020.
5. Khanal, S.; Ghimire, P.; Dhamoon, A.S. The repertoire of adenovirus in human disease: The innocuous to the deadly. *Biomedicines* **2018**, *6*, 30. [CrossRef] [PubMed]
6. Lion, T. Adenovirus infections in immunocompetent and immunocompromised patients. *Clin. Microbiol. Rev.* **2014**, *27*, 441–462. [CrossRef] [PubMed]

7. Yu, C.-F.; Peng, W.-M.; Oldenburg, J.; Hoch, J.; Bieber, T.; Limmer, A.; Hartmann, G.; Barchet, W.; Eis-Hubinger, A.M.; Novak, N. Human Plasmacytoid Dendritic Cells Support Th17 Cell Effector Function in Response to TLR7 Ligation. *J. Immunol.* **2010**, *184*, 1159–1167. [CrossRef]
8. Tran, T.T.P.; Eichholz, K.; Amelio, P.; Moyer, C.; Nemerow, G.R.; Perreau, M.; Mennechet, F.J.D.; Kremer, E.J. Humoral immune response to adenovirus induce tolerogenic bystander dendritic cells that promote generation of regulatory T cells. *PLoS Pathog.* **2018**, *14*, 1–31. [CrossRef] [PubMed]
9. Kremer, E.J.; Boutin, S.; Chillon, M.; Danos, O. Canine adenovirus vectors: An alternative for adenovirus-mediated gene transfer. *J. Virol.* **2000**, *74*, 505–512. [CrossRef]
10. Moyer, C.L.; Wiethoff, C.M.; Maier, O.; Smith, J.G.; Nemerow, G.R. Functional genetic and biophysical analyses of membrane disruption by human adenovirus. *J. Virol.* **2011**, *85*, 2631–2641. [CrossRef]
11. Soudais, C.; Boutin, S.; Hong, S.S.; Chillon, M.; Danos, O.; Bergelson, J.M.; Boulanger, P.; Kremer, E.J. Canine adenovirus type 2 attachment and internalization: Coxsackievirus-adenovirus receptor, alternative receptors, and an RGD-independent pathway. *J. Virol.* **2000**, *74*, 10639–10649. [CrossRef]
12. Imelli, N.; Ruzsics, Z.; Puntener, D.; Gastaldelli, M.; Greber, U.F. Genetic reconstitution of the human adenovirus type 2 temperature-sensitive 1 mutant defective in endosomal escape. *Virol. J.* **2009**, *6*, 174. [CrossRef]
13. Eichholz, K.; Bru, T.; Tran, T.T.T.P.; Fernandes, P.; Welles, H.; Mennechet, F.F.J.D.; Manel, N.; Alves, P.; Perreau, M.; Kremer, E.J.; et al. Immune-complexed adenovirus induce AIM2-mediated pyroptosis in human dendritic cells. *PLoS Pathog.* **2016**, *12*, e1005871. [CrossRef]
14. Mittereder, N.; Yei, S.; Bachurski, C.; Cuppoletti, J.; Whitsett, J.A.; Tolstoshev, P.; Trapnell, B.C. Evaluation of the efficacy and safety of in vitro, adenovirus-mediated transfer of the human cystic fibrosis transmembrane conductance regulator cDNA. *Hum. Gene Ther.* **1994**, *5*, 717–729. [CrossRef]
15. Lore, K.; Adams, W.C.; Havenga, M.J.; Precopio, M.L.; Holterman, L.; Goudsmit, J.; Koup, R.A. Myeloid and plasmacytoid dendritic cells are susceptible to recombinant adenovirus vectors and stimulate polyfunctional memory T cell responses. *J. Immunol.* **2007**, *179*, 1721–1729. [CrossRef] [PubMed]
16. Chéneau, C.; Eichholz, K.; Tran, T.H.; Tran, T.T.P.; Paris, O.; Henriquet, C.; Bajramovic, J.J.; Pugniere, M.; Kremer, E.J. Lactoferrin Retargets Human Adenoviruses to TLR4 to Induce an Abortive NLRP3-Associated Pyroptotic Response in Human Phagocytes. *Front. Immunol.* **2021**, *12*, 1894. [CrossRef] [PubMed]
17. Perreau, M.; Mennechet, F.; Serratrice, N.; Glasgow, J.N.; Curiel, D.T.; Wodrich, H.; Kremer, E.J. Contrasting Effects of Human, Canine, and Hybrid Adenovirus Vectors on the Phenotypical and Functional Maturation of Human Dendritic Cells: Implications for Clinical Efficacy. *J. Virol.* **2007**, *81*, 3272–3284. [CrossRef]
18. Perreau, M.; Pantaleo, G.; Kremer, E.J. Activation of a dendritic cell–T cell axis by Ad5 immune complexes creates an improved environment for replication of HIV in T cells. *J. Exp. Med.* **2008**, *205*, 2717–2725. [CrossRef] [PubMed]
19. Mathan, T.S.M.; Figdor, C.G.; Buschow, S.I. Human plasmacytoid dendritic cells: From molecules to intercellular communication network. *Front. Immunol.* **2013**, *4*, 372. [CrossRef] [PubMed]
20. Swiecki, M.; Colonna, M. The multifaceted biology of plasmacytoid dendritic cells. *Nat. Rev. Immunol.* **2015**, *15*, 471–485. [CrossRef]
21. Isnard, S.; Hatton, E.X.; Iannetta, M.; Guillerme, J.-B.; Hosmalin, A. Cell-Associated HIV Cross-Presentation by Plasmacytoid Dendritic Cells Is Potentiated by Noncognate CD8 + T Cell Preactivation. *J. Immunol.* **2021**, *207*, 15–22. [CrossRef]
22. Wiethoff, C.M.; Wodrich, H.; Gerace, L.; Nemerow, G.R. Adenovirus protein VI mediates membrane disruption following capsid disassembly. *J. Virol.* **2005**, *79*, 1992–2000. [CrossRef]
23. Guiducci, C.; Ott, G.; Chan, J.H.; Damon, E.; Calacsan, C.; Matray, T.; Lee, K.-D.D.; Coffman, R.L.; Barrat, F.J. Properties regulating the nature of the plasmacytoid dendritic cell response to Toll-like receptor 9 activation. *J. Exp. Med.* **2006**, *203*, 1999–2008. [CrossRef]
24. Lund, J.; Sato, A.; Akira, S.; Medzhitov, R.; Iwasaki, A. Toll-like receptor 9-mediated recognition of Herpes simplex virus-2 by plasmacytoid dendritic cells. *J. Exp. Med.* **2003**, *198*, 513–520. [CrossRef] [PubMed]
25. Perreau, M.; Welles, H.C.; Pellaton, C.; Gjoksi, B.; Potin, L.; Martin, R.; Harari, A.; Bett, A.; Casimiro, D.; Gall, J.; et al. The Number of Toll-Like Receptor 9-Agonist Motifs in the Adenovirus Genome Correlates with Induction of Dendritic Cell Maturation by Adenovirus Immune Complexes. *J. Virol.* **2012**, *86*, 6279–6285. [CrossRef] [PubMed]
26. Perreau, M.; Kremer, E.J. The conundrum between immunological memory to adenovirus and their use as vectors in clinical gene therapy. *Mol. Biotechnol.* **2006**, *34*, 247–256. [CrossRef]
27. Martinez, R.; Schellenberger, P.; Vasishtan, D.; Aknin, C.; Austin, S.; Dacheux, D.; Rayne, F.; Siebert, A.; Ruzsics, Z.; Gruenewald, K.; et al. The amphipathic helix of adenovirus capsid protein VI contributes to penton release and postentry sorting. *J. Virol.* **2015**, *89*, 2121–2135. [CrossRef]
28. Moyer, C.L.; Nemerow, G.R. Disulfide-bond formation by a single cysteine mutation in adenovirus protein VI impairs capsid release and membrane lysis. *Virology* **2012**, *428*, 41–47. [CrossRef] [PubMed]
29. Krammer, P.H.; Weyd, H. Life, death and tolerance. *Biochem. Biophys. Res. Commun.* **2017**, *482*, 470–472. [CrossRef]
30. Alculumbre, S.G.; Saint-André, V.; Di Domizio, J.; Vargas, P.; Sirven, P.; Bost, P.; Maurin, M.; Maiuri, P.; Wery, M.; Roman, M.S.; et al. Diversification of human plasmacytoid predendritic cells in response to a single stimulus article. *Nat. Immunol.* **2018**, *19*, 63–75. [CrossRef]

31. Reizis, B. Plasmacytoid Dendritic Cells: Development, Regulation, and Function. *Immunity* **2019**, *50*, 37–50. [CrossRef]
32. Ghosh, H.S.; Cisse, B.; Bunin, A.; Lewis, K.L.; Reizis, B. Continuous Expression of the Transcription Factor E2-2 Maintains the Cell Fate of Mature Plasmacytoid Dendritic Cells. *Immunity* **2010**, *33*, 905–916. [CrossRef] [PubMed]
33. Arnberg, N. Adenovirus receptors: Implications for targeting of viral vectors. *Trends Pharmacol. Sci.* **2012**, *33*, 442–448. [CrossRef] [PubMed]
34. Mennechet, F.J.D.; Paris, O.; Ouoba, A.R.; Salazar Arenas, S.; Sirima, S.B.; Takoudjou Dzomo, G.R.; Diarra, A.; Traore, I.T.; Kania, D.; Eichholz, K.; et al. A review of 65 years of human adenovirus seroprevalence. *Expert Rev. Vaccines* **2019**, *18*, 597–613. [CrossRef] [PubMed]
35. Baker, A.H.; Herzog, R.W. Did Dendritic Cell Activation, Induced by Adenovirus-Antibody Complexes, Play a Role in the Death of Jesse Gelsinger? *Mol. Ther.* **2020**, *28*, 704–706. [CrossRef]
36. Lion, T. Adenovirus persistence, reactivation, and clinical management. *FEBS Lett.* **2019**, *593*, 3571–3582. [CrossRef] [PubMed]
37. Zheng, Y.; Stamminger, T.; Hearing, P. E2F/Rb Family Proteins Mediate Interferon Induced Repression of Adenovirus Immediate Early Transcription to Promote Persistent Viral Infection. *PLoS Pathog.* **2016**, *12*, 1–24. [CrossRef] [PubMed]
38. Takahashi, K.; Asabe, S.; Wieland, S.; Garaigorta, U.; Gastaminza, P.; Isogawa, M.; Chisari, F.V. Plasmacytoid dendritic cells sense hepatitis C virus-infected cells, produce interferon, and inhibit infection. *Proc. Natl. Acad. Sci. USA* **2010**, *107*, 7431–7436. [CrossRef]
39. Decembre, E.; Assil, S.; Hillaire, M.L.B.; Dejnirattisai, W.; Mongkolsapaya, J.; Screaton, G.R.; Davidson, A.D.; Dreux, M. Sensing of Immature Particles Produced by Dengue Virus Infected Cells Induces an Antiviral Response by Plasmacytoid Dendritic Cells. *PLoS Pathog.* **2014**, *10*, e1004434. [CrossRef]
40. Baglio, S.R.; Van Eijndhoven, M.A.J.; Koppers-Lalic, D.; Berenguer, J.; Lougheed, S.M.; Gibbs, S.; Léveillé, N.; Rinkel, R.N.P.M.; Hopmans, E.S.; Swaminathan, S.; et al. Sensing of latent EBV infection through exosomal transfer of 5′pppRNA. *Proc. Natl. Acad. Sci. USA* **2016**, *113*, E587–E596. [CrossRef]
41. Vande Walle, L.; Lamkanfi, M. Pyroptosis. *Curr. Biol.* **2016**, *26*, R568–R572. [CrossRef]

viruses

MDPI

Article

In Vitro and In Vivo Evaluation of Human Adenovirus Type 49 as a Vector for Therapeutic Applications

Emily A. Bates [1,†], John R. Counsell [2,3,†], Sophie Alizert [4], Alexander T. Baker [1,5], Natalie Suff [6], Ashley Boyle [7], Angela C. Bradshaw [4], Simon N. Waddington [7,8], Stuart A. Nicklin [4], Andrew H. Baker [4,9,*] and Alan L. Parker [1,*]

1 Division of Cancer and Genetics, School of Medicine, Cardiff University, Cardiff CF14 4XN, UK; BatesE@Cardiff.ac.uk (E.A.B.); Baker.Alexander@mayo.edu (A.T.B.)
2 Genetics and Genomic Medicine Research and Teaching Department, UCL Great Ormond Street Institute of Child Health, London WC1N 1EH, UK; j.counsell@ucl.ac.uk
3 NIHR Great Ormond Street Hospital Biomedical Research Centre, 30 Guilford Street, London WC1N 1EH, UK
4 Institute of Cardiovascular and Medical Sciences, BHF Glasgow Cardiovascular Research Centre, University of Glasgow, Glasgow G12 8TA, UK; sophie.alizert84@gmail.com (S.A.); Angela.Bradshaw@glasgow.ac.uk (A.C.B.); Stuart.Nicklin@Glasgow.ac.uk (S.A.N.)
5 Center for Individualized Medicine, Mayo Clinic, Scottsdale, AZ 85259, USA
6 Department of Women and Children's Health, King's College London, St Thomas' Hospital, Westminster Bridge Road, London SE1 7EH, UK; natalie.suff@kcl.ac.uk
7 Gene Transfer Technology Group, EGA Institute for Women's Health, University College London, 86-96 Chenies Mews, London WC1E 6BT, UK; ashley.boyle@ucl.ac.uk (A.B.); s.waddington@ucl.ac.uk (S.N.W.)
8 MRC Antiviral Gene Therapy Research Unit, Faculty of Health Sciences, University of the Witswatersrand, Johannesburg 2193, South Africa
9 Queen's Medical Research Institute, University of Edinburgh, 47 Little France Crescent, Edinburgh EH16 4TJ, UK
* Correspondence: andy.baker@ed.ac.uk (A.H.B.); parkeral@cardiff.ac.uk (A.L.P.)
† These authors contributed equally to this work.

Abstract: The human adenovirus phylogenetic tree is split across seven species (A–G). Species D adenoviruses offer potential advantages for gene therapy applications, with low rates of pre-existing immunity detected across screened populations. However, many aspects of the basic virology of species D—such as their cellular tropism, receptor usage, and in vivo biodistribution profile—remain unknown. Here, we have characterized human adenovirus type 49 (HAdV-D49)—a relatively understudied species D member. We report that HAdV-D49 does not appear to use a single pathway to gain cell entry, but appears able to interact with various surface molecules for entry. As such, HAdV-D49 can transduce a broad range of cell types in vitro, with variable engagement of blood coagulation FX. Interestingly, when comparing in vivo biodistribution to adenovirus type 5, HAdV-D49 vectors show reduced liver targeting, whilst maintaining transduction of lung and spleen. Overall, this presents HAdV-D49 as a robust viral vector platform for ex vivo manipulation of human cells, and for in vivo applications where the therapeutic goal is to target the lung or gain access to immune cells in the spleen, whilst avoiding liver interactions, such as intravascular vaccine applications.

Keywords: adenovirus; viral vector; gene therapy; vaccines

1. Introduction

Human adenoviruses are non-enveloped, icosahedral viruses, divided across seven species (A–G) [1]. They have emerged as popular gene therapy vectors for therapeutic purposes, where they can be grown to high titers and efficiently transduce a range of cell types in vivo and in vitro [2]. Their diverse genetic background enables flexibility when selecting capsid serotypes with unique cell targeting profiles and host interaction characteristics, depending on the therapeutic goal of gene transfer.

Adenoviruses have been used clinically for gene supplementation, vaccination, and as oncolytic virotherapy [3]. To date, there have been over 100 adenovirus types described (http://hadvwg.gmu.edu/, accessed on 10 June 2021). When selecting an adenoviral vector for in vivo gene therapy applications, it is advantageous to select a viral platform with low levels of seroprevalence in the population. Such pre-existing immunity would otherwise hamper the efficacy of adenovirus vectors, due to rapid neutralization by the reticuloendothelial system. This is particularly relevant to species C adenoviruses, such as human adenovirus type 5 (HAdV-C5), for which neutralizing antibodies have been detected at rates of 30–90% in human populations [4,5], which result in the rapid sequestration and elimination of therapeutics based on HAdV-C5 [6]. For intravenous applications, HAdV-C5 is also hampered by interactions with host proteins, which result in efficient elimination by the liver and spleen (reviewed in [7,8]). Critical amongst such interactions is the high-affinity, Ca^{2+}-dependent interaction between the major adenoviral capsid protein—the hexon—and circulating blood clotting factor X (FX), which results in efficient, heparan sulphate proteoglycan (HSPG)-dependent transduction of liver hepatocytes [9–11]. Circumventing such interactions in the blood has required extensive genetic engineering approaches to develop heavily modified HAdV-C5-based vectors better suited to targeted intravenous approaches [12,13]. An alternative approach is to develop viral platforms with alternative receptor usage [14] and more limited interactions with blood clotting factors, such as those derived from species D [10].

Several groups have attempted to circumvent this restriction by selecting adenovirus capsids with naturally low seroprevalence rates, such as the chimpanzee adenovirus platform developed by The Jenner Institute for vaccination against SARS-CoV-2 [15,16]. However, it has been reported that some populations do harbour pre-existing immunity to chimpanzee adenoviruses, as observed in a Chinese cohort [17]. Further attempts to mitigate pre-existing immunity against adenoviruses has focused on those derived from species B or D, due to their comparative rarity in human populations [4,18,19]. This includes the species D member HAdV-D26, which is the basis of the Janssen Pharmaceuticals vaccine against SARS-CoV-2 [20], and the species B variant enadenotucirev (ColoAd1), which was developed by directed evolution of a panel of different adenovirus strains [21].

Adenovirus type 49 (HAdV-D49) is a species D member, which has also been evaluated for potential for gene therapy applications in vitro and ex vivo, owing to its particularly low seroprevalence rates. A study of a Scottish cohort failed to detect any neutralizing antibodies to HAdV-D49 at all [5], although low levels of 1–2% were detected in further screening studies in Europe [22,23]. Previous studies have highlighted HAdV-D49 as a potential vaccine vector [24], and as an effective agent for ex vivo cardiovascular gene delivery due to its efficiency in rapidly transducing endothelial and vascular smooth muscle cells [5]. Previous studies into the basic virology of HAdV-D49 suggest that it may engage CD46 as a cellular receptor [24], although its exact entry mechanism remains unclear, and a recent study indicated that the highly charged fibre knob protein may provide a novel mechanism for cell entry [25]. Overall, there remain limited investigations into the basic biology of HAdV-D49 and its human cell transduction characteristics, despite its described advantages.

Here, we have evaluated the transduction characteristics and tropism of HAdV-D49 for in vitro and in vivo manipulation of cells, to better evaluate its potential use and therapeutic exploitation.

2. Materials and Methods

2.1. Cells and Tissues

HEK 293T cells (human embryonic kidney: ATCC CRL-1573) were used for viral production, cultured in Dulbecco's modified Eagle's medium (DMEM; Invitrogen, Grand

Island, NY, USA) supplemented with 2 mM L-glutamine (Invitrogen) and 10% foetal bovine serum (FBS; PAA Laboratories, Cölbe, Germany). HepG2 (hepatocellular carcinoma: ATCC HB-8065), A549 (human lung carcinoma: ATCC CCL-185), SKOV3 (human ovarian carcinoma: ATCC HTB-77), HeLa (cervical adenocarcinoma cells: ATCC CCL-2), HT-29 (colon adenocarcinoma cells: ATCC HTB-38) and MDA435 (melanocyte: ATCC HTB-129) cells were cultured in DMEM, minimal essential medium (MEM), or RPMI-1640 medium (Invitrogen with 2 mM L-glutamine, 10% FCS, and 1 mM sodium pyruvate (Sigma-Aldrich, St. Louis, MO, USA)). Cells were maintained at 37 °C and 5% CO_2.

2.2. Adenovirus Culture

HAdV-C5 and HAdV-D49 are replication-incompetent E1/E3-deleted vectors constructed as described previously [14]. HAdV-C5/49 fibre (49F) and fibre knob (49K) pseudotypes and HVR mutant versions were generated using previously described recombineering methods [23]. Viruses were propagated in HEK 293T, E1-complementing cell lines and purified using CsCl gradients. Viral recovery was quantified by micro-BCA assay (Thermo Fisher, Loughborough, UK), assuming that 1 g protein = 4×10^9 viral particles (vp), and confirmed by NanoSight measurement (NanoSight, Malvern, UK). Infectious units (pfu) were quantified by end-point dilution plaque assay.

2.3. Luciferase Transduction Assay

Assay was performed using a commercially available luciferase assay kit (Promega). Cells were seeded into 96-well cell plates at a density of 2×10^4 cells/well in 200 µL of cell culture media and left to adhere overnight. Viruses were administered in relevant media and incubated for 3 h, before replenishing with complete media and culturing for a further 45 h. Cells were then lysed, with lysates assayed for luciferase activity according to the kit manufacturer's instructions. Luciferase activity was measured in relative light units (RLUs) using a plate reader (CLARIOstar, BMG Labtech, Aylesbury, UK). Total protein concentration was determined in the lysate using the Pierce BCA protein assay kit (Thermo Fisher, Loughborough, UK) according to the manufacturer's protocol. Transduction efficiency was expressed as RLU/mg of lysate protein.

2.4. Heparinase and Neuraminidase Transduction Assay

Cells were seeded at a density of 5×10^4 cells/well in 96-well plates. Neuraminidase (from Vibrio Cholera, Merck, Darmstadt, Germany) was added at a concentration of 50 mU/mL, whilst heparinase III (from Flavobacterium heparinum, Merck, Darmstadt, Germany) was added at a concentration of 1 U/mL, diluted in serum-free media. Adenoviruses carrying a luciferase transgene were added in serum-free media and incubated on ice for 1 h. Cells were then washed with cold PBS and replenished with complete media, before returning to incubation at 37 °C for 48 h. Cells were analysed for luciferase activity as described above.

2.5. Hemagglutination Assay

Erythrocytes were extracted from blood cells derived from a human donor, who gave informed consent. Then, 50 µL of 1% erythrocyte suspension was layered in each well of a 96-well plate, before adding 50 µL of relevant virus (1×10^9 VP) and gently mixing into the erythrocyte suspension. Haemagglutination was assessed visually.

2.6. Generation of Recombinant Fibre Knob Proteins

Recombinant fibre knob proteins were produced as described previously [14]. pQE-30 vectors, containing the coding sequence spanning 13 amino acids upstream of the TLW motif to the stop codon, were transformed into SG13009-competent cells harbouring the pREP-4 plasmid. Then, 1 L of bacterial cells was grown to OD0.6, and protein expression was induced with a final concentration of 0.5 mM IPTG. Cells were pelleted by centrifugation and resuspended in 50 mL lysis buffer (50 mM Tris, pH 8.0, 300 mM NaCl,

1% (*v*/*v*) NP40, 1 mg/mL Lysozyme, 1 mM β-mercaptoethanol). Cell lysate was then loaded to a HisTrap FF crude column and eluted with imidazole. Fractions determined to contain proteins of interest were then concentrated to <1 mL total volume and purified by size exclusion chromatography using a Superdex 200 10/300 GL increase column. Validation of recombinant knob trimerisation was performed with 5 μg and 10 μg using Coomassie staining.

2.7. Blocking of Virus Infection with Recombinant Fibre Knob Protein

Cells were seeded in 96-well plates at a density of 2×10^4 cells/well. The relevant adenovirus fibre knob was added to wells in 200 μL of cold PBS and incubated at 4 °C for 1 h. Media were then removed and adenovirus vectors containing luciferase payloads were introduced at the required dose in culture media before incubating at 4 °C for 1 h. Media were then removed, replaced with complete media, and cells incubated at 37 °C for 48 h. Transduction efficiency was measured by luciferase assay, as described above.

2.8. Animals

For in vivo studies, animal procedures were performed in strict accordance with UK Home Office guidelines. These studies were approved by the University of Glasgow Animal Procedures and Ethics Committee and performed under UK Home Office license PPL 60/4429, or by the ethical review committee of University College London under UK Home Office License PPL 70/6014. All efforts were made to minimize suffering.

For intravenous studies, outbred CD1 mice (Charles River Laboratories International, USA) were injected intravenously with the stated doses of adenoviral vectors and blinded during the course of in vivo investigations. Foetal intracranial injection was performed as previously described [26,27]. Briefly, pregnant mice carrying pups at 16 days of gestation were anaesthetized using isoflurane inhalation anaesthesia, and a midline laparotomy was performed to expose the uterus. Five microliters of vector was administered to each foetus via a transuterine injection targeting the anterior horn of the lateral ventricle of the left hemisphere of the brain. The laparotomy was sutured and mice were provided topical and systemic analgesia and allowed to recover in a warm chamber.

2.9. In Vivo Bioluminescent Imaging

Images and bioluminescence data were gathered as described previously [28]. Firefly D-luciferin (150 mg/kg) was administered to mice by intraperitoneal injection 5-min before imaging with a cooled charge-coupled device (CCD) camera (IVIS Lumina II, PerkinElmer). Detection of bioluminescence in visceral organs was performed using the auto region of interest (ROI) quantification function in Living Image 4.4 (PerkinElmer). Signal intensities were expressed as photons per second per centimeter^2 per steradian.

2.10. Luminex Quantification of Cytokine and Chemokine Levels

HAdV-C5 or HAdV-D49 (1×10^{11} vp) vectors encoding luciferase were administered by intravenous injection into outbred CD1 mice. Then, 50 μL of blood was extracted by venesection and collected in capillary tubes 6 h following virus administration. Serum was prepared by allowing blood to coagulate at room temperature for 30 min, then centrifuging at 10,000× *g* rpm for 15 min. Cytokine and chemokine analysis of sera was performed using a mouse cytokine 20-plex Luminex panel according to the manufacturer's instructions (Invitrogen, Paisley, UK), quantifying levels of basic fibroblast growth factor (bFGF), granulocyte macrophage colony-stimulating factor (GM-CSF), interferon-gamma (IFN-γ), interleukin (IL)-1α, IL-1β, IL-2, IL-4, IL-5, IL-6, IL-10, IL-12(p40/p70), IL-13, IL-17, IFN-induced protein (IP10), keratinocyte-derived cytokine (KC), monocyte chemoattractant protein (MCP-1), monokine induced by gamma interferon (MIG), macrophage inflammatory protein-1alpha (MIP-1α), tumour necrosis factor-alpha (TNF-α), and vascular endothelial growth factor (VEGF). Data were analysed using Bio-Plex manager software with 5PL curve fitting.

2.11. Calculation of Vector Genome Copies in Mouse Organs

Genomic DNA was extracted from the organ of interest following tissue harvest after termination of in vivo experiments. DNA was extracted from tissue using the Qiagen DNeasy blood and tissue kit. Vector genome copies were quantified by qPCR, targeted to the adenoviral vector genome, and expressed per 50 ng of extracted DNA.

2.12. Statistical Analysis

Data presented are derived from a minimum of two experimental replicates per group, unless otherwise stated. Transduction graphs are displayed with a log scale, and for these experiments data were log transformed before analysis. Statistical significance was calculated using ANOVA or Student's *t*-test; $p < 0.05$ was considered statistically significant.

3. Results

Our experiments focused on the in vitro and in vivo transduction characteristics of HAdV-D49 vectors, with relevance to gene therapy applications.

3.1. HAdV-D49 Transduces a Broad Range of Cell Types In Vitro

We initially profiled the transduction efficiency of HAdV-D49 in a range of cell types to gauge its potential use for cellular gene transfer in vitro. To evaluate the efficiency of transduction, we performed assays at two doses of HAdV—a standard dose of 1000 vp/cell (Figure 1A), and a higher dose of 10,000 vp/cell (Figure 1B). Expression of luciferase transgene highlighted the efficient transduction of several immortalized cell types mediated by HAdV-D49, including alveolar basal endothelial cells (A549 cells), colon adenocarcinoma cells (HT29), cervical adenocarcinoma cells (Hela), melanocytes (MDA435), and ovarian carcinoma cells (SKOV3). In the majority of cases, HAdV-D49 demonstrated a significantly increased transduction efficiency compared to HAdV-C5.

In light of the above findings, we investigated whether FX binding could further enhance HAdV-D49-mediated transduction. A previous study indicated that HAdV-D49 forms an unstable complex with FX [10] and, therefore, it remained unclear how this unstable complex might impact on cellular transduction efficiency. Cells were transduced with either HAdV-D49 or HAdV-C5 vectors expressing luciferase in serum-free media, in the presence or absence of physiological concentrations of FX (10 μg/mL), with expression determined by bioluminescent signal from the luciferase transgene (Figure 1C). HAdV-D49 transduction was only significantly increased by the presence of FX in MDA 435 cells, possibly reflecting high expression of heparan sulphate proteoglycan (HSPGs)—known to mediate FX-mediated viral cell entry—in MDA-435 cells. FX consistently enhanced the transduction of HAdV-C5, with a significant increase for three cell lines (A549, MDA 435, and SKOV3). This indicates that FX plays a variable role in HAdV-D49 infection; however, the presence of FX is not a requirement for cell entry, and nor does its presence enhance transduction mediated by HAdV-D49 in all cell lines tested.

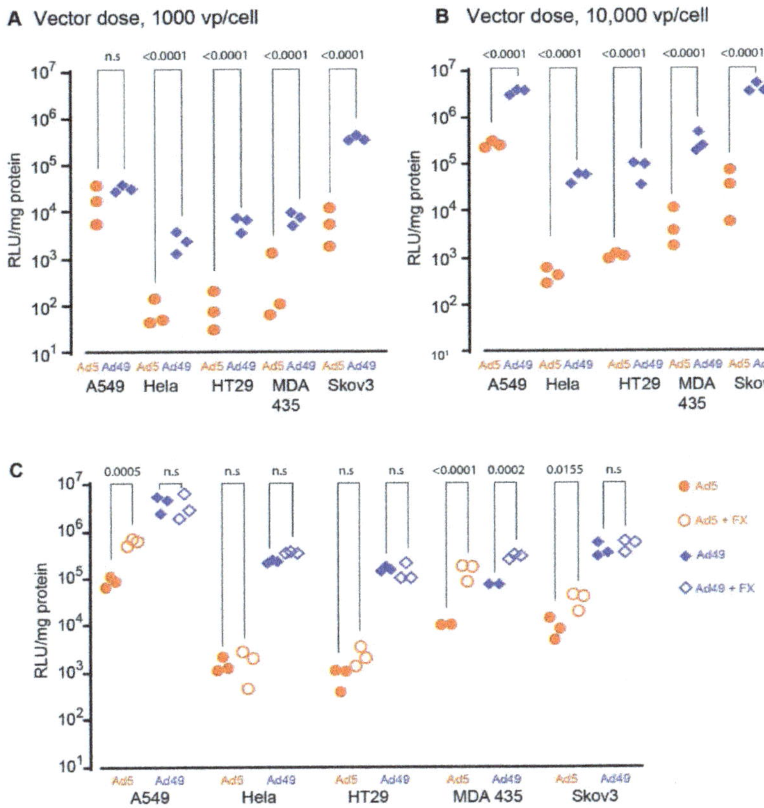

Figure 1. Profiling of HAdV-D49 transduction in cancer cell lines in vitro. Transduction of HAdV-C5 (shown as Ad5) and HAdV-D49 (shown as Ad49) in A549, HeLa, HT20, MDA 435, and SKOV3 cell lines at (**A**) 1000 vp/cell and (**B**) 10,000 vp/cell. Luciferase expression was measured 48 h post-infection. (**C**) Transduction of the same cell lines using 5000 vp/cell, in either the presence or absence of recombinant FX. n.s: not statistically significant, $p > 0.05$.

3.2. In Vivo Biodistribution Profiling Shows That FX Interaction Does Not Confer Increased Liver Targeting

We next investigated the in vivo biodistribution of HAdV-D49 vectors in mice, to determine which visceral organs were targeted by systemic delivery. We delivered HAdV-D49 vectors carrying a luciferase transgene to mice via intravenous injection, and quantified vector genome copy numbers and luciferase expression in a range of target organs 48 h later, comparing their distribution to that of HAdV-C5 (Figure 2). As expected, HAdV-C5 vectors showed strong liver tropism (Figure 2A). Despite a possible interaction with FX, HAdV-D49 did not appear to transduce the liver (Figure 2B), but instead demonstrated increased uptake in the spleen (Figure 2C). Quantification of vector genomes recovered from the lung, liver, spleen, heart, kidney, and pancreas support these findings (Figure 2D). HAdV-D49 shows the highest uptake in the lung and spleen, with significantly lower uptake in the liver compared to HAdV-C5. We then considered the effect of significantly higher levels of HAdV-D49 in the spleen on cytokine production (Figure 2E and Figure S1). Evaluation of cytokine profiles from mouse plasma shows that HAdV-D49 treatment significantly increases a panel of inflammatory cytokines, including FGF, MCP1, MIG, IFN-gamma, IL1-beta, IL-2, IL-5, IL-6, IL-12, and IP-10.

Increased cytokine activation coupled with reduced liver targeting suggests that HAdV-D49 possesses advantages over HAdV-C5 for use in gene therapy or vaccine applications.

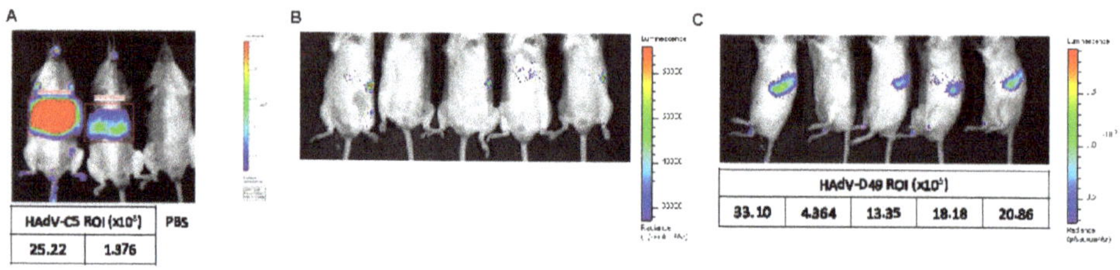

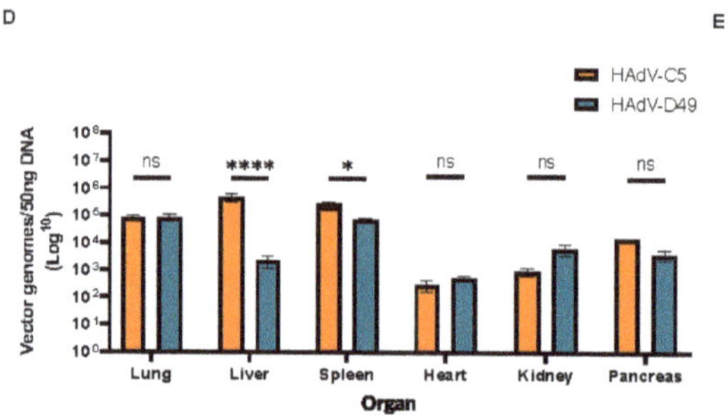

E

Cytokine	HAdV-C5 (pg/ml)	HAdV-D49 (pg/ml)	Fold Increase	Significance
VEGF	1.33	0.45	0.34	ns
FGF	227.61	254.34	1.12	*
MCP1	1125.64	2798.23	2.49	**
MIG	46.44	2606.41	56.12	*
TNFα	10.60	22.28	2.10	ns
MIP1α	63.16	63.87	1.01	ns
GM-CSF	13.40	14.60	1.09	ns
IFN-γ	53.81	895.53	16.64	*
IL-1α	35.23	50.34	1.43	ns
IL-1β	138.44	385.37	2.78	*
IL-2	32.52	42.68	1.31	*
IL-4	39.36	40.25	1.02	ns
IL-5	129.95	260.85	2.01	**
IL-6	46.67	229.01	4.91	**
IL-12	107.05	277.47	2.59	***
IP-10	1181.08	2223.06	1.88	***

Figure 2. In vivo biodistribution analysis of HAdV-D49 vectors compared to HAdV-C5 vectors. IVIS imaging system used for in vivo quantification of (**A**) HAdV-C5 (ROI $\times 10^8$) liver transduction with PBS control; (**B**) IVIS images of HAdV-D49 (shown as Ad49) transduction with mice imaged on their backs, and (**C**) IVIS images of HAdV-D49 (Ad49) transduction when mice are imaged on their sides. Uptake in the spleen was quantified using ROI ($\times 10^5$). (**D**) Quantification of biodistribution using a dose of 1×10^{11} virus particles. qPCR to evaluate vector genomes in the lung, liver, spleen, heart, kidney, and pancreas. (**E**) Cytokine concentrations, gauged by Luminex 20-plex cytokine analysis in mouse plasma after HAdV-C5 (Ad5) or HAdV-D49 (Ad49) administration (shown as pg/mL), and demonstrated as fold increase of HAdV-D49 in comparison to HAdV-C5. Individual plots are shown in Figure S1. ns, not statistically significant, $p > 0.05$; *, $p < 0.05$; ** $p < 0.01$; ***, $p < 0.001$; ****, $p < 0.0001$.

3.3. HAdV-D49 Fibre Alone Does Not Mediate Liver Transduction In Vivo

To further dissect the mechanistic basis of the decreased hepatic transduction and increased splenic transduction observed following intravascular administration of HAdV-D49 compared to HAdV-C5, we utilised a panel of HAdV-C5/49-pseudotyped vectors (Figure 3). To assess whether this altered distribution results from the interaction between the HAdV-D49 primary receptors and the fibre knob protein, we developed HAdV-C5 vectors pseudotyped with either the fibre knob protein alone from HAdV-D49 (HAdV-C5/K49), or with the whole fibre (HAdV-C5/F49). Since hepatic transduction of HAdV-C5 is known to be mediated by the high-affinity interaction between the HAdV-C5 hexon protein and FX, we also generated mutants with substitutions of critical FX binding residues in HVR7 to abrogate FX binding and, thus, limit liver transduction. As expected, HAdV-C5 vectors with a WT hexon showed higher transduction of the liver compared to those with ablation of the FX binding (HVR mutant). Interestingly, when the whole HAdV-C5 fibre was replaced by HAdV-D49 whole fibre, no uptake in the liver was observed for either the WT hexon or the HVR mutant. However, liver transduction was restored when only the HAdV-D49 fibre knob was swapped. No liver transduction was observed in HAdV-C5/K49 when HVR was mutated; this indicates that the fibre alone does not mediate splenic transduction, and HAdV-D49 most likely requires additional components, beyond its fibre, to define its tropism in vivo. In addition, this data suggests that the HAdV-D49 fibre knob does not significantly alter HAdV-C5 biodistribution when administered intravenously, and is able to mediate transduction in the liver.

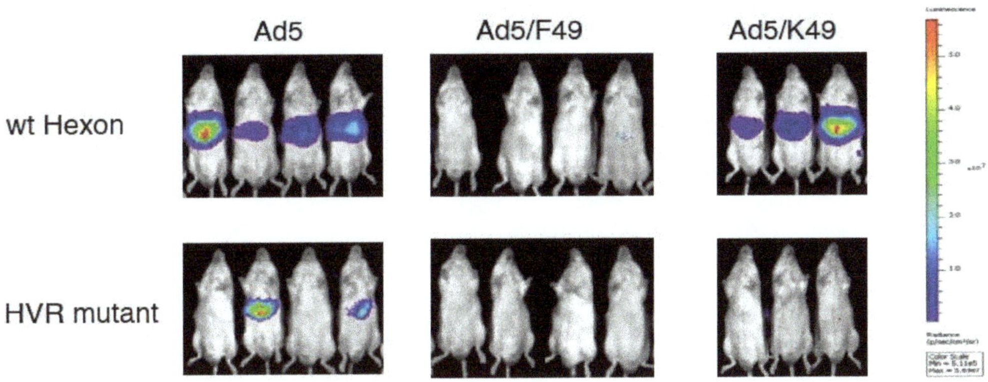

	HAdV-C5 ROI (x10^7)			
wt hexon	201	11.3	50.3	73.7
HVR mutant	0.095	1.33	0.081	0.65

	HAdV-C5/F49 ROI (x10^5)			
wt hexon	19.6	2.76	2.23	11.0
HVR mutant	1.21	1.13	2.05	1.29

	HAdV-C5/K49 ROI (x10^7)		
wt hexon	8.82	32.3	209
HVR mutant	0.0177	0.0198	0.233

Figure 3. Evaluation of the HAdV-D49 fibre and fibre knob in vivo. Transduction of HAdV-C5 and HAdV-C5/49 fibre (F49) and fibre knob (49K) pseudotypes in vivo, with and without the hexon mutation (HVR mutant), to reduce liver uptake with ROI indicated in tables.

3.4. HAdV-D49 Show Similar Levels of Intracranial Transduction to HAdV-C5

In order to evaluate the potential of HAdV-D49 as a platform for neurological gene transfer applications, MF1 mice were injected intracranially in utero with 1×10^9 vp of either HAdV-C5- or HAdV-D49-expressing luciferase. At three months of age, mice were IVIS imaged for luciferase activity (Figure 4). Mice treated with HAdV-D49.Luc mediated readily detectable levels of transduction in the brain, although the levels achieved were lower than those treated with HAdV-C5.

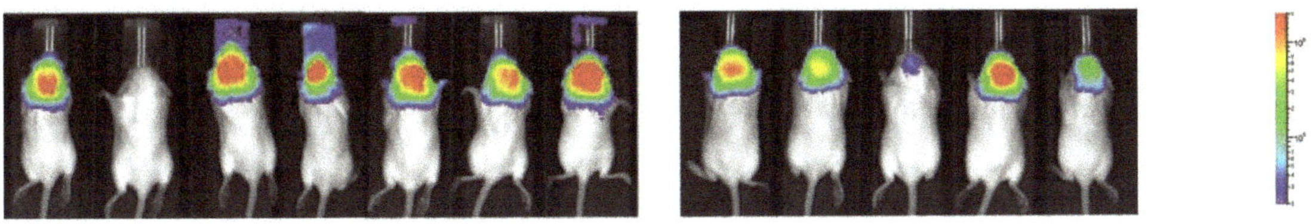

HAdV-C5 Total Flux (x10^5 p/s)						
185	1.04	320	146	269	164	512

HAdV-D49 Total Flux (x10^5 p/s)				
101	37.8	3.29	232	17.0

Figure 4. In vivo bioluminescence imaging of luciferase expression mediated by HAdV-C5 (Ad5) vectors compared to HAdV-D49 (Ad49) vectors after foetal intracranial injection. IVIS imaging system was performed at three months of age. Quantification of bioluminescence is shown in table (total flux $\times 10^5$ p/s).

3.5. HAdV-D49 Infection of Target Cells Does Not Appear to Be Attributed to a Single Cell Surface Marker

We have demonstrated that HAdV-D49 exhibits a broad tropism both in vitro and in vivo; we therefore aimed to further our understanding of the mechanism underpinning HAdV-D49 receptor usage. We initially investigated the extent to which HAdV-D49 can engage CAR for transduction of hepatocytes (HepG2). We first demonstrated that we were able to produce HAdV-C5 and HAdV-D49 recombinant knob proteins in a trimeric state (Figure 5A). For these studies, CAR binding was blocked using recombinant HAdV-C5 knob protein prior to transduction (Figure 5B). As expected, HAdV-C5 knob treatment clearly blocked transduction mediated by HAdV-C5; however, the same treatment had no effect on HAdV-D49 transduction. In addition, recombinant HAdV-D49 knob was also used to block cellular receptors. Interestingly, recombinant HAdV-D49 knob was able to inhibit transduction mediated by HAdV-C5, indicating that HAdV-D49 knob protein can bind and block CAR—albeit less efficiently that recombinant HAdV-C5. Interestingly, recombinant HAdV-D49 knob protein was not able to inhibit HAdV-D49-mediated cell infectivity, except poorly at the highest concentration. This indicates that HAdV-D49 is not solely dependent on binding of its fibre knob protein to cells for transduction of cells, and that other capsid proteins may be critical.

We observed that HAdV-D49 haemagglutinates human serum at high concentrations (10^9 vp/well), indicating that it may bind to CAR or sialic acid, like other species D adenoviruses [29] (Figure 5C). We investigated this potential targeting of sialic acid as well as heparan sulphate proteoglycans (HSPGs), given that these have been identified as interacting partners for adenovirus transduction [25,29–33]. To probe this, hepatocytes (HepG2 cells) were transduced following heparinase and neuraminidase treatment to cleave HSPGs and sialic acid, respectively. Whilst the removal of HSPGs by heparinase demonstrated a small but significant reduction in HAdV-D49-mediated transduction (Figure 5D), no effect was observed following treatment with neuraminidase, indicating that sialic acid is not essential for transduction (Figure 5E).

Finally, we investigated CD46 interaction, as it has previously been described as having a potential involvement in HAdV-D49 infection [15,23]. CD46 was blocked on A549, HeLa, and SKOV3 cells using a specific monoclonal antibody (MEM-258) prior to transduction with HAdV-C5, HAdV-D49, and HAdV-B35 (Figure 5F). HAdV-D49 infection was reduced significantly in all three cell lines when CD46 was blocked. HAdV-B35 saw significant decreases in transduction in all three cell lines, whereas blocking CD46 had a minimal effect on HAdV-C5 infection. These data suggest that CD46 is a possible receptor for HAdV-D49, consistent with the previous data; however, it is not likely to be the sole receptor, as HAdV-D49 remains able to readily transduce cells even when CD46 is blocked.

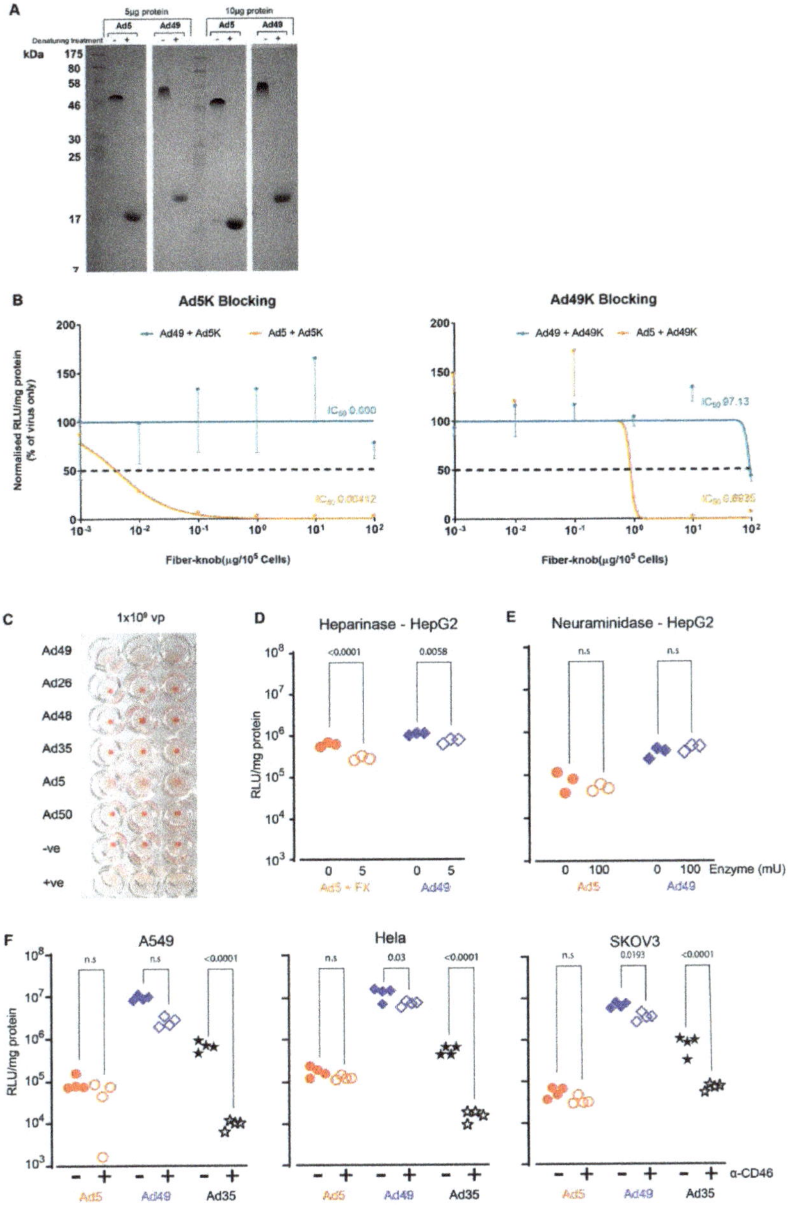

Figure 5. HAdV-D49 can interact with multiple receptors to transduce target cells. (**A**) Recombinant HAdV-C5K and HAdV-D49K with or without treatment using a denaturing agent were stained with Coomassie. (**B**) Blocking of HAdV-C5 (Ad5) and HAdV-D49 (Ad49) transduction in hepatocytes (HepG2) with dilution of HAdV-C5 or HAdV-D49 fibre knob (100–0.001 μg/10^5 cells). (**C**) Haemagglutination profiling of several adenovirus serotypes, including HAdV-C5 and HAdV-D49. Negative (-ve, PBS) and positive (+ve, PBS containing 0.2% Tween-20) were also used as relevant controls; three wells are shown reflecting technical replicates. (**D**) HAdV-C5 + FX and HAdV-D49 transduction in HepG2 cells after treatment with heparinase. (**E**) HAdV-C5 (shown as Ad5) and HAdV-D49 transduction in HepG2 cells after treatment with neuraminidase. (**F**) Transduction of HAdV-C5, HAdV-D49, and HAdV-B35 (Ad35) in A549, HeLa, and SKOV3 cells following blocking with 258-MEM, anti-CD46 antibody. n.s; not statistically significant, $p > 0.05$.

4. Discussion

Adenoviruses have been widely used in gene therapy for a variety of applications, owing to their diversity and efficiency in delivering therapeutic DNA cassettes to the nuclei of cells in vitro and in vivo. Common applications of adenoviruses include oncolytic virotherapy [1], ex vivo cell manipulation [5], and vaccine delivery [16,20,34,35], whilst recently they have been used pre-clinically to mediate in utero gene editing of disease-causing mutations in foetal mice [36]. Clinical translation of adenovirus therapies could be accelerated by greater understanding of their biology and interactions with human proteins. Additionally, many of the current adenovirus platforms are restricted by the prevalence of neutralizing antibodies against adenoviral capsid proteins, which restrict their delivery to the desired cell types in vivo.

Here, we have shown that HAdV-D49 vectors can be used to deliver genes to a wide range of target cells, both in vitro and in vivo. We identified that HAdV-D49 showed enhanced transduction of hepatocytes in vitro in the presence of human serum, which was linked to a potential interaction with human FX. Given previous reports of HAdV-C5 utilizing FX as a partner for enhanced interaction with its CAR receptor [10,37], we expected a similar concept for HAdV-D49, although this was not an absolute requirement. We observed that FX significantly increased HAdV-D49 transduction in MDA-435 cells alone; we consider whether this can be attributed to varying levels of HSPGs on the cells' surfaces. However, as there was no significant increase observed in other cell lines tested, we conclude that FX is not essential in mediating HAdV-D49 transduction. Furthermore, the fibre knob domain of HAdV-D49 did not block HAdV-D49 entry into hepatocytes, suggesting that further interacting partners may be involved. This reaffirms the apparent complexity of HAdV-D49 tropism and interactions with human cells during transduction.

Evaluation of HAdV-D49 biodistribution was efficient in mediating gene transfer, predominantly to the spleen, in vivo, with surprisingly low levels of vector genomes detected in the liver of treated mice considering the potential for FX engagement. The high uptake of HAdV-D49 by the spleen was correlated with an increase in inflammatory cytokine release. Overall, this indicates that HAdV-D49 is a versatile vector with potential usage for a range of in vivo and ex vivo gene therapy and vaccine applications.

We extended our studies deeper into the basic biology of HAdV-D49 vectors, by probing the potential mechanism for its transduction of target cells. Previous studies have suggested that HAdV-D49 utilizes CD46 as its primary cellular receptor [24]. Our data suggest that, although CD46 is interacted with, it is not the exclusive receptor. A recent study describes a number of species D serotypes that are able to engage CD46 directly via the hexon [30]. Although this has not yet been described for HAdV-D49, it should be considered as a mechanism for CD46 interaction. Furthermore, we did not see evidence of a single cell adhesion molecule being absolutely required for HAdV-D49 transduction, instead identifying multiple interacting partners that may be utilized for cell entry. Future studies to unpick the complex tropism of HAdV-D49 will require combinations of inhibitors to fully delineate whether HAdV-D49 uses multiple previously described receptors and/or a completely novel means of cell entry. Evidence of potential interactions with CAR, CD46, HSPGs, and FX were all evident in our data, but no single factor was deemed to be necessary and sufficient for transduction. The broad transduction profile of HAdV-D49 is therefore potentially explained by its apparent promiscuous interaction profile, maintaining efficient transduction across a range of cell types by establishing a repertoire of interaction mechanisms, for greater flexibility over other adenovirus types.

A particularly intriguing aspect of our study was the paradoxical cell tropism of HAdV-D49 in vitro versus in vivo, where it showed efficient transduction of many cell types in vitro, including liver cells, whereas in vivo its biodistribution was largely restricted to lung and spleen, with minimal transduction of liver compared to HAdV-C5. Further studies will be required to identify which specific subsets of cells within the spleen are efficiently transduced by HAdV-D49, but it is likely that increased uptake by immune cells within the spleen correlates with the enhanced innate immune responses we noted following intravascular delivery of HAdV-D49 compared to HAdV-C5. Given that all in vitro studies were conducted in human cells in the presence of human serum or human FX protein, it is reasonable to conclude that efficient HAdV-D49 transduction shows species dependency in its interactions with cell receptors and co-receptors. In order to maximize HAdV-D49 utilization in in vivo gene therapy applications, further work should be aimed at interrogating the species requirements of its cellular interactions, in order to identify the crucial determinants for entry. Indeed, in the context of ex vivo gene therapy, HAdV-D49 clearly shows promise as a versatile and robust gene transfer agent, with exciting potential in pre-clinical and clinical applications.

Supplementary Materials: The following are available online at https://www.mdpi.com/1999-4915/13/8/1483/s1, Figure S1: Luminex quantification of cytokine and chemokine levels.

Author Contributions: Conceptualization, A.L.P. and A.H.B.; methodology, E.A.B., A.C.B., S.N.W., S.A.N., A.L.P., and A.H.B.; investigation, E.A.B., J.R.C., S.A., A.T.B., N.S., A.C.B., S.N.W., S.A.N., and A.L.P.; data curation, E.A.B., J.R.C., S.N.W., A.H.B., A.B. and A.L.P.; formal analysis, E.A.B., J.R.C., A.C.B., S.N.W., S.A.N., A.L.P., and A.H.B.; writing—original draft preparation, E.A.B., J.R.C., A.H.B., and A.L.P.; writing—review and editing, S.A., A.T.B., A.C.B., S.N.W., S.A.N., A.L.P., and A.H.B.; funding acquisition, S.N.W., A.H.B., and A.L.P. All authors have read and agreed to the published version of the manuscript.

Funding: E.A.B. is supported by a Cardiff University School of Medicine PhD studentship AC1170AP02 (to ALP) and by Cancer Research UK Experimental Cancer Medicine Centre funding (C7838/A25173). J.R.C. received funding from NIHR GOSH BRC grant 17BX23. A.T.B. was funded by a Tenovus Cancer Care Ph.D. studentship (reference PhD2015/L13) to A.L.P. A.C.B. was supported by a Personal Research Fellowship from the Royal Society of Edinburgh (RSE/33457). N.S. received funding from the Wellbeing of Women Research training fellowship grant RT414 and the Priory foundation and Action Medical Research grant GN2647. ALP received funding from Cancer Research UK (C52915/A29104).

Institutional Review Board Statement: All in vivo animal studies were performed in strict accordance with UK Home Office guidelines under University of Glasgow Animal Procedures and Ethics Committee and performed under UK Home Office license PPL 60/4429 (awarded from 24th October 2012), or under UK Home Office regulations and approved by the ethical review committee of University College London under UK Home Office License PPL 70/6014 (dated 19th February 2014). Every effort was made to minimize suffering.

Informed Consent Statement: Not applicable.

Data Availability Statement: The datasets generated during and/or analysed during the current study are available from the corresponding authors on reasonable request.

Acknowledgments: We thank Nicola Britton and Gregor Aitchison of the University of Glasgow for technical support. We are grateful to Ryan Ritchie and the Institute of Infection, Immunity, and Inflammation (University of Glasgow) imaging facility for assistance in providing support accessing IVIS imaging facilities and data.

Conflicts of Interest: The authors declare no conflict of interest. The funders had no role in the design of the study, in the collection, analyses, or interpretation of data, in the writing of the manuscript, or in the decision to publish the results.

References

1. Baker, A.T.; Aguirre-Hernández, C.; Halldén, G.; Parker, A.L. Designer Oncolytic Adenovirus: Coming of Age. *Cancers* **2018**, *10*, 201. [CrossRef]
2. Lee, C.S.; Bishop, E.S.; Zhang, R.; Yu, X.; Farina, E.M.; Yan, S.; Zhao, C.; Zeng, Z.; Shu, Y.; Wu, X.; et al. Adenovirus-Mediated Gene Delivery: Potential Applications for Gene and Cell-Based Therapies in the New Era of Personalized Medicine. *Genes Dis.* **2017**, *4*, 43–63. [CrossRef]
3. Wold, W.S.M.; Toth, K. Adenovirus Vectors for Gene Therapy, Vaccination and Cancer Gene Therapy. *Curr. Gene Ther.* **2013**, *13*, 421–433. [CrossRef] [PubMed]

4. Abbink, P.; Lemckert, A.A.C.; Ewald, B.A.; Lynch, D.M.; Denholtz, M.; Smits, S.; Holterman, L.; Damen, I.; Vogels, R.; Thorner, A.R.; et al. Comparative Seroprevalence and Immunogenicity of Six Rare Serotype Recombinant Adenovirus Vaccine Vectors from Subgroups B and D. *J. Virol.* **2007**, *81*, 4654 LP–4663. [CrossRef] [PubMed]
5. Dakin, R.S.; Parker, A.L.; Delles, C.; Nicklin, S.A.; Baker, A.H. Efficient Transduction of Primary Vascular Cells by the Rare Adenovirus Serotype 49 Vector. *Hum. Gene Ther.* **2015**, *26*, 312–319. [CrossRef] [PubMed]
6. Parker, A.L.; Waddington, S.N.; Buckley, S.M.K.; Custers, J.; Havenga, M.J.E.; van Rooijen, N.; Goudsmit, J.; McVey, J.H.; Nicklin, S.A.; Baker, A.H. Effect of Neutralizing Sera on Factor X-Mediated Adenovirus Serotype 5 Gene Transfer. *J. Virol.* **2009**, *83*, 479–483. [CrossRef]
7. Cunliffe, T.G.; Bates, E.A.; Parker, A.L. Hitting the Target but Missing the Point: Recent Progress towards Adenovirus-Based Precision Virotherapies. *Cancers* **2020**, *12*, E3327. [CrossRef]
8. Parker, A.L.; Nicklin, S.A.; Baker, A.H. Interactions of Adenovirus Vectors with Blood: Implications for Intravascular Gene Therapy Applications. *Curr. Opin. Mol.* **2008**, *10*, 439–448.
9. Parker, A.L.; Waddington, S.N.; Nicol, C.G.; Shayakhmetov, D.M.; Buckley, S.M.; Denby, L.; Kemball-Cook, G.; Ni, S.; Lieber, A.; McVey, J.H.; et al. Multiple Vitamin K-Dependent Coagulation Zymogens Promote Adenovirus-Mediated Gene Delivery to Hepatocytes. *Blood* **2006**, *108*, 2554–2561. [CrossRef]
10. Waddington, S.N.; McVey, J.H.; Bhella, D.; Parker, A.L.; Barker, K.; Atoda, H.; Pink, R.; Buckley, S.M.K.; Greig, J.A.; Denby, L.; et al. Adenovirus Serotype 5 Hexon Mediates Liver Gene Transfer. *Cell* **2008**, *132*, 397–409. [CrossRef]
11. Kalyuzhniy, O.; Di Paolo, N.C.; Silvestry, M.; Hofherr, S.E.; Barry, M.A.; Stewart, P.L.; Shayakhmetov, D.M. Adenovirus Serotype 5 Hexon Is Critical for Virus Infection of Hepatocytes In Vivo. *Proc. Natl. Acad. Sci. USA* **2008**, *105*, 5483–5488. [CrossRef] [PubMed]
12. Uusi-Kerttula, H.; Davies, J.A.; Thompson, J.M.; Wongthida, P.; Evgin, L.; Shim, K.G.; Bradshaw, A.; Baker, A.T.; Rizkallah, P.J.; Jones, R.; et al. Ad5NULL-A20: A Tropism-Modified, Avβ6 Integrin-Selective Oncolytic Adenovirus for Epithelial Ovarian Cancer Therapies. *Clin. Cancer Res.* **2018**, *24*, 4215–4224. [CrossRef] [PubMed]
13. Davies, J.A.; Marlow, G.; Uusi-Kerttula, H.K.; Seaton, G.; Piggott, L.; Badder, L.M.; Clarkson, R.W.E.; Chester, J.D.; Parker, A.L. Efficient Intravenous Tumor Targeting Using the Avβ6 Integrin-Selective Precision Virotherapy Ad5NULL-A20. *Viruses* **2021**, *13*, 864. [CrossRef]
14. Baker, A.T.; Greenshields-Watson, A.; Coughlan, L.; Davies, J.A.; Uusi-Kerttula, H.; Cole, D.K.; Rizkallah, P.J.; Parker, A.L. Diversity within the Adenovirus Fiber Knob Hypervariable Loops Influences Primary Receptor Interactions. *Nat. Commun.* **2019**, *10*, 741. [CrossRef] [PubMed]
15. Dicks, M.D.J.; Spencer, A.J.; Edwards, N.J.; Wadell, G.; Bojang, K.; Gilbert, S.C.; Hill, A.V.S.; Cottingham, M.G. A Novel Chimpanzee Adenovirus Vector with Low Human Seroprevalence: Improved Systems for Vector Derivation and Comparative Immunogenicity. *PLoS ONE* **2012**, *7*, e40385. [CrossRef]
16. Voysey, M.; Clemens, S.A.C.; Madhi, S.A.; Weckx, L.Y.; Folegatti, P.M.; Aley, P.K.; Angus, B.; Baillie, V.L.; Barnabas, S.L.; Bhorat, Q.E.; et al. Safety and Efficacy of the ChAdOx1 NCoV-19 Vaccine (AZD1222) against SARS-CoV-2: An Interim Analysis of Four Randomised Controlled Trials in Brazil, South Africa, and the UK. *Lancet* **2021**, *397*, 99–111. [CrossRef]
17. Zhao, H.; Xu, C.; Luo, X.; Wei, F.; Wang, N.; Shi, H.; Ren, X. Seroprevalence of Neutralizing Antibodies against Human Adenovirus Type-5 and Chimpanzee Adenovirus Type-68 in Cancer Patients. *Front. Immunol.* **2018**, *9*, 335. [CrossRef]
18. Teigler, J.E.; Iampietro, M.J.; Barouch, D.H. Vaccination with Adenovirus Serotypes 35, 26, and 48 Elicits Higher Levels of Innate Cytokine Responses than Adenovirus Serotype 5 in Rhesus Monkeys. *J. Virol.* **2012**, *86*, 9590–9598. [CrossRef]
19. Weaver, E.A.; Barry, M.A. Low Seroprevalent Species D Adenovirus Vectors as Influenza Vaccines. *PLoS ONE* **2013**, *8*, e73313. [CrossRef]
20. Sadoff, J.; Gray, G.; Vandebosch, A.; Cárdenas, V.; Shukarev, G.; Grinsztejn, B.; Goepfert, P.A.; Truyers, C.; Fennema, H.; Spiessens, B.; et al. Safety and Efficacy of Single-Dose Ad26.COV2.S Vaccine against Covid-19. *N. Engl. J. Med.* **2021**, *384*, 2187–2201. [CrossRef]
21. Dyer, A.; Di, Y.; Calderon, H.; Illingworth, S.; Kueberuwa, G.; Tedcastle, A.; Jakeman, P.; Chia, S.L.; Brown, A.; Silva, M.A.; et al. Oncolytic Group B Adenovirus Enadenotucirev Mediates Non-Apoptotic Cell Death with Membrane Disruption and Release of Inflammatory Mediators. *Mol. Ther. Oncolytics* **2016**, *4*, 18–30. [CrossRef]
22. De Jong, J.C.; Wermenbol, A.G.; Verweij-Uijterwaal, M.W.; Slaterus, K.W.; Wertheim-Van Dillen, P.; Van Doornum, G.J.J.; Khoo, S.H.; Hierholzer, J.C. Adenoviruses from Human Immunodeficiency Virus-Infected Individuals, Including Two Strains That Represent New Candidate Serotypes Ad50 and Ad51 of Species B1 and D, Respectively. *J. Clin. Microbiol.* **1999**, *37*, 3940–3945. [CrossRef]
23. Vogels, R.; Zuijdgeest, D.; van Rijnsoever, R.; Hartkoorn, E.; Damen, I.; de Béthune, M.-P.; Kostense, S.; Penders, G.; Helmus, N.; Koudstaal, W.; et al. Replication-Deficient Human Adenovirus Type 35 Vectors for Gene Transfer and Vaccination: Efficient Human Cell Infection and Bypass of Preexisting Adenovirus Immunity. *J. Virol.* **2003**, *77*, 8263–8271. [CrossRef]
24. Lemckert, A.A.C.; Grimbergen, J.; Smits, S.; Hartkoorn, E.; Holterman, L.; Berkhout, B.; Barouch, D.H.; Vogels, R.; Quax, P.; Goudsmit, J.; et al. Generation of a Novel Replication-Incompetent Adenoviral Vector Derived from Human Adenovirus Type 49: Manufacture on PER.C6 Cells, Tropism and Immunogenicity. *J. Gen. Virol.* **2006**, *87*, 2891–2899. [CrossRef] [PubMed]
25. Baker, A.T.; Davies, J.A.; Bates, E.A.; Moses, E.; Mundy, R.M.; Marlow, G.; Cole, D.K.; Bliss, C.M.; Rizkallah, P.J.; Parker, A.L. The Fiber Knob Protein of Human Adenovirus Type 49 Mediates Highly Efficient and Promiscuous Infection of Cancer Cell Lines Using a Novel Cell Entry Mechanism. *J. Virol.* **2021**, *95*, e01849-20. [CrossRef] [PubMed]

26. Massaro, G.; Mattar, C.N.Z.; Wong, A.M.S.; Sirka, E.; Buckley, S.M.K.; Herbert, B.R.; Karlsson, S.; Perocheau, D.P.; Burke, D.; Heales, S.; et al. Fetal Gene Therapy for Neurodegenerative Disease of Infants. *Nat. Med.* **2018**, *24*, 1317–1323. [CrossRef]
27. Rahim, A.A.; Wong, A.M.; Ahmadi, S.; Hoefer, K.; Buckley, S.M.K.; Hughes, D.A.; Nathwani, A.N.; Baker, A.H.; McVey, J.H.; Cooper, J.D.; et al. In Utero Administration of Ad5 and AAV Pseudotypes to the Fetal Brain Leads to Efficient, Widespread and Long-Term Gene Expression. *Gene* **2012**, *19*, 936–946. [CrossRef] [PubMed]
28. Buckley, S.M.K.; Delhove, J.M.K.M.; Perocheau, D.P.; Karda, R.; Rahim, A.A.; Howe, S.J.; Ward, N.J.; Birrell, M.A.; Belvisi, M.G.; Arbuthnot, P.; et al. In Vivo Bioimaging with Tissue-Specific Transcription Factor Activated Luciferase Reporters. *Sci. Rep.* **2015**, *5*, 11842. [CrossRef]
29. Baker, A.T.; Mundy, R.M.; Davies, J.A.; Rizkallah, P.J.; Parker, A.L. Human Adenovirus Type 26 Uses Sialic Acid–Bearing Glycans as a Primary Cell Entry Receptor. *Sci. Adv.* **2019**, *5*, eaax3567. [CrossRef]
30. Arnberg, N.; Edlund, K.; Kidd, A.H.; Wadell, G. Adenovirus Type 37 Uses Sialic Acid as a Cellular Receptor. *J. Virol.* **2000**, *74*, 42–48. [CrossRef]
31. Dechecchi, M.C.; Tamanini, A.; Bonizzato, A.; Cabrini, G. Heparan Sulfate Glycosaminoglycans Are Involved in Adenovirus Type 5 and 2-Host Cell Interactions. *Virology* **2000**, *268*, 382–390. [CrossRef] [PubMed]
32. Dechecchi, M.C.; Melotti, P.; Bonizzato, A.; Santacatterina, M.; Chilosi, M.; Cabrini, G. Heparan Sulfate Glycosaminoglycans Are Receptors Sufficient to Mediate the Initial Binding of Adenovirus Types 2 and 5. *J. Virol.* **2001**, *75*, 8772 LP–8780. [CrossRef] [PubMed]
33. Tuve, S.; Wang, H.; Jacobs, J.D.; Yumul, R.C.; Smith, D.F.; Lieber, A. Role of Cellular Heparan Sulfate Proteoglycans in Infection of Human Adenovirus Serotype 3 and 35. *PLoS Pathog.* **2008**, *4*, e1000189. [CrossRef] [PubMed]
34. Zhu, F.-C.; Li, Y.-H.; Guan, X.-H.; Hou, L.-H.; Wang, W.-J.; Li, J.-X.; Wu, S.-P.; Wang, B.-S.; Wang, Z.; Wang, L.; et al. Safety, Tolerability, and Immunogenicity of a Recombinant Adenovirus Type-5 Vectored COVID-19 Vaccine: A Dose-Escalation, Open-Label, Non-Randomised, First-in-Human Trial. *Lancet* **2020**, *395*, 1845–1854. [CrossRef]
35. Logunov, D.Y.; Dolzhikova, I.V.; Shcheblyakov, D.V.; Tukhvatulin, A.I.; Zubkova, O.V.; Dzharullaeva, A.S.; Kovyrshina, A.V.; Lubenets, N.L.; Grousova, D.M.; Erokhova, A.S.; et al. Safety and Efficacy of an RAd26 and RAd5 Vector-Based Heterologous Prime-Boost COVID-19 Vaccine: An Interim Analysis of a Randomised Controlled Phase 3 Trial in Russia. *Lancet* **2021**, *397*, 671–681. [CrossRef]
36. Rossidis, A.C.; Stratigis, J.D.; Chadwick, A.C.; Hartman, H.A.; Ahn, N.J.; Li, H.; Singh, K.; Coons, B.E.; Li, L.; Lv, W.; et al. In Utero CRISPR-Mediated Therapeutic Editing of Metabolic Genes. *Nat. Med.* **2018**, *24*, 1513–1518. [CrossRef] [PubMed]
37. Greig, J.A.; Buckley, S.M.K.; Waddington, S.N.; Parker, A.L.; Bhella, D.; Pink, R.; Rahim, A.A.; Morita, T.; Nicklin, S.A.; McVey, J.H.; et al. Influence of Coagulation Factor X on In Vitro and In Vivo Gene Delivery by Adenovirus (Ad) 5, Ad35, and Chimeric Ad5/Ad35 Vectors. *Mol. Ther.* **2009**, *17*, 1683–1691. [CrossRef] [PubMed]

viruses

MDPI

Communication

Intracellular Sequestration of the NKG2D Ligand MIC B by Species F Adenovirus

Edson R. A. Oliveira, Lenong Li and Marlene Bouvier *

Department of Microbiology and Immunology, University of Illinois at Chicago, 909 S Wolcott Avenue, Chicago, IL 60612, USA; edsonrao@gmail.com (E.R.A.O.); Lenong@uic.edu (L.L.)
* Correspondence: mbouvier@uic.edu

Abstract: The enteric human adenoviruses of species F (HAdVs-F), which comprise HAdV-F40 and HAdV-F41, are significant pathogens that cause acute gastroenteritis in children worldwide. The early transcription unit 3 (E3) of HAdVs-F is markedly different from that of all other HAdV species. To date, the E3 proteins unique to HAdVs-F have not been characterized and the mechanism by which HAdVs-F evade immune defenses in the gastrointestinal (GI) tract is poorly understood. Here, we show that HAdV-F41 infection of human intestinal HCT116 cells upregulated the expression of MHC class I-related chain A (MIC A) and MIC B relative to uninfected cells. Our results also showed that, for MIC B, this response did not however result in a significant increase of MIC B on the cell surface. Instead, MIC B was largely sequestered intracellularly. Thus, although HAdV-F41 infection of HCT116 cells upregulated MIC B expression, the ligand remained inside infected cells. A similar observation could not be made for MIC A in these cells. Our preliminary findings represent a novel function of HAdVs-F that may enable these viruses to evade immune surveillance by natural killer (NK) cells in the infected gut, thereby paving the way for the future investigation of their unique E3 proteins.

Keywords: adenoviruses; adenovirus species F; viral tropism; gut immune system; enteric viruses; immune evasion; NK cells; MIC A and MIC B

1. Introduction

HAdVs represents a large family of genetically diverse pathogens. To date, more than 100 different HAdVs have been identified and classified into seven species, A to G (http://hadvwg.gmu.edu/ (accessed on 22 November 2019)). HAdVs cause partially overlapping, species-specific diseases associated with infections of the respiratory (species B, C, and E), urinary (species B), gastrointestinal (species A and F), and ocular (species D) systems [1]. HAdVs are highly contagious and can cause severe local outbreaks. Although healthy adults can generally control the virus, HAdV infections in children and immunocompromised individuals can be fatal [2–5]. HAdV devotes a considerable portion of its genome to modulation of host immune functions, which presumably enables some species to establish and maintain lifelong asymptomatic infections. The vast majority of HAdV genes involved in the modulation of host immune functions are grouped in the E3 region [6,7]. The E3 region is not essential for HAdV replication in cultured cells, but the fact that this transcription unit is always maintained in natural isolates strongly suggests that E3 gene products are critical for natural infections in humans [6,7]. Notably, E3 is one of the most divergent gene regions between species (see Figure 1). This genetic variability is not well understood, but it strongly suggests that E3 proteins play a role in the manifestation of species-specific tissue tropism and diseases [6]. The 19K protein of HAdV-C binds to and retains MHC class I molecules in the endoplasmic reticulum, thereby rendering HAdV-C-infected cells less efficient at presenting viral antigens and less sensitive to lysis by CD8+ T cells [8–22]. That this 19K gene is maintained in HAdV-B, -D, and -E (see Figure 1) underscores the critical need for these species to retain a

MHC I-binding function. The x-ray crystal structures of AdV-C2 and AdV-E4 E3-19K bound to HLA-A2 provided insights into the mechanism of immune modulation [23,24]. Other E3 proteins were shown to function by inhibiting tumor necrosis factor activities and interfering with apoptotic cell death processes and leukocyte activation, ultimately suppressing the lysis of infected cells by NK cells [25] and references therein.

The enteric HAdV-F, which comprises HAdV-F40 and HAdV-F41, shows a narrow tropism for epithelial cells of the GI tract and is a leading etiologic agent of acute gastroenteritis in infants and young children worldwide [26–29]. Notably, there are marked differences in the E3 region of HAdV-F compared to the other species (see Figure 1): HAdV-F lacks the common E3-19K gene and has two genes, 19.4K and 31.6K, that are unique to that species [30–32]. To date, E3-19.4K and E3-31.6K proteins have not been characterized and we have no knowledge of their functions. The study of HAdV-F thus offers a unique opportunity to study how interactions between E3 proteins and the immune microenvironment at the site of infection contributes to viral tropism and pathogenicity. We suggest that the selective pressure on HAdV-F40 and HAdV-F41 by the immune system of the GI tract has led these viruses to adapt in order to replicate effectively in gut cells, and that E3-19.4K and/or E3-31.6K proteins play an important role in this process. More specifically, we suggest that HAdVs-F have evolved a function directed at suppressing the expression of MIC A and MIC B molecules on infected cells. MIC A and MIC B are stress-inducible surface ligands that are recognized by the NKG2D activating receptor expressed on NK cells to eliminate stressed cells. Given that the distribution of MIC activating ligands is largely restricted to intestinal epithelial cells under normal conditions, and that HAdVs-F are exquisitely adapted to replicate in the intestinal epithelium [33] and references therein, it might not be surprising that these viruses interfere with MIC A and MIC B to suppress immune surveillance by NK cells.

To advance our understanding of HAdVs-F, and given the significance of these viruses as pathogens, we have initiated a study to examine the effects of HAdV-F infection on cell surface expression of MIC ligands. We have established an in vitro culture system based on infection of human intestinal HCT116 cells with HAdVs-F from which we show that HAdV-F41 causes the intracellular sequestration of MIC B. These preliminary results support the hypothesis that interferences with NKG2D MIC ligands is a mechanism used by HAdVs-F to evade immune surveillance in the gut and may be a determinant of viral tropism.

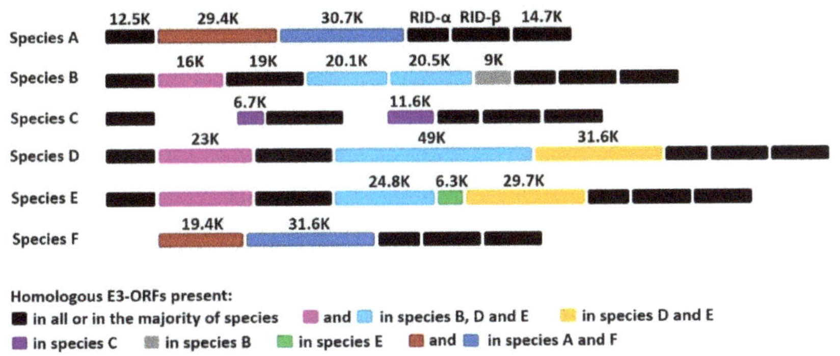

Figure 1. Sequence alignment showing the coding potential of E3 regions of the most common HAdVs-A, -B, -C, -D, -E, and -F. The expected molecular mass of each gene product is indicated. Proteins with amino acid sequence homology, generally ~35%, have the same shade coding: 19.4K and 31.6K are unique to HAdV-F.

2. Materials and Methods

2.1. Virus Growth and Cells

HAdV-F41 (ATCC® VR-930™) was grown in 50–60% confluent HEK-293 cells (ATCC® CRL-1573™) in DMEM (ATCC® 30-2002) supplemented with 1–2% FBS (ATCC® 30-2020™). Infection was done with virus at passage five at an MOI = 1. After infection, when cells show clear cytopathic effect (round up with increased nucleus size), cultures were harvested with a cell scraper and transferred to falcon tubes. Cell suspensions were centrifuged at 700× g, 4 °C for 10 min, and cells were resuspended in culture medium discharging the supernatant. Samples were subjected to three freeze/thaw cycles (−80 °C and 37 °C), then centrifuged at 1500× g, 4 °C for 10 min. Supernatants were aliquoted in small volumes and kept at −80 °C until use. To determine viral titers, an aliquot of the virus preparation was used for titration in HEK-293 cells via immunohistochemistry using the QuickTiter™ Adenovirus Quantitation Kit (Cell Biolabs, Catalog no. VPK-109, San Diego, CA, USA), following instructions by the manufacturer. Two cell types were used for the in vitro infection models: human colorectal carcinoma HCT116 cells (ATCC® CCL-247, Manassas, VI, USA). HCT116 cells were grown in McCoy's 5A medium (ATCC® 30-2007™, Manassas, VI, USA) supplemented with 10% FBS.

2.2. Immunofluorescence Staining

For immunofluorescence (IF) staining, cells were grown on sterile glass coverslips placed on 12-well plates prior to infection with HAdV-F41 (MOI 0.5). After 2 days, cells were fixed in 4% PFA for 10 min, permeabilized with 0.1% triton x-100 for 20 min, and blocked with 1% PBS/BSA for 30 min. For virus staining, a rabbit anti-pVIII polyclonal Ab (provided by Dr. W. Wold, St-Louis University, St. Louis, MO, USA) was used. Cells were washed and stained for 1 h with a mixture of donkey anti-rabbit secondary Ab conjugated with rhodamine (Invitrogen, Catalog no. 31685, Waltham, MA, USA), and Phalloidin-iFluor 488 (Abcam, Catalog no. ab176753, Cambridge, UK) to stain actin fibers. MIC A and MIC B staining were done using primary mouse anti-MIC A and mouse anti-MIC B Abs. A goat anti-mouse-FITC was used as the secondary Ab. Coverslips were mounted on slides using ProLong™ Diamond Antifade with DAPI (Invitrogen, Catalog no. P36962, Waltham, MA, USA) and cured at 4 °C for 24 h in the dark. Samples were analyzed under an Olympus BX51 IF microscope coupled with a CCD camera to acquire individual channels of DAPI, alexa fluor 488 or rhodamine. Acquired channels were merged using ImageJ software v1.53a. Uninfected cells, and secondary Abs alone, produced no relevant signals in the rhodamine channel.

2.3. Flow Cytometry

HCT116 cells were infected with HAdV-F41 (MOI 0.5) and expression levels of MIC A and MIC B were determined on the cell surface and intracellularly by flow cytometry on days 2 and 4 post-infection. Infection was assessed based on the expression of intracellular hexon protein. At the harvest time, cells were scraped, washed in PBS by centrifugation at 700× g for 10 min, incubated with Zombie Violet Fixable Viability Kit (Biolegend, Catalog no. 423114, San Diego, CA, USA) at 1:500 for 30 min in the dark for discriminating live versus dead cells, washed, and fixed in 4% PFA for 20 min on ice. Cells were then washed and incubated with a mixture of anti-MIC A-phycoerythrin (PE) (Sino Biological Catalog no. 12302-MM04-P, Beijing, China) and anti-MIC B-allophycocyanin (APC) (Sino Biological Catalog no. 10759-MM12-A, Beijing, China) Abs for 40 min on ice. Isotype Abs recommended by the manufacturer, as well as uninfected HCT116 cells, were used as negative controls. In the case of samples prepared for extra- and intra-cellular staining, cells were incubated with Ab cocktail for surface staining prior to permeabilization with 0.1% triton x-100 for 10 min at RT. Hexon staining was carried out using a 2Hx-2 monoclonal anti-hexon Ab (provided by Dr. W. Wold, St-Louis University, St. Louis, MO, USA) [34] with further detection using a secondary anti-mouse-FITC Ab. After staining, cells were washed 2 times in PBS, resuspended in 300 μL PBS, and data were acquired on a Gallios flow instrument (Beckman & Coulter, Brea, CA, USA). Samples were analyzed offline using FlowJo software v10 considering only live cells, after exclusion of cell debris and aggregates.

2.4. Protein Sequence Analysis

The GenBank accession number for HAdV-F41 E3 region is M85254 [32]. Bioinformatics software and servers used for protein sequence analysis were: SignalP v. 5.0 (DTU Bioinformatics, Denmark) [35], SMART (EMBL, Heidelberg, Germany) [36], TMHMM v. 2.0 (DTU Bioinformatics, Denmark) [37], and PHYRE v. 2.0 (Imperial College, London, UK) [38].

3. Results

3.1. E3 Region

E3 is one of the most divergent gene regions between species (Figure 1): while some E3 proteins are found in all species, others have counterparts in only a few species, and some E3 proteins are unique to a given species. Remarkably, HAdV-F lacks the common E3-19K protein and instead expresses two proteins, 19.4K and 31.6K, that are unique to this species (Figure 1). To date, the functions of these proteins are unknown.

3.2. In Vitro Models of HAdV-F Infection

That HAdVs-F are one of the least characterized species may be due to the difficulty of propagating these viruses in most common human cell culture systems that permit replication of all other HAdVs. We established an in vitro culture system for infection of human intestinal HCT116 with HAdV-F41 (Figure 2). To characterize this cell system, we used IF staining of HAdV structural proteins hexon and pVIII as a way to track infected cells. HCT116 cells were infected with HAdV-F41 (MOI 0.5) and results shows a clear nuclear staining of pVIII on day 2 post-infection, consistent with the permissiveness of these cells for HAdV-F41 infection. The results establish new conditions for HAdV-F41 infection in HCT116 cells.

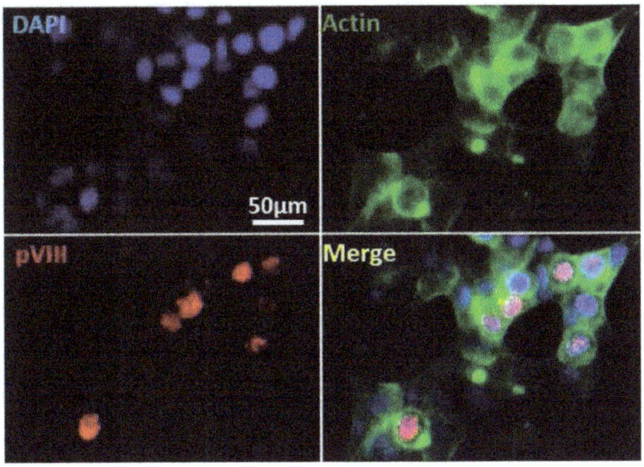

Figure 2. Immunofluorescence staining of HAdV pVIII protein in HAdV-F41-infected HCT116 cells. Cells infected with HAdV-F41 (MOI 0.5) at day 2 post-infection showing nuclear localization of the viral structural pVIII protein (red). Actin fibers and cell chromatin are presented in green and blue, respectively. Samples were analyzed under an Olympus BX51 IF microscope coupled with a CCD camera Acquired channels were merged using ImageJ software v1.53a. Uninfected cells or secondary Ab alone yielded no relevant signals.

3.3. HAdV-F41 Interferes with Cell Surface Expression of MIC B

We examined if HAdV-F41 impairs the cell surface expression of MIC A and MIC B in HCT116 cells by flow cytometry and IF. We first characterized the basal expression levels of MIC ligands in uninfected HCT116 cells over four days. Results show that for both MIC A and MIC B, expression levels are higher intracellularly than on the cell surface (Figure 3a). Furthermore, MIC B is more abundant overall than MIC A (Figure 3a,b), and MIC A is negligibly expressed on HCT116 cells (Figure 3a). Finally, it is important to note that, in uninfected HCT116 cells, MIC B cell surface expression levels decreased slightly from day 2 to day 4 (Figure 3a). This may be due to the proteolytic shedding of MIC B from the cell surface, a process that occurs during normal cell growth and the expression of MIC proteins [39].

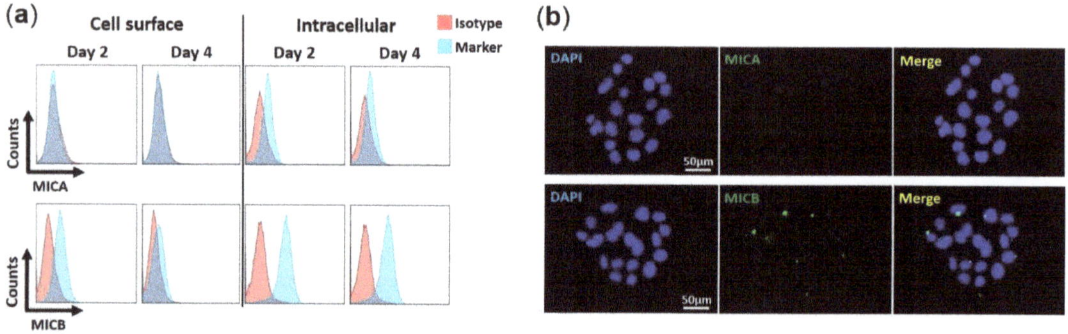

Figure 3. Expression of MIC ligands in uninfected HCT116 cells. (**a**) Flow cytometry histograms showing levels of MIC A and MIC B on the surface and in the intracellular environment of uninfected HCT116 cells. Cells were harvested at day 2 and 4 in culture. Isotype Abs recommended by the manufacturer were used as negative controls. Sample were analyzed on a Gallios (Beckman & Coulter, Brea, CA, USA) flow instrument and analysis was done offline using FlowJo software v10. Histograms are gated on live cells, after exclusion of cell debris and aggregates. (**b**) Immunofluorescence staining of MIC A and MIC B in uninfected HCT116 cells at day 2 in culture, with DAPI in blue and MICs in green. Data are representative of three independent experiments.

We then examined if HAdV-F41 modulates the expression levels of MIC ligands in infected HCT116 cells over four days using flow cytometry. To account for the natural proteolytic shedding of MICs, the controls consisted of uninfected HCT116 cells collected at each time point. HAdV-F41-infected cells were detected using a monoclonal anti-hexon Ab (Figure 4a). Results show that MIC A expression levels, whether on the cell surface or intracellularly, are consistently higher in hexon$^+$ cells than hexon$^-$ cells (Figure 4b). The same observation was made for MIC B (Figure 4b). Thus, the expression of MIC ligands is upregulated by HCT116 cells infected with HAdV-F41. The data in Figure 4b were further analyzed by considering changes in median fluorescence intensity (MFI) of MIC expression levels in hexon$^+$ cells relative to hexon$^-$ cells and the values plotted as "fold increase" (Figure 4c, see legend). The analysis revealed that hexon$^+$ cells expressed significantly more MIC B in intracellular compartments relative to hexon$^-$ cells, 18-fold increase on day 2 versus 15-fold increase on day 4. In contrast, there was only a small increase of MIC B expression on the cell surface, 1.5-fold on day 2 versus 3.7-fold on day 4 (Figure 4c). Thus, although HAdV-F41 cause an upregulation of MIC B in HCT116 cells, this did not lead to increased expression of the ligand on the cell surface suggesting that MIC B is largely sequestered intracellularly in infected cells. A similar trend for MIC A could not be observed in these cells, instead similar relative changes in intracellular and cell surface expression levels of MIC A were determined (Figure 4c). Finally, consistent with the flow cytometry results showing higher levels of MIC B expression in hexon$^+$ cells (Figure 4b), IF analysis showed that fluorescence intensity signals are stronger in

HAdV-F41-infected cells (shown as pVIII⁺ cells) (Figure 5b) relative to uninfected cells (pVIII⁻ cells) (Figure 5a).

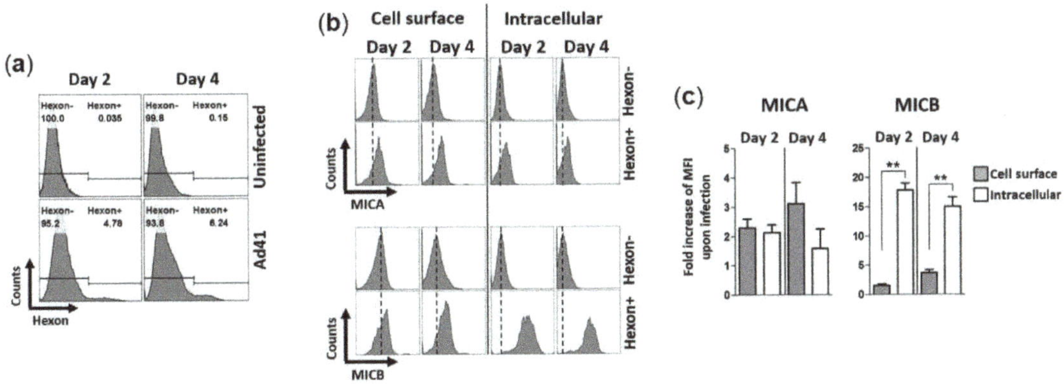

Figure 4. Expression of MIC ligands in HAdV-F41-infected HCT116 cells. HCT116 cells were infected with HAdV-F41 (MOI 0.5) and levels of MIC A and MIC B were assessed on the cell surface and intracellularly by flow cytometry, on day 2 and 4 post-infection. (**a**) Flow cytometry histograms showing staining of HAd41-infected cells (Hexon⁺) using a 2Hx-2 monoclonal anti-hexon Ab [34] with further detection with a secondary anti-mouse-FITC Ab. (**b**) Flow cytometry histogram showing expression levels of MIC A and MIC B on the cell surface and intracellularly. Hexon⁻ populations were gated from uninfected samples. Dashed lines represent the MFI levels of MIC A or MIC B on uninfected samples. (**c**) Fold increases in expression of MIC A and MIC B upon HAdV-F41 infection were calculated as MFI^{hexon+} / MFI^{hexon-}. ** $p < 0.01$ defined by t student test.

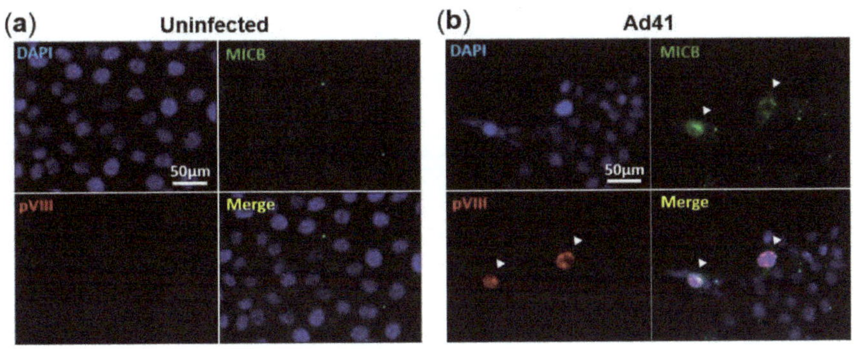

Figure 5. Immunofluorescence assay of MIC B in HAdV-F41-infected HCT116 cells. IF assay showing (**a**) uninfected and (**b**) HAdV-F41-infected HCT116 cells (MOI 0.5; day 2). HAdV-F41 was traced using a rabbit polyclonal anti-pVIII Ab and a secondary goat anti-rabbit-Rhodamine. MIC B detection was performed using a mouse anti-MIC B Ab followed by incubation with a goat anti-mouse-FITC. Cell nuclei were counterstained with DAPI. Arrows show HAdV-F41-infected cells exhibiting stronger signals of MIC B compared to uninfected cells.

3.4. E3-19.4K and E3-31.6K Proteins

Given the uniqueness of E3-19.4K and E3-31.6K to HAdV-F, and the strong possibility that these proteins participate in immune evasion functions in the gut, we characterized their basic properties. HAdV-F41 19.4K has 173 residues (Supplementary Figure S1) and its amino acid sequence is 99% identical to its counterpart in HAdV-F40 except for a single amino acid change at residue 144, which is an isoleucine in HAdV-F41 and asparagine in HAdV-F40 [31,32]. An analysis of HAdV-F41 19.4K sequence using the

Bioinformatics software SignalP-5.0, SMART, and TMHMM [35–38] predicts that the protein is a type I transmembrane protein with the signal sequence comprising the first 15 or 18 N-terminal residues (depending on the software used) and transmembrane domain spanning residues 144 to 166 (Supplementary Figure S1). Because residue 144 is predicted to be membrane-localized, the ectodomains of E3-19.4K are expected to be identical between HAdV-F40 and HAdV-F41. Secondary structure predictions using the Phyre2 package indicated that E3-19.4K has 13% α-helix, 58% β-strand, and 20% disordered regions, with the putative transmembrane domain correctly identified as an α-helix. The accuracy of these predictions awaits a determination of the three-dimensional structure of E3-19.4K. A BLAST analysis indicated that the amino acid sequence of HAdV-F41 E3-19.4K shows no homology to the common E3-19K immunomodulatory proteins of species B, C, D, and E. The BLAST analysis also revealed that HAdV-F41 E3-19.4K is 50% homologous to the E3 CR1-α1 protein of HAdV-G52 [40]. The homology comprises residues 39 to 122 (underlined in Supplementary Figure S1) and represents a conserved region within E3 named "adenoE3CR1rpt" (also known as 6.7K) [41,42]. HAdV-G52 has not been extensively studied and the function of its CR1-α1 protein is unknown. However, it was shown that HAdV-C2 CR1-α1 directs E3-19K to the ER [43] and can also cooperate with the RID proteins to evade TNF-α-induced NK-κB activation [44]. The lack of general information about E3 CR1-α1 proteins makes it difficult to draw any conclusions on the role of this motif in E3-19.4K.

HAdV-F41 E3-31.6K has 276 residues (Supplementary Figure S2) and its amino acid sequence is 99% homologous to that of HAdV-F41 E3-31.6K, except for two amino acid changes at residues 192 and 264 which are both glutamic acid in Ad41 and lysine in Ad40 [31,32]. E3-31.6K is predicted to be a type I transmembrane protein with the signal sequence comprising the first 16 N-terminal residues and the transmembrane domain spanning residues 234 to 256 (Supplementary Figure S2). Secondary structure predictions using Phyre^2 indicated that E3-31.6K has 17% α-helix, 50% β-strand, and 11% disordered regions, with the putative transmembrane domain correctly identified as an α-helix. A BLAST analysis indicated that the amino acid sequence of HAdV-F40 E3-31.6K shows no homology to E3-19K proteins. No other significant sequence homology was identified.

4. Discussion

HAdVs-F replicate preferentially in the GI tract and cause severe gastroenteritis in children. The selective pressure on HAdVs-F by the microenvironment of the gut has led these viruses to eliminate the E3-19K gene, and presumably the MHC I-binding function, and to evolve two new E3 genes that are conspicuously absent in all other species. This raises intriguing questions on the underlying mechanisms by which HAdVs-F are exquisitely adapted to the GI tract. Given that E3 proteins modulate various host immune functions [6,7], it is reasonable to assume that the distinct pathogenicity of HAdVs-F stems in some way from interactions of their E3 proteins with the immune system of the gut. On that basis, we have investigated the effect of HAdV-F infection on stress-induced MIC A and B molecules. MIC A and MIC B are normally expressed at low levels almost exclusively on intestinal epithelial cells and engage with activating receptors on NK cells as part of host immunosurveillance of stressed cells.

HAdVs-F are notoriously difficult to grow in most cell culture systems [45–47] and whether this characteristic arises from a feature unique to this species, namely that they contain two different fiber proteins (long and short) and unique penton base proteins, is unclear [48–50]. We have developed optimal cell culture conditions for infection of intestinal HCT 116 cells with HAdV-F41. Our results showed that HAdV-F41 infection of HCT116 cells upregulated the expression of MIC A and MIC B relative to uninfected cells, on the cell surface as well as intracellularly. These results are consistent with the role of MIC A and MIC B as stress-inducible ligands and underline a possible role for the NKG2D pathway in HAdV-F infection. Our results also showed that for MIC B, this response did not however lead to a significant increase of the ligand on the cell surface. Instead, MIC B was largely sequestered intracellularly. Thus, although HAdV-F41 infection upregulates the expression levels of MIC B in HCT116 cells, the ligand remained inside infected cells. A similar trend for MIC A could not be observed in HCT116 cells and therefore it remains to be further evaluated if HAdVs-F selectively target MIC B-the selective targeting of MIC ligands (MIC A or MIC B) has been reported previously for several human viruses [51–55]. Taken together, we showed for the first time that HAdV-F41 infection of HCT116 cells led to the intracellular sequestration of the NKG2D activating ligand MIC B. Whether our findings represent a viral escape mechanism to prevent recognition and elimination of HAdV-F41-infected cells in the gut by NK cells requires further investigation.

Our results raise important questions concerning the mechanism by which HAdV-F sequesters MIC B inside cells, and the viral factor responsible for this effect. The E3-19.4K and E3-31.6K proteins are highly conserved in HAdVs-F, 99% amino acid sequence identity between HAdV-F40 and HAdV-F41, which suggests that these proteins are critical for viral tropism or virulence in the gut. Interestingly, it was shown previously that infection of human fibroblasts by HAdV-C2 and HAdV-C5 led to the sequestration of MIC A and MIC B inside infected cells [56], an effect that was attributed to the E3-19K protein [56,57]. HAdVs-C are tropic for epithelial cells of the lungs and cause respiratory illnesses. However, these viruses can also cause GI symptoms, as part of a systemic infection with accompanying respiratory disorders, and are persistently detected in stools of healthy and infected individuals. Furthermore, tumorigenic HAdV-A12 of species A, which is also associated with gastroenteritis, was shown to suppress the expression of NKG2D activating ligands on transformed mouse and rat cells through the transcriptional repression of these ligands [58]. The protein responsible for this effect has not yet been identified. Taken together, interferences with NKG2D activating ligands may be an important mechanism by which HAdVs mediate immune evasion in the GI tract, most especially for species F. In this context, others have shown, using a model system of human enteroids, that HAdV-F41 was resistant to the activity of enteric alpha-defensin 5 while in contrast HAdV-C5 was neutralized [59,60]. These results show that a host factor such as alpha-defensin 5, an innate defense peptide expressed in the crypts of the intestine, could also modulate the tropism of enteric species F.

In conclusion, we showed that HAdV-F41 sequesters MIC B inside infected cells. This represents a novel function of HAdVs-F. HAdVs-F have received considerably less attention than the other HAdVs, despite being significant pathogens, and our findings cast a new light on how these viruses, under immune pressure in the GI tract, have remarkably adapted to this site. Future investigations will provide more details about this function and the viral protein responsible for it, and will reveal if the suppression of MIC ligands impairs the recognition of HAdV-F-infected cells by NK cells.

Supplementary Materials: The following are available online at https://www.mdpi.com/1999-4915/13/7/1289/s1, Figure S1: The amino acid sequence of HAdV-F41 E3-19.4K protein (residues 1-173), Figure S2: The amino acid sequences of HAdV-F41 E3-31.6K protein (residues 1- 276).

Author Contributions: Conceptualization of the study and methodology, M.B. and E.R.A.O.; experiments, E.R.A.O. and L.L.; data analysis, all authors; writing–original draft preparation, M.B.; writing–review and editing, M.B., E.R.A.O. and L.L.; supervision, M.B.; funding acquisition, M.B. All authors have read and agreed to the published version of the manuscript.

Funding: This work was supported in whole or in part by grants from the US National Institute of Allergy and Infectious Diseases R01 AI114467 (M.B.), R01 AI108546 (M.B.), and R21 AI135917 (M.B.).

Institutional Review Board Statement: Not applicable.

Informed Consent Statement: Not applicable.

Data Availability Statement: Not applicable.

Acknowledgments: We thank Abhishek Kumar and Hui Deng for their initial efforts working on HAdV-F. We are grateful to W. Wold and Ann Tollefson (St-Louis University, St. Louis, MO, USA) for gifts of reagents and discussion.

Conflicts of Interest: All authors declare no conflict of interests.

References

1. Berk, A. Adenoviridae. In *Fields Virology*, 6th ed.; Fields, B.N., Knipe, D.M., Howley, P.M., Eds.; Wolters Kluwer Health/Lippincott Williams & Wilkins: Philadelphia, PA, USA, 2013; pp. 1704–1731.
2. Lion, T. Adenovirus infections in immunocompromised patients. *Clin. Microbiol. Rev.* **2014**, *27*, 441–462. [CrossRef]
3. Hoffman, J.A. Adenovirus infections in solid organ transplant recipients. *Curr. Opin. Organ. Transplant.* **2009**, *14*, 625–633. [CrossRef] [PubMed]
4. Feghoul, L.; Chevret, S.; Cuinet, A.; Dalle, J.H.; Ouachee, M.; Yacouben, K.; Fahd, M.; Guerin-El Khourouj, V.; Roupret-Serzec, J.; Sterkers, G.; et al. Adenovirus infection and disease in paediatric haematopoietic stem cell transplant patients: Clues for antiviral preemptive treatment. *Clin. Microbiol. Infect.* **2015**, *21*, 701–709. [CrossRef]
5. Fisher, B.T.; Boge, C.L.K.; Petersen, H.; Seif, A.E.; Bryan, M.; Hodinka, R.L.; Cardenas, A.M.; Purdy, D.R.; Loudon, B.; Kajon, A.E. Outcomes of human adenovirus infection and disease in a retrospective cohort of pediatric hematopoietic cell transplant patients. *J. Pediatr. Infect. Dis. Soc.* **2018**, *8*, 317–324. [CrossRef] [PubMed]
6. Burgert, H.; Blusch, J.H. Immunomodulatory functions encoded by the E3 transcription unit of adenoviruses. *Virus Genes* **2000**, *21*, 13–25. [CrossRef] [PubMed]
7. Wold, W.S.M.; Gooding, L.R. Region E3 of adenovirus: A cassette of genes involved in host immunosurveillance and virus-cell interactions. *Virology* **1991**, *184*, 1–8. [CrossRef]
8. Burgert, H.-G.; Kvist, S. The E3/19K protein of adenovirus type 2 binds to the domains of histocompatibility antigens required for CTL recognition. *EMBO J.* **1987**, *6*, 2019–2026. [CrossRef]
9. Burgert, H.-G.; Kvist, S. An adenovirus type 2 glycoprotein blocks cell surface expression of human histocompatibility class I antigens. *Cell* **1985**, *41*, 987–997. [CrossRef]
10. Kvist, S.; Ostberg, L.; Persson, H.; Philipson, L.; Peterson, P.A. Molecular association between transplantation antigens and cell surface antigen in adenovirus-transformed cell line. *Proc. Natl. Acad. Sci. USA* **1978**, *75*, 5674–5678. [CrossRef]
11. Paabo, S.; Bhat, B.M.; Wold, W.S.M.; Peterson, P.A. A short sequence in the COOH terminus makes an adenovirus membrane glycoprotein a resident of the endoplasmic reticulum. *Cell* **1987**, *50*, 311–317. [CrossRef]
12. Sester, M.; Ruszics, Z.; Mackley, E.; Burgert, H.-G. The transmembrane domain of the adenovirus E3/19K protein acts as an endoplasmic reticulum retention signal and contribute to intracellular sequestration of major histocompatibility complex class I molecules. *J. Virol.* **2013**, *87*, 6104–6117. [CrossRef]
13. Cox, J.H.; Bennink, J.R.; Yewdell, J.W. Retention of adenovirus E19 glycoprotein in the endoplasmic reticulum is essential to its ability to block antigen presentation. *J. Exp. Med.* **1991**, *174*, 1629–1637. [CrossRef]
14. Liu, H.; Stafford, W.F.; Bouvier, M. The endoplasmic reticulum lumenal domain of the adenovirus type 2 E3-19K binds to peptide-filled and peptide-deficient HLA-A*1101 molecules. *J. Virol.* **2005**, *79*, 13317–13325. [CrossRef]
15. Liu, H.; Fu, J.; Bouvier, M. Allele- and locus-specific recognition of class I MHC molecules by the immunomodulatory E3-19K protein from adenovirus. *J. Immunol.* **2007**, *178*, 4567–4575. [CrossRef]
16. Fu, J.; Li, L.; Bouvier, M. Adenovirus E3-19K proteins of different serotypes and subgroups have similar, yet distinct, immunomodulatory functions towards major histocompatibility class I molecules. *J. Biol. Chem.* **2011**, *286*, 17631–17639. [CrossRef] [PubMed]
17. Fu, J.; Bouvier, M. Determinants of the endoplasmic reticulum (ER) lumenal-domain of the Adenovirus serotype 2 E3-19K protein for association with and ER-retention of major histocompatibility complex class I molecules. *Mol. Immunol.* **2011**, *48*, 532–538. [CrossRef] [PubMed]
18. Flomenberg, P.; Piaskowski, V.; Truitt, R.L.; Casper, J.T. Human adenovirus-specific CD8+ T-cell responses are not inhibited by E3-19K in the presence of gamma interferon. *J. Virol.* **1996**, *70*, 6314–6322. [CrossRef]
19. Andersson, M.; McMichael, A.; Peterson, P.A. Reduced allorecognition of adenovirus-2 infected cells. *J. Immunol.* **1987**, *138*, 3960–3966. [PubMed]
20. Burgert, H.-G.; Maryanski, J.L.; Kvist, S. "E3/19K" protein of adenovirus type 2 inhibits lysis of cytolytic T lymphocytes by blocking cell surface expression of histocompatibility class I antigens. *Proc. Natl. Acad. Sci. USA* **1987**, *84*, 1356–1360. [CrossRef] [PubMed]
21. Tanaka, Y.; Tevethia, S.S. Differential effect of adenovirus 2 E3/19K glycoprotein on the expression of H-2Kb and H-2Db class I antigens and H-2Kb- and H-2Db-restricted SV40-specific CTL-mediated lysis. *Virology* **1988**, *165*, 357–366. [CrossRef]
22. Rawle, F.C.; Tollefson, A.E.; Wold, W.S.M.; Gooding, L.R. Mouse anti-adenovirus cytoxic T lymphocytes. *J. Immunol.* **1989**, *138*, 3960–3966.
23. Li, L.; Muzahim, Y.; Bouvier, M. Crystal structure of adenovirus E3-19K bound to HLA-A2 reveals mechanism for immunomodulation. *Nat. Struc. Mol. Biol.* **2012**, *19*, 1176–1181. [CrossRef]
24. Li, L.; Santarserio, B.D.; Bouvier, M. Structure of the Adenovirus Type 4 (Species E) E3-19K/HLA-A2 complex reveals species-specific features in MHC I recognition. *J. Immunol.* **2016**, *197*, 1399–1407. [CrossRef] [PubMed]

25. Oliveira, E.R.A.; Bouvier, M. Immune evasion by adenoviruses: A window into host-virus adaptation. *FEBS Lett.* **2019**, *593*, 3496–3503. [CrossRef] [PubMed]
26. Lee, J.I.; Lee, G.-C.; Chung, J.Y.; Han, T.H.; Lee, Y.K.; Kim, M.S.; Lee, C.H. Detection and molecular characterization of adenoviruses in Korean children hospitalized with acute gastroenteretitis. *Microbiol. Immunol.* **2012**, *56*, 523–528. [CrossRef] [PubMed]
27. LaRosa, G.; Libera, S.D.; Petricca, S.; Iaconelli, M.; Donia, D.; Saccucci, P.; Cenko, F.; Xhelilaj, G.; Divizia, M. Genetic diversity of human adenovírus in children with acute gastreoenteretitis, Albania, 2013–2015. *Biomed. Res. Intern.* **2015**, *2015*, 142912.
28. Afrad, M.H.; Avzun, T.; Haque, J.; Haguw, W.; Hossain, M.E.; Rahman, A.F.M.R.; Ahmed, S.; Faruque, A.S.G.; Rahman, M.Z.; Rahman, M. Detection of enteric- and non-enteric adenoviruses in gastreoenteretitis patients, Bangladesh, 2012–2015. *J. Med. Virol.* **2017**. [CrossRef]
29. Kumthip, K.; Khamrin, P.; Ushijima, H.; Maneekarn, N. Enteric- and non-enteric adenoviruses associated with acute gastreoenteretitis in pediatric patients in Thailand, 2011 to 2017. *PLoS ONE* **2019**, *14*, e0220263. [CrossRef]
30. Davison, A.J.; Telford, A.R.; Watson, M.S.; McBride, K.; Mautner, V. The DNA sequence of adenovirus type 40. *J. Mol. Biol.* **1993**, *234*, 1308–1316. [CrossRef]
31. Bailey, A.; Mautner, V. Phylogenetic relationships among adenovirus serotypes. *Virology* **1994**, *205*, 438–452. [CrossRef]
32. Yeh, H.-Y.; Pieniazek, N.; Pieniazek, D.; Luftig, R. Genetic organization, size, and complete sequence of early region 3 genes of human adenovirus type 41. *J. Virol.* **1996**, *70*, 2658–2663. [CrossRef]
33. Kosulin, K. Intestinal HAdV infection: Tissue specificity, persistence, and implications for antiviral therapy. *Viruses* **2019**, *11*, 804. [CrossRef]
34. Cepko, C.L.; Whetsone, C.A.; Sharp, P.A. Adenovirus hexon monoclonal antibody that is group specific and potentially useful as a diagnostic reagent. *J. Clin. Microbiol.* **1983**, *17*, 360–364. [CrossRef] [PubMed]
35. Armenteros, J.J.A.; Tsirigos, K.D.; Sønderby, C.K.; Petersen, T.N.; Winther, O.; Brunak, S.; von Heijne, G.; Nielsen, H. SignalP 5.0 improves signal peptide predictions using deep neural networks. *Nat. Biotech.* **2019**, *37*, 420–423. [CrossRef]
36. Letunic, I.; Bork, P. SMART: Recent updates, new developments and status in 2020. *Nucleic Acid Res.* **2020**, *49*, D458–D460. [CrossRef] [PubMed]
37. Krogh, A.; Larsson, B.; von Heijne, G.; Sonnhammer, E.L.L. Predicting transmembrane protein topology with a hidden Markov model: Application to complete genomes. *J. Mol. Biol.* **2011**, *305*, 567–580. [CrossRef] [PubMed]
38. Kelley, L.; Jefferys, B. *Phyre2: Protein Homology/Analogy Recognition Engine V 2.0*; Structural Bioinformatics Group, Imperial College: London, UK, 2011.
39. Moncaya, G.; Lin, D.; McCarthy, M.T.; Watson, A.A.; O'Calaghan, C.A. MICA expression is regulated by cell adhesion and contact in a FAK/Src-dependent manner. *Front. Immunol.* **2017**, *7*. [CrossRef] [PubMed]
40. Jones, M.S.; Harrach, B.; Ganac, R.D.; Gozum, M.M.A.; de la Cruz, W.P.; Riedel, B.; Pan, C.; Delwart, E.L.; Schnurr, D.P. New adenoviruses species found in a patient presenting with gastroenteritis. *J. Virol.* **2007**, *81*, 5978–5984. [CrossRef]
41. Davison, A.J.; Benko, M.; Harrach, B. Genetic content and evolution of adenoviruses. *J. Gen. Virol.* **2003**, *84*, 2895–2908. [CrossRef] [PubMed]
42. Deryckere, F.; Burgert, H.-G. Early region 3 adenovirus type 19 (subgroup D) encodes an HLA-binding protein distinct from that of subgroups B and C. *J. Virol.* **1996**, *70*, 2832–2841. [CrossRef]
43. Rawls-Wilson, J.; Deutscher, S.L.; Wold, W.S.M. The signal-anchor domain of adenovirus E3-6.7K, a type III integral membrane protein can direct adenovirus E3-gp19K, a type I integral membrane protein, into the membrane of the endoplasmic reticulum. *Virology* **1994**, *201*, 66–76. [CrossRef] [PubMed]
44. Benedict, C.A.; Norris, P.S.; Prigozy, T.I.; Bodmer, J.-L.; Mahr, J.A.; Garnett, C.T.; Martinnon, F.; Tschopp, J.; Gooding, L.R.; Ware, C.F. Three adenovirus E3 proteins cooperate to evade apoptosis by tumor necrosis factor-related apoptosis-inducing ligand receptor-1 and -2. *J. Biol. Chem.* **2001**, *276*, 3270–3278. [CrossRef] [PubMed]
45. Tiemessen, C.T.; Kidd, A.H. Adenoviruses type 40 and 41 growth in vitro: Host range diversity reflected by differences in patterns of DNA replication. *J. Virol.* **1994**, *68*, 1239–1244. [CrossRef] [PubMed]
46. de Jong, J.C. Candidate adenoviruses 40 and 41: Fastidious adenoviruses from human infant stool. *J. Med. Virol.* **1983**, *11*, 215–231. [CrossRef] [PubMed]
47. Witt, D.J.; Bousquet, E.B. Comparison of enteric adenovirus infection in various human cell lines. *J. Virol. Methods* **1988**, *20*, 295–308. [CrossRef]
48. Kidd, A.H.; Chroboczek, J.; Ruigrok, R.W. Adenovirus type 40 virions contain two distinct fibers. *Virology* **1983**, *192*, 73–84. [CrossRef] [PubMed]
49. Yeh, H.Y.; Pieniazek, N.; Pieniazek, D.; Gelderblom, H.; Luftig, R.B. Human adenovirus type 41 contains two fibers. *Virus Res.* **1994**, *33*, 179–198. [CrossRef]
50. Albinsson, B.; Kidd, A.H. Adenovirus type 41 lacks an RGD alpha(v)-integrin binding motif on the penton base and undergoes delayed uptake in A549 cells. *Virus Res.* **1999**, *64*, 125–136. [CrossRef]
51. Jonjic, S.; Babic, M.; Polic, B.; Krmpotic, A. Immune evasion of natural killer cells by viruses. *Curr. Opin. Immunol.* **2008**, *20*, 30–38. [CrossRef]
52. Cosman, D.; Müllberg, J.; Sutherland, C.L.; Chin, W.; Armitage, R.; Fanslow, W.; Kubin, M.; Chalupny, N.J. ULPBs, novel MHC class I-related molecules, bind to CMV glycoprotein UL16 and stimulates NK cytotoxicity through NKG2D receptor. *Immunity* **2001**, *14*, 123–133. [CrossRef]

53. Ashiru, O.; Bennett, N.J.; Boyle, L.H.; Thomas, M.; Trowsdale, J.; Wills, M.R. NKG2D ligand MICA is retained in the cis-Golgi apparatus by human cytomegalovirus protein UL142. *J. Virol.* **2009**, *83*, 12345–12354. [CrossRef]
54. Schneider, C.L.; Hudson, A.W. The human herpesvirus-7 (HHV-7) U21 immunoevasin subverts NK-mediated cytotoxicity through modulation of MICA and MICB. *PLoS Pathog.* **2011**, *7*, e1002362. [CrossRef]
55. Thomas, M.; Boname, J.M.; Field, S.; Nejentsev, S.; Salio, M.; Cerundolo, V.; Wills, M.; Lehner, P.J. Down-regulation of NKG2D and NKp80 ligands by Kaposi's sarcoma-associated herpesvirus K5 protects against NK cell cytotoxicity. *Proc. Natl. Acad. Sci. USA* **2008**, *105*, 1656–1661. [CrossRef]
56. McSharry, B.P.; Burgert, H.G.; Owen, D.P.; Stanton, R.J.; Prod'homme, V.; Sester, M.; Koebernick, K.; Groh, V.; Spies, T.; Cox, S.; et al. Adenovirus E3/19K promotes evasion of NK cell recognition by intracellular sequestration of the NKG2D ligands major histocompatibility complex class I chain-related proteins A and B. *J. Virol.* **2008**, *82*, 4585–4594. [CrossRef] [PubMed]
57. Sester, M.; Ruszics, Z.; Mackley, E.; Burget, H.-G. Conserved amino acids within the adenovirus 2 E3/19K protein differentially affect downregulation of MHC class I and MICA/B proteins. *J. Immunol.* **2010**, *184*, 255–257. [CrossRef]
58. Heyward, C.Y.; Patel, R.; Mace, E.M.; Grier, J.T.; Guan, H.; Makrygiannis, A.P.; Orange, J.S.; Ricciardi, R.P. Tumorigenic adenovirus 12 cells evade NK cell lysis by reducing the expression of NKG2D ligands. *Immunol. Lett.* **2012**, *4*, 16–23. [CrossRef]
59. Smith, J.G.; Silvestry, M.; Lindert, S.; Lu, W.; Nemerow, G.R.; Stewart, P.L. Insights into the mechanisms of adenovirus capsid disassembly from studies of defensin neutralization. *PLoS Pathog.* **2010**, *6*, e1000959. [CrossRef] [PubMed]
60. Holly, M.K.; Smith, J.G. Adenovirus infections of human enteroids reveals interferon sensitivity and preferential infection of goblet cells. *J. Virol.* **2018**, *92*, e00250-18. [CrossRef] [PubMed]

viruses

MDPI

Communication

Differential Regulation of Cellular FAM111B by Human Adenovirus C Type 5 E1 Oncogenes

Wing-Hang Ip *, Britta Wilkens, Anastasia Solomatina, Judith Martin, Michael Melling, Paloma Hidalgo, Luca D. Bertzbach, Thomas Speiseder and Thomas Dobner *

Department of Viral Transformation, Leibniz Institute for Experimental Virology (HPI), 20251 Hamburg, Germany; britta.wilkens@leibniz-hpi.de (B.W.); a.solomatina@uni-bonn.de (A.S.); judith_martin@gmx.net (J.M.); mickamell@outlook.de (M.M.); paloma.hidalgo@leibniz-hpi.de (P.H.); luca.bertzbach@leibniz-hpi.de (L.D.B.); thomas.speiseder@gmx.net (T.S.)

* Correspondence: winghang.ip@leibniz-hpi.de (W.-H.I.); thomas.dobner@leibniz-hpi.de (T.D.)

Abstract: The adenovirus type 5 (HAdV-C5) E1 transcription unit encodes regulatory proteins that are essential for viral replication and transformation. Among these, E1A and E1B-55K act as key multifunctional HAdV-C5 proteins involved in various steps of the viral replication cycle and in virus-induced cell transformation. In this context, HAdV-C5-mediated dysregulations of cellular factors such as the tumor suppressors p53 and pRB have been intensively investigated. However, cellular components of downstream events that could affect infection and viral transformation are widely unknown. We recently observed that cellular FAM111B is highly regulated in an E1A-dependent fashion. Intriguingly, previous reports suggest that FAM111B might play roles in tumorigenesis, but its exact functions are not known to date. Here, we set out to investigate the role of FAM111B in HAdV-C5 infections. We found that (i) FAM111B levels are upregulated early and downregulated late during infection, that (ii) FAM111B expression is differentially regulated, that (iii) FAM111B expression levels depend on the presence of E1B-55K and E4orf6 and that (iv) a FAM111B knockdown increases HAdV-C5 replication. Our data indicate that FAM111B acts as an anti-adenoviral host factor that is involved in host cell defense mechanisms in productive HAdV-C5 infection. Moreover, these findings suggest that FAM111B might play an important role in the host antiviral immune response that is counteracted by HAdV-C5 E1B-55K and E4orf6 oncoproteins.

Keywords: human adenovirus; FAM111 trypsin-like peptidase B; adenovirus early region 1 (E1), E1A; E1B-55K; E4orf6; viral replication; antiviral host factor

1. Introduction

Human adenoviruses (HAdVs) are non-enveloped, linear dsDNA viruses that mostly cause asymptomatic or mild disease in younger and immunocompetent individuals. However, HAdV infections of immunocompromised patients or patients with pre-existing respiratory or cardiac conditions can cause severe disease courses and are serious health issues [1]. Therefore, it is important to thoroughly study basic mechanisms of HAdV infection as well as the interplay between HAdV gene products and host immunity.

In a yet unpublished transcriptome analysis, we found that cellular FAM111B was strongly upregulated upon transduction with the two HAdV-C5 oncogenes E1A and E1B. The exact function of FAM111B, the second and last member of the "family with sequence similarity 111" gene family, is widely unknown. Interestingly though, a recent study demonstrated that FAM111B co-precipitates with HAdV-C5 E1B-55K [2]. FAM111B contains a predicted trypsin-like cysteine/serine peptidase domain, comprising of two subdomains connected by a linker region. Notably, three specific alterations within the linker region were previously found to induce hereditary fibrosing poikiloderma [3]. Mercier et al. showed that FAM111B is important for fibrosis development, which is a key pathological process in a variety of human diseases [3]. The trypsin-like cysteine/serine peptidase domain of FAM111B shares 45% homology with FAM111A. Mutations in this region were identified to induce Kenny–Caffey syndrome and osteocraniostenosis, which are characterized by impaired skeletal development [4,5]. Intriguingly, the 11q12 locus (and the genes encoding FAM111A and FAM111B in this region, specifically) is associated with prostate cancer susceptibility and serves as a signature to discriminate the risk of metastases [6]. In line with this, more and more recent evidence continues to point towards an involvement of FAM111B in tumorigenesis [7–9]. Finally, the SV40 virus large T antigen interacts with FAM111A to inhibit viral replication in restrictive cells and FAM111A restricts vaccinia virus (VACV) replication [10–13]. However, it was also found that SV40 virus still replicates in a restrictive cell line upon FAM111A depletion [10].

In this report, we set out to investigate the role of FAM111B in the context of HAdV-C5 infection and the regulation of this putative cellular oncogene by HAdV-C5 E1 genes.

2. Materials and Methods

2.1. Cells and Culture Conditions

H1299 (ATCC no. CRL-5803: ATCC Global Bioresource Center, Manassas, Virginia, VA, USA) and A549 (ACC 107; DSMZ-German Collection of Microorganisms and Cell Cultures, Braunschweig, Germany) were grown in DMEM (Gibco; Carlsbad, CA, USA) supplemented with 10% FCS, 100 U/mL penicillin and 100 µg/mL streptomycin. All cells were kept in incubators at 37 °C in 5% CO_2. FAM111B knockdown cell lines were generated by lentiviral transduction of A549 cells. Transduced cells were maintained under 2 µM puromycin selection. All cells were tested for mycoplasma contamination on a regular basis.

2.2. Plasmids and Transient Transfections

HAdV-C5 E1B, HA-E1A (wild type (wt), RB-, CBP- and RB/CBP-binding deficient) and E2F-1 were expressed from pcDNA plasmids (Invitrogen, Carlsbad, CA, USA) under CMV IE1-control. For transient transfections, subconfluent H1299 cells were incubated with DNA and 25-kDa linear polyethylenimine (PEI) as described previously [14]. For the short hairpin RNA (shRNA)-mediated FAM111B knockdown, we purchased shRNAs against FAM111B targeting the coding strand sequence 5′-GCCTGCCTAGTGATTCTCATT-3′ (mission shRNA, clone ID NM_198947.1-638s1c1; Sigma-Aldrich, St. Louis, MO, USA) based on the viral pLKO.1-puro vector. The shFAM111B plasmids were co-transfected with lentiviral packaging plasmids by calcium phosphate transfection according to the manufacturer's protocol (ProFection Mammalian Transfection System; Promega, Madison, WI, USA).

2.3. Luciferase Reporter Assays

For dual luciferase assays, subconfluent H1299 cells were transfected with 0.2 µg reporter (pLightSwitch-FAM111B prom; Product ID: S707175; SwitchGear Genomics, Carlsbad, CA, USA), 0.1 µg pFirefly-TK (expresses firefly luciferase under the control of the HSV-TK promoter) and 0.6 µg of effector plasmids (HAdV-C5 HA-E1A (wt, RB-, CBP-, RB/CBP-binding deficient), E1B-55K and E2F-1 as well as LeGO-iBLB2 E1B-55K and LeGO-iVLN2 E1A [15]) by PEI transfection as described above. Cell extracts were prepared 24 h post transfection (h p.t.) and renilla activity was recorded with a luminometer (Lumat LB9510, Berthold Technologies GmbH & Co.KG, Bad Wildbad, Germany).

2.4. Viruses

H5*pg*4100 harboring deletions in the E3-coding region served as the wt HAdV-C5 virus [16]. The mutant virus H5*pm*4149 (4x) harbors four stop codons within the E1B coding region and H5*pm*4154 (E4orf6-) has a stop codon at P66 of E4orf6 to abrogate translation of the respective proteins [17,18]. Viruses were propagated, titrated and used for infections as described previously [19].

Virus yield was analyzed at indicated time points by quantitative E2A immunofluorescence staining. Viral titers were determined as described before and are represented as the number of fluorescence-forming units (FFU)/µL [20].

2.5. Antibodies and Protein Analysis

Primary monoclonal and polyclonal antibodies used to detect viral proteins included mouse monoclonal E1A (M73), E1B-55K (2A6), E4orf6 (RSA3), E2A (B6-8), rat monoclonal L4-100K (6B10) and HAdV-C5 rabbit polyclonal serum L133 [21]. Primary monoclonal and polyclonal antibodies specific for cellular and ectopically expressed proteins included rabbit polyclonal FAM111B (HPA038637; Sigma-Aldrich) and mouse monoclonal β-actin (AC-15; Sigma-Aldrich). Horseradish peroxidase (HRP)-conjugated secondary antibodies anti-rabbit IgG, anti-mouse IgG and anti-rat IgG for detection of proteins by immunoblotting were obtained from Jackson/Dianova (Hamburg, Germany).

All protein extracts were incubated in radioimmunoprecipitation assay (RIPA) lysis buffer on ice for 30 min. Total cell lysates were sonicated in a high-intensity cup horn (Branson Ultrasonics, Brookfield, CT, USA) for 45 s (40 pulses; output, 0.6 and 0.8 impulses/second) before insoluble debris was removed by centrifugation (11,000 rpm, 3 min, 4 °C). Protein concentration was measured photometrically with Bradford reagent (Bio-Rad, Hercules, CA, USA). Equal amounts of total protein were separated on 10% SDS-polyacrylamide gels after denaturation (5× SDS sample buffer, 95 °C, 3 min) and subjected to immunoblotting exactly as previously described [21].

2.6. In Vitro Translation and Pulldown Assays

FLAG-tagged FAM111B was in vitro translated with the TNT Coupled Wheat Germ Systems (VWR, Radnor, PA, USA) according to the manufacturer's instruction. In vitro-translated FLAG-tagged FAM111B was purified by immunoprecipitation (IP) with an ANTI-FLAG M2 Affinity Gel (Sigma-Aldrich). Briefly, FLAG-tagged FAM111B was added to the supplied beads together with TBS buffer and protease inhibitors (1% phenylmethylsulfonyl fluoride (PMSF), 0.1% aprotinin, 1 µg/mL leupeptin, 1 µg/mL pepstatin, 25 mM iodacetamide and 25 mM N-ethylmaleimide) and incubated overnight at 4 °C. An aliquot of FLAG-tagged FAM111B was eluted from the beads using 2× Laemmli buffer as controls. The remaining beads coupled with FLAG-tagged FAM111B were incubated with protein lysates prepared in RIPA buffer as described in the previous section for 2 h at 4 °C. Bound proteins were eluted with 2× Laemmli buffer at 95 °C for 5 min.

2.7. Indirect Immunofluorescence

For indirect immunofluorescence analyses, 1.5×10^5 A549 cells were grown on glass coverslips in 6-well cell dishes one day prior to infection. Cells were fixed with paraformaldehyde (PFA; 4% (v/v) in PBS) at room temperature (RT) for 20 min. PFA-fixed cells were permeabilized with phosphate-buffered saline (PBS) supplemented with 0.5% Triton X-100 for 10 min at room temperature and blocked with 1.5 mL Tris-buffered saline-BG (TBS-BG; BG represents 5% (w/v) BSA and 5% (w/v) glycine) for 1 h at RT. Next, coverslips were incubated with the indicated primary antibody diluted in PBS in a humidity chamber for 30 min. The corresponding secondary antibody (Alexa 488-(Invitrogen), Cy3 (Jackson, West Grove, PA, USA)-conjugated secondary antibodies) diluted in PBS was added afterwards. Finally, nuclei were stained with DAPI (4,6-diaminidino-2-phenylindole) in PBS (1:1000, (v/v) from 1 mg/mL stock) for 5 min, before the cover slips were mounted in mounting solution (Energene, Regensburg, Germany). Images were acquired using a confocal spinning-disk microscope (Nikon Eclipse Ti-E stand (Nikon, Tokyo, Japan); Yokogawa CSU-W1 spinning disk (Yokogawa, Tokyo, Japan; 2× Andor888 EM-CCD camera (Oxford Instruments, Abingdon, UK); Nikon 100× NA 1.49 objective (Nikon)).

2.8. Isolation and Quantification of Nucleic Acids

RNAs were isolated by incubating fresh cell pellets (4 × 10^6 cells) in 600 µL TRIzol (ThermoFisher, Waltham, MA, USA) and 60 µL of 1-bromo-3-chloropropane for 10 min at RT. Samples were then shaken and centrifuged at 12,000× *g* for 15 min at 4 °C. The aqueous phase with nucleic acids was precipitated with 500 µL of isopropanol (12,000× *g*, 10 min, 4 °C), washed with 1 mL ethanol (75%, *v*/*v*) by vortexing and pelleted (7500× *g*, 5 min, 4 °C). The resulting pellets were air-dried for 5–10 min and resuspended in 80 µL RDD buffer containing 10 µL RNase-free DNase I (Qiagen, Hilden, Germany) to digest traces of the remaining DNA for 30 min at RT. DNase I was heat-inactivated (75 °C, 5 min) followed by RNA precipitation with RNase-free LiCl solution (final concentration, 2.5 M; Applied Biosystems, Foster City, CA, USA) for 30 min at −20 °C. Samples were then centrifuged (16,000× *g* for 20 min at 4 °C) and RNAs were washed with ice-cold ethanol (75%, *v*/*v*). After air-drying for 5–10 min, the RNA was dissolved for 10 min in 5–50 µL nuclease-free water (Qiagen). RNA was transcribed into cDNA using a reverse transcription system (Applied Biosystems) according to the manufacturer's protocol using random primers. cDNAs were quantified by real-time quantitative PCR (qPCR) using SYBR green reagents (SensiMix Plus SYBR; Quantace, London, UK), FAM111B and GAPDH-specific primers (FAM111B forward: 5′-GCCCTTGAAATGCAGAATCCA-3′; reverse: 5′-GCTGTAAACACACTACGGTCTAA-3′ and GAPDH forward: 5′-ACCACAGTCCATGCCATCAC-3′; reverse: 5′-TCCACCACCCTG TTGCTGTA-3′) and a Rotor-Gene cycler (Corbett Life Sciences, Sydney, Australia). FAM111B mRNA levels were calculated relative to the level of cellular GAPDH RNA.

3. Results and Discussion

3.1. FAM111B RNA and Protein Levels Are Upregulated in HAdV-C5-Infections

To analyze whether FAM111B transcript and protein levels are differentially regulated during HAdV-C5 infection, we infected the human lung adenocarcinoma cell line A549, that is widely used in HAdV research, with wt HAdV-C5 [16] and first investigated mRNA levels of FAM111B during infection at different time points post infection (Figure 1A). FAM111B transcript levels increased at 16–24 h post infection (h p.i.) compared to mock (Figure 1A) but interestingly dropped at 48 h p.i. (Figure 1A). These results indicate that FAM111B is regulated at the transcriptional level in response to HAdV-C5 infection, implying that FAM111B might belong to the family of immediate early response genes activated upon HAdV infection.

To check whether this is also reflected in FAM111B protein levels, we analyzed FAM111B protein expression by immunofluorescence. We stained HAdV- and mock-infected A549 cells with specific antibodies detecting FAM111B and the viral protein DBP (E2A), which formed nuclear spherical structures representing the sites of viral replication [22]. FAM111B showed diffuse nuclear staining in mock-infected cells excluding the nucleoli and a weak cytoplasmic signal, while FAM111B intensity is reduced in HAdV-C5-infected cells (Figure 1B). However, these findings are in contrast to the elevated FAM111B transcript levels at 24 h p.i. (Figure 1A). We therefore performed a time-resolved analysis to investigate FAM111B protein expression levels during a course of 8–72 h p.i. by immunoblotting (Figure 1C). Here, FAM111B levels increased early, but decreased later during infection—observations that are consistent with our previous mRNA analyses. The steady increase in FAM111B protein levels at 8–24 h was followed by a decrease at 24–72 h (Figure 1C).

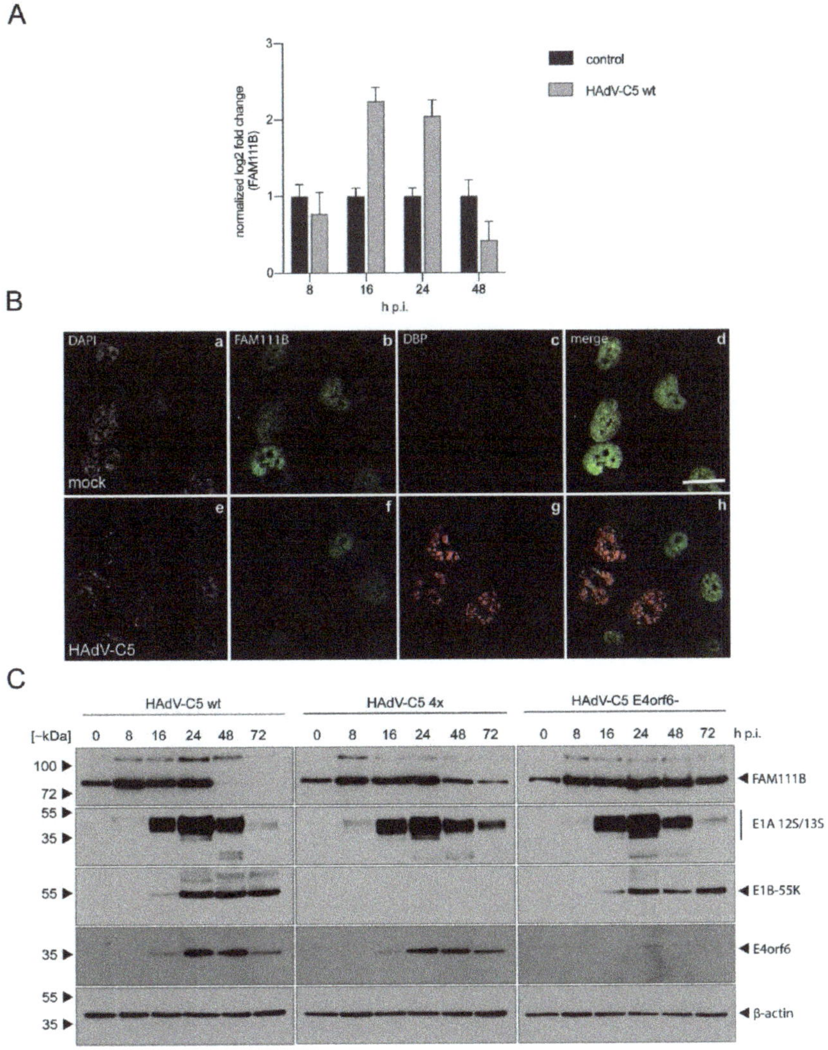

Figure 1. FAM111B transcript and protein levels are differentially regulated during infection. Subconfluent (**A**,**B**) A549 cells or (**C**) H1299 cells were left uninfected (mock) or infected with HAdV-C5 (MOI 20) and harvested at indicated time points. (**A**) Total RNA was extracted and reverse transcribed and resulting cDNA was amplified with primer pairs specific for FAM111B or cellular GAPDH by RT-PCR (ΔΔCt). The experiment was done in two triplicates. Representative FAM111B mRNA amounts are shown normalized to GAPDH. Error bars indicate standard error of mean. (**B**) Cells were fixed with 4% PFA at 24 h p.i. DBP was visualized as infection control with a mouse mAb B6-8 (α-DBP). FAM111B was visualized using the rabbit FAM111B pAb HPA038637. Primary antibodies were detected with secondary antibodies conjugated to Alexa 488 (green) or Alexa 555 (red), and nuclei were stained with DAPI. Representative staining patterns of >30 analyzed cells and overlays of the single images are shown. The scale bar in (d) corresponds to 10 μm. (**C**) H1299 cells were infected with wt HAdV-C5, E1B-55K-deleted (4×) and E4orf6-deleted (6-) viruses (MOI 20) and harvested at indicated time points. Total cell lysates were resolved by SDS-PAGE and visualized by immunoblotting. Proteins were detected using pAb HPA038637 (α-FAM111B), mAb M73 (α-E1A), mAb 2A6 (α-E1B-55K), mAb RSA3 (α-E4orf6) and mAb AC-15 (β-actin). Molecular masses in kilodaltons (kDa) are indicated on the left, while corresponding proteins are indicated on the right. Mock is indicated as 0 h p.i.

These results indicate that FAM111B expression is upregulated early in infection. In contrast, FAM111B protein levels are downregulated during the late course of infection. Complete abrogation of FAM111B protein expression late during wt HAdV-C5 infection could point towards potent post-translational regulations that remain to be elucidated. Taken together, these data show that wt HAdV-C5 infection strongly induced mRNA and protein levels of FAM111B at early time points, which decreased in the late phase of the infection.

3.2. FAM111B Expression Levels Are Differentially Regulated by E1A, E1B-55K and E4orf6

Adenovirus E4orf6 and E1B-55K proteins form a functional cullin-based E3 ubiquitin ligase complex to target a number of cellular proteins for proteasomal degradation [18,23,24]. Within the complex, E4orf6 associates via multiple BC boxes with cellular elongins B and C to enable binding of either Cul5 or Cul2 and further components to build up the core ligase complex, and E1B-55K serves as a substrate recognition factor. This complex is formed to degrade cellular proteins, which would otherwise counteract viral replication [25–27]. To investigate whether E1B and E4orf6 mediate the degradation of FAM111B during late times of infection, we used HAdV-C5 mutants that do not express E1B-55K or E4orf6 [17,18] (Figure 1C). As infection control and control for the virus mutants, we stained for E1A, E1B-55K and E4orf6 and could detect viral proteins starting at 16 h p.i. While FAM111B protein levels of wt-infected cells were not detectable at 48–72 h p.i., they were substantially reduced in cells infected with virus mutants lacking E1B-55K or E4orf6, most likely due to decreased transcript levels at these stages of infection (Figure 1C). These data suggest that the E3 ubiquitin ligase complex formed by E1B and E4orf6 together with cellular proteins targets FAM111B at late time points of infection.

Since E1B-55K serves as a substrate recognition factor to target cellular antiviral proteins for proteasomal degradation, we then tested whether E1B-55K interacts with FAM111B. We first in vitro translated FLAG-FAM111B (Figure 2A) and immobilized FAM111B using FLAG antibody-bound agarose beads. Although a considerable amount of FAM111B was still bound on the agarose beads, the eluate was positive for FAM111B after staining with a FAM111B-specific monoclonal antibody (mAb) (Figure 2B). For IPs, HAdV-C5-infected or mock-infected A549 whole-cell lysates were incubated with immobilized FLAG-tagged FAM111B. Immunoblot analysis of E1A 12S/13S and E1B-55K proteins after co-IP with the FAM111B-specific mAb revealed that FAM111B interacts only with E1B-55K but not with E1A 12S/13S (Figure 2C), supporting the idea that FAM111B interacts with E1B-55K and serves as a novel substrate of the viral E3 ubiquitin ligase complex. These data are supported by work from Hung et al., who have previously identified FAM111B as a binding partner of the viral E1B-55K protein [2].

To analyze transcriptional regulation of FAM111B by E1A and E1B, we performed luciferase reporter gene assays under control of the FAM111B promoter. Activity was measured after co-transfection with the respective constructs, demonstrating that FAM111B promoter activity is enhanced upon co-transfection with the E1A- but not the E1B-containing plasmid (Figure 3A,B). E1A is known to interact with pRB and its family members to activate E2F-dependent transcription and cell cycle entry, and transcriptional regulation occurs through CBP/p300-dependent genes by E1A [28,29]. Moreover, the FAM111B promoter is responsive to E2F-1, E2F-4 and CREB [30]. To investigate whether E1A activates the FAM111B promoter via the pRB/E2F pathway, we first analyzed responsiveness of the FAM111B promoter to E2F-1 (Figure 3C). Additionally, E1A was co-transfected to investigate whether it has an additive effect on FAM111B promoter activation (Figure 3D). We found that E2F-1 stimulated transcription of the FAM111B promoter approximately 4-fold higher compared to the FAM111B promoter alone (Figure 3C). Intriguingly, co-transfection of E2F-1 plus E1A further increased FAM111B promoter activity (Figure 3D).

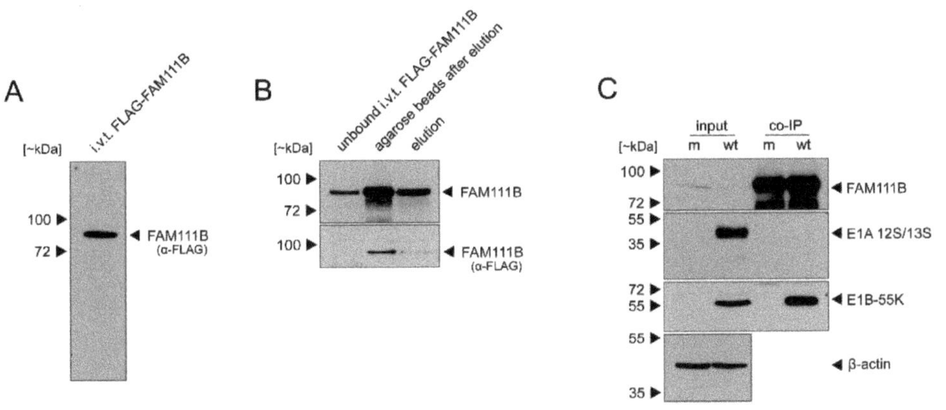

Figure 2. In vitro-translated FAM111B interacts with E1B-55K. (**A**) FLAG-tagged FAM111B was produced by in vitro transcription/translation (i.v.t.) with wheat germ extract. (**B**) Samples were coupled with anti-FLAG M2 affinity gels. (**C**) Subconfluent A549 cells were infected with wt HAdV-C5 (MOI 5) and total cell lysates were prepared at 48 h p.i. and incubated with FLAG-tagged FAM111B beads. Input cell lysates and FAM111B co-immunoprecipitated proteins were resolved by SDS-PAGE and visualized by immunoblotting. Proteins were detected using pAb HPA038637 (α-FAM111B), mAb M73 (α-E1A), mAb 2A6 (α-E1B-55K) and mAb AC-15 (β-actin). Molecular masses in kilodaltons (kDa) are indicated on the left and corresponding proteins are indicated on the right.

Finally, we performed luciferase assays with E1A mutants deficient in binding to RB or CBP or both (Figure 3E). As expected, the FAM111B promoter was activated about 4-fold after co-transfection with wt E1A (Figure 3E), while functional inactivation of E1A-binding sites to RB, CBP or both clearly diminished E1A's transcriptional stimulation of the FAM111B promoter (Figure 3E). To control for the expression levels of the different E1A mutants, whole cell lysates were immuno-stained with an E1A-specific antibody (Figure 3F). In summary, these results show that binding of E1A to p300/CBP and pRB is important for transcriptional regulation of FAM111B. While E1A seems to modulate FAM111B transcription, E1B might regulate protein levels.

3.3. FAM111B Knockdown Increases HAdV-C5 Gene Expression

Finally, we set out to analyze the role of FAM111B during HAdV-C5 infection in more detail. Therefore, we used shRNA knockdowns (KD) of FAM111B in A549 cells, in comparison to control cells that express scrambled shRNA (Figure 4A). First, the KD was confirmed by immunoblotting as FAM111B shRNA reduced steady-state levels of the FAM111B protein, while no effect in FAM111B levels was observed with the control shRNA (Figure 4A). Next, cell growth analyses revealed that A549s replicate slightly better with depleted FAM111B, although this was not statistically significant (Figure 4B). Finally, we infected A549-derived FAM111B KD cells and the respective control cell line with wt HAdV-C5 (MOI 5 and 20) and performed virus yield analysis at time points 24 h, 48 and 72 h p.i. (Figure 4C,D). The FAM111B KD enhanced progeny virus production by up to 3-fold at 48 h p.i. These results suggest that FAM111B may function as a host restriction factor that affects virus replication and virus progeny production.

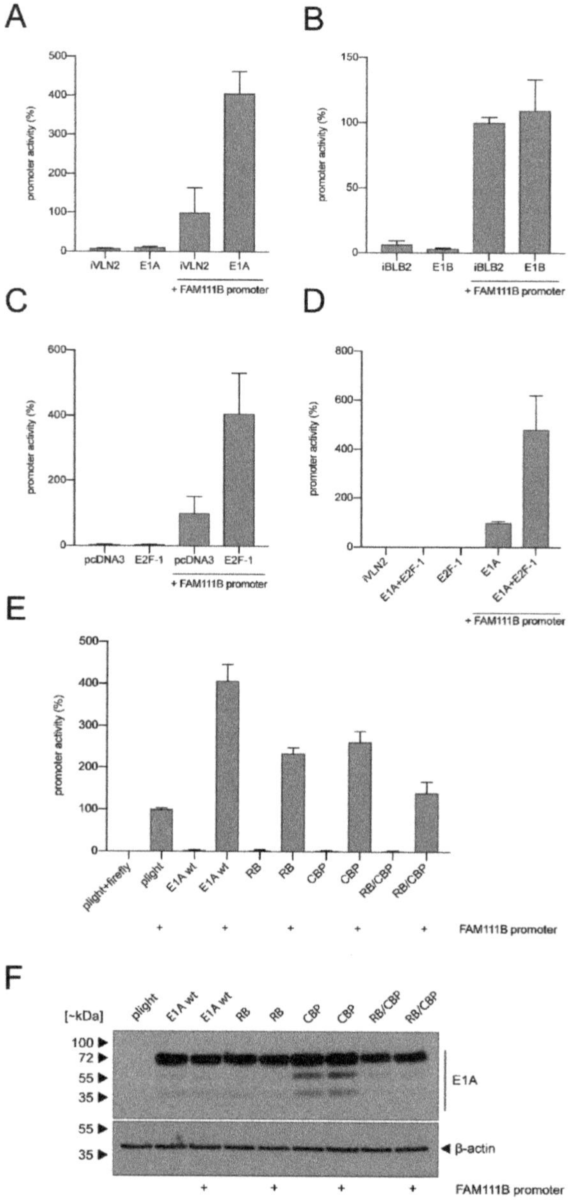

Figure 3. E1A activates FAM111B transcription that is enhanced by E2F-1 and depends on its RB/CBP motif. H1299 cells were transfected with 0.5 μg of pFirefly-TK, 0.5 μg of FAM111B promoter. (**A**) LeGO-iVLN2 E1A, (**B**) LeGO-iBLB2 E1B-55K, (**C**) E2F-1, (**D**) LeGO-iVLN2 E1A and E2F-1, (E) LeGO-iVLN2 wt E1A, RB, CBP single- or RB/CBP double-binding mutants. (F) Input levels of total cell lysates of (**E**) were detected using mouse mAb M-58 (α-E1A) and mAb AC-15 (β-actin). Luciferase activities were determined at 24 h p.t. with total cell extracts. Firefly luciferase activities are shown in percentages. (**A–E**) Representative means and standard deviations are shown. Experiments were done in two triplicates.

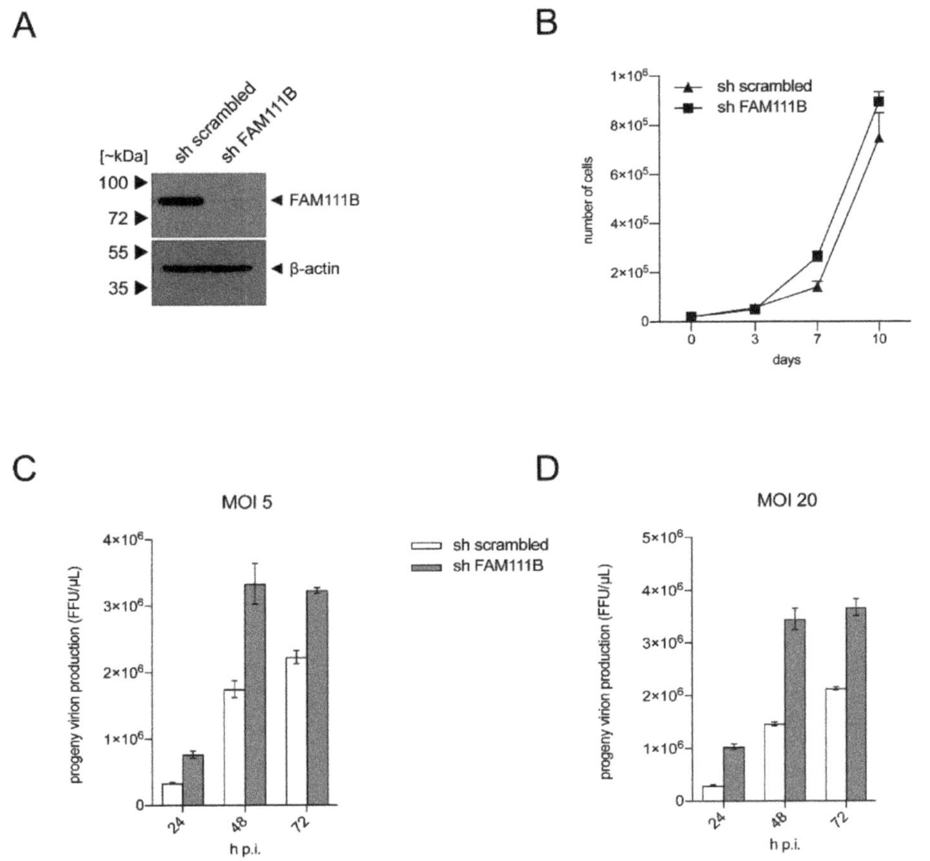

Figure 4. FAM111B depletion enhances viral progeny production. A549 cells were transduced with lentiviral particles harboring an shRNA against FAM111B and scrambled shRNA as control. Transduced cells were selected with puromycin from 48 h p.t. for 7 days. (**A**) An aliquot of transduced and positively selected cells were subjected to total cell extract preparation. Proteins were separated by SDS-PAGE and visualized by immunoblotting by applying pAb HPA038637 (α-FAM111B) and mAb AC-15 (β-actin). Molecular masses in kilodaltons (kDa) are indicated on the left and corresponding proteins on the right. (**B**) A total of 1×10^4 cells were cultivated and cell growth was monitored by determining the absolute cell numbers at the indicated time points. Values show the mean of triplicate measurements of a representative experiment. The experiment was performed twice. Error bars represent the corresponding standard error of mean. (**C**) Cells were infected with HAdV-C5 at (**C**) MOI 5 and (**D**) MOI 20 and viral particles were harvested at the indicated time points after infection. Virus yield was determined by quantitative B6-8 (α-DBP) immunofluorescence staining on HEK 293 cells after methanol fixation. The experiments were performed twice in triplicate. A representative experiment is shown for each MOI. Error bars indicate the standard error of the mean.

4. Conclusions

This study demonstrates that FAM111B expression is transcriptionally activated at early times after HAdV-C5 infection. This upregulation is mediated by E1A on a transcriptional level, resulting in elevated FAM111B expression dependent on E1A's CBP/RB-binding motifs in an E2F-dependent manner. While FAM111B protein levels increase early during infection, they are downregulated at later time points, likely mediated by the E3 ubiquitin ligase complex composed of E1B and E4orf6. Taken together, our data strongly indicate that FAM111B is a novel member of a growing list of cellular restriction factors that are activated in response to HAdV infection during the immediate early phase of the infection. This, together with the findings that the related FAM111A protein restricts SV40 and VACV replication, suggests that members of nuclear trypsin-like proteases act as restriction factors to antagonize replication of DNA viruses. Future investigations will aim to further reveal its impact on HAdV replication and also possible contributions of FAM111B to viral transformation. These data could be applied to other viral oncogenes and viral infections in general that manipulate key cellular pathways. Finally, deeper insights into the exact functions of antiviral restriction factors might potentially contribute to the identification of new therapeutic antiviral strategies.

Author Contributions: Conceptualization, T.S. and T.D.; methodology, T.S.; formal analysis, W.-H.I. and T.S.; investigation, W.-H.I., B.W., A.S., J.M., M.M. and T.S.; writing—original draft preparation, W.-H.I. and L.D.B.; writing—review and editing, W.-H.I., A.S., P.H., L.D.B. and T.D.; visualization, W.-H.I., A.S., J.M., P.H., L.D.B. and T.S.; project administration, T.D.; funding acquisition, T.D. All authors have read and agreed to the published version of the manuscript.

Funding: The Leibniz Institute for Experimental Virology (HPI) is supported by the Freie und Hansestadt Hamburg and the Bundesministerium für Gesundheit. This work was supported by Wilhelm Sander-Stiftung and Deutsche Forschungsgemeinschaft (Do 343/7-1) grants.

Acknowledgments: We thank Stefanie Wernick for excellent technical assistance.

Conflicts of Interest: The authors declare no conflict of interest.

References

1. Lion, T. Adenovirus infections in immunocompetent and immunocompromised patients. *Clin. Microbiol. Rev.* **2014**, *27*, 441–462. [CrossRef] [PubMed]
2. Hung, G.; Flint, S.J. Normal human cell proteins that interact with the adenovirus type 5 E1B 55kDa protein. *Virology* **2017**, *504*, 12–24. [CrossRef]
3. Mercier, S.; Kury, S.; Shaboodien, G.; Houniet, D.T.; Khumalo, N.P.; Bou-Hanna, C.; Bodak, N.; Cormier-Daire, V.; David, A.; Faivre, L.; et al. Mutations in FAM111B cause hereditary fibrosing poikiloderma with tendon contracture, myopathy, and pulmonary fibrosis. *Am. J. Hum. Genet.* **2013**, *93*, 1100–1107. [CrossRef]
4. Abraham, M.B.; Li, D.; Tang, D.; O'Connell, S.M.; McKenzie, F.; Lim, E.M.; Hakonarson, H.; Levine, M.A.; Choong, C.S. Short stature and hypoparathyroidism in a child with Kenny-Caffey syndrome type 2 due to a novel mutation in FAM111A gene. *Int. J. Pediatr. Endocrinol.* **2017**, *2017*, 1. [CrossRef]
5. Nikkel, S.M.; Ahmed, A.; Smith, A.; Marcadier, J.; Bulman, D.E.; Boycott, K.M. Mother-to-daughter transmission of Kenny-Caffey syndrome associated with the recurrent, dominant FAM111A mutation p.Arg569His. *Clin. Genet.* **2014**, *86*, 394–395. [CrossRef] [PubMed]
6. Akamatsu, S.; Takata, R.; Haiman, C.A.; Takahashi, A.; Inoue, T.; Kubo, M.; Furihata, M.; Kamatani, N.; Inazawa, J.; Chen, G.K.; et al. Common variants at 11q12, 10q26 and 3p11.2 are associated with prostate cancer susceptibility in Japanese. *Nat. Genet.* **2012**, *44*, 426–429. [CrossRef] [PubMed]
7. Kawasaki, K.; Nojima, S.; Hijiki, S.; Tahara, S.; Ohshima, K.; Matsui, T.; Hori, Y.; Kurashige, M.; Umeda, D.; Kiyokawa, H.; et al. FAM111B enhances proliferation of KRAS-driven lung adenocarcinoma by degrading p16. *Cancer Sci.* **2020**, *111*, 2635–2646. [CrossRef]
8. Sun, H.; Liu, K.; Huang, J.; Sun, Q.; Shao, C.; Luo, J.; Xu, L.; Shen, Y.; Ren, B. FAM111B, a direct target of p53, promotes the malignant process of lung adenocarcinoma. *Onco Targets Ther.* **2019**, *12*, 2829–2842. [CrossRef] [PubMed]
9. Xiong, W.; Wu, X.; Starnes, S.; Johnson, S.K.; Haessler, J.; Wang, S.; Chen, L.; Barlogie, B.; Shaughnessy, J.D., Jr.; Zhan, F. An analysis of the clinical and biologic significance of TP53 loss and the identification of potential novel transcriptional targets of TP53 in multiple myeloma. *Blood* **2008**, *112*, 4235–4246. [CrossRef] [PubMed]
10. Fine, D.A.; Rozenblatt-Rosen, O.; Padi, M.; Korkhin, A.; James, R.L.; Adelmant, G.; Yoon, R.; Guo, L.; Berrios, C.; Zhang, Y.; et al. Identification of FAM111A as an SV40 host range restriction and adenovirus helper factor. *PLoS Pathog.* **2012**, *8*, e1002949. [CrossRef] [PubMed]
11. Nie, M.; Oravcová, M.; Jami-Alahmadi, Y.; Wohlschlegel, J.A.; Lazzerini-Denchi, E.; Boddy, M.N. Activation of FAM111A Protease Induces Defects in Nuclear Function that Likely Underlie its Roles in Disease and Viral Restriction. *bioRxiv* **2020**. [CrossRef]
12. Panda, D.; Fernandez, D.J.; Lal, M.; Buehler, E.; Moss, B. Triad of human cellular proteins, IRF2, FAM111A, and RFC3, restrict replication of orthopoxvirus SPI-1 host-range mutants. *Proc. Natl. Acad. Sci. USA* **2017**, *114*, 3720–3725. [CrossRef]
13. Tarnita, R.M.; Wilkie, A.R.; DeCaprio, J.A. Contribution of DNA Replication to the FAM111A-Mediated Simian Virus 40 Host Range Phenotype. *J. Virol.* **2019**, *93*. [CrossRef]
14. Schreiner, S.; Wimmer, P.; Sirma, H.; Everett, R.D.; Blanchette, P.; Groitl, P.; Dobner, T. Proteasome-dependent degradation of Daxx by the viral E1B-55K protein in human adenovirus-infected cells. *J. Virol.* **2010**, *84*, 7029–7038. [CrossRef]

15. Speiseder, T.; Hofmann-Sieber, H.; Rodriguez, E.; Schellenberg, A.; Akyuz, N.; Dierlamm, J.; Spruss, T.; Lange, C.; Dobner, T. Efficient Transformation of Primary Human Mesenchymal Stromal Cells by Adenovirus Early Region 1 Oncogenes. *J. Virol.* **2017**, *91*, e01782-16. [CrossRef] [PubMed]
16. Groitl, P.; Dobner, T. Construction of adenovirus type 5 early region 1 and 4 virus mutants. *Methods Mol. Med.* **2007**, *130*, 29–39. [CrossRef]
17. Kindsmüller, K.; Groitl, P.; Härtl, B.; Blanchette, P.; Hauber, J.; Dobner, T. Intranuclear targeting and nuclear export of the adenovirus E1B-55K protein are regulated by SUMO1 conjugation. *Proc. Natl. Acad. Sci. USA* **2007**, *104*, 6684–6689. [CrossRef] [PubMed]
18. Blanchette, P.; Kindsmüller, K.; Groitl, P.; Dallaire, F.; Speiseder, T.; Branton, P.E.; Dobner, T. Control of mRNA export by adenovirus E4orf6 and E1B55K proteins during productive infection requires E4orf6 ubiquitin ligase activity. *J. Virol.* **2008**, *82*, 2642–2651. [CrossRef] [PubMed]
19. Kindsmüller, K.; Schreiner, S.; Leinenkugel, F.; Groitl, P.; Kremmer, E.; Dobner, T. A 49-kilodalton isoform of the adenovirus type 5 early region 1B 55-kilodalton protein is sufficient to support virus replication. *J. Virol.* **2009**, *83*, 9045–9056. [CrossRef]
20. Bürck, C.; Mund, A.; Berscheminski, J.; Kieweg, L.; Müncheberg, S.; Dobner, T.; Schreiner, S. KAP1 Is a Host Restriction Factor That Promotes Human Adenovirus E1B-55K SUMO Modification. *J. Virol.* **2016**, *90*, 930–946. [CrossRef] [PubMed]
21. Freudenberger, N.; Meyer, T.; Groitl, P.; Dobner, T.; Schreiner, S. Human Adenovirus Core Protein V Is Targeted by the Host SUMOylation Machinery To Limit Essential Viral Functions. *J. Virol.* **2018**, *92*. [CrossRef]
22. Puvion-Dutilleul, F.; Pedron, J.; Cajean-Feroldi, C. Identification of intranuclear structures containing the 72K DNA-binding protein of human adenovirus type 5. *Eur. J. Cell Biol.* **1984**, *34*, 313–322.
23. Cheng, C.Y.; Blanchette, P.; Branton, P.E. The adenovirus E4orf6 E3 ubiquitin ligase complex assembles in a novel fashion. *Virology* **2007**, *364*, 36–44. [CrossRef]
24. Schreiner, S.; Wimmer, P.; Dobner, T. Adenovirus degradation of cellular proteins. *Future Microbiol.* **2012**, *7*, 211–225. [CrossRef]
25. Blanchette, P.; Cheng, C.Y.; Yan, Q.; Ketner, G.; Ornelles, D.A.; Dobner, T.; Conaway, R.C.; Conaway, J.W.; Branton, P.E. Both BC-box motifs of adenovirus protein E4orf6 are required to efficiently assemble an E3 ligase complex that degrades p53. *Mol. Cell. Biol.* **2004**, *24*, 9619–9629. [CrossRef]
26. Cheng, C.Y.; Gilson, T.; Wimmer, P.; Schreiner, S.; Ketner, G.; Dobner, T.; Branton, P.E.; Blanchette, P. Role of E1B55K in E4orf6/E1B55K E3 ligase complexes formed by different human adenovirus serotypes. *J. Virol.* **2013**, *87*, 6232–6245. [CrossRef] [PubMed]
27. Querido, E.; Blanchette, P.; Yan, Q.; Kamura, T.; Morrison, M.; Boivin, D.; Kaelin, W.G.; Conaway, R.C.; Conaway, J.W.; Branton, P.E. Degradation of p53 by adenovirus E4orf6 and E1B55K proteins occurs via a novel mechanism involving a Cullin-containing complex. *Genes Dev.* **2001**, *15*, 3104–3117. [CrossRef] [PubMed]
28. Pelka, P.; Miller, M.S.; Cecchini, M.; Yousef, A.F.; Bowdish, D.M.; Dick, F.; Whyte, P.; Mymryk, J.S. Adenovirus E1A directly targets the E2F/DP-1 complex. *J. Virol.* **2011**, *85*, 8841–8851. [CrossRef] [PubMed]
29. Pelka, P.; Ablack, J.N.; Torchia, J.; Turnell, A.S.; Grand, R.J.; Mymryk, J.S. Transcriptional control by adenovirus E1A conserved region 3 via p300/CBP. *Nucleic Acids Res.* **2009**, *37*, 1095–1106. [CrossRef]
30. Rouillard, A.D.; Gundersen, G.W.; Fernandez, N.F.; Wang, Z.; Monteiro, C.D.; McDermott, M.G.; Ma'ayan, A. The harmonizome: A collection of processed datasets gathered to serve and mine knowledge about genes and proteins. *Database* **2016**, *2016*. [CrossRef] [PubMed]

viruses

MDPI

Article

Heparan Sulfate Is a Cellular Receptor for Enteric Human Adenoviruses

Anandi Rajan [1,2,3,*], Elin Palm [1,2], Fredrik Trulsson [1,2,4], Sarah Mundigl [1,2], Miriam Becker [1,2,5,†], B. David Persson [1,2,6], Lars Frängsmyr [1,2] and Annasara Lenman [1,5,†]

1 Section of Virology, Department of Clinical Microbiology, Umeå University, 90185 Umeå, Sweden; elin.palm@umu.se (E.P.); c.f.b.trulsson@lumc.nl (F.T.); sarah.mundigl@outlook.de (S.M.); miriam.becker@umu.se (M.B.); david.persson@sva.se (B.D.P.); lars.frangsmyr@umu.se (L.F.); anna-sara.lenman@umu.se (A.L.)
2 The Laboratory for Molecular Infection Medicine Sweden (MIMS), Umeå University, 90185 Umeå, Sweden
3 Department of Medical Biochemistry and Cell Biology, University of Gothenburg, 41390 Gothenburg, Sweden
4 Department of Cell and Chemical Biology, Leiden University Medical Center, 2333 ZC Leiden, The Netherlands
5 Institute for Experimental Virology, TWINCORE, Centre for Experimental and Clinical Infection Research, 30625 Hannover, Germany
6 National Veterinary Institute, SVA, 75189 Uppsala, Sweden
* Correspondence: anandi.rajan@gu.se
† A joint venture between the Medical School Hannover and the Helmholtz Centre for Infection Research.

Abstract: Human adenovirus (HAdV)-F40 and -F41 are leading causes of diarrhea and diarrhea-associated mortality in children under the age of five, but the mechanisms by which they infect host cells are poorly understood. HAdVs initiate infection through interactions between the knob domain of the fiber capsid protein and host cell receptors. Unlike most other HAdVs, HAdV-F40 and -F41 possess two different fiber proteins—a long fiber and a short fiber. Whereas the long fiber binds to the Coxsackievirus and adenovirus receptor (CAR), no binding partners have been identified for the short fiber. In this study, we identified heparan sulfate (HS) as an interaction partner for the short fiber of enteric HAdVs. We demonstrate that exposure to acidic pH, which mimics the environment of the stomach, inactivates the interaction of enteric adenovirus with CAR. However, the short fiber:HS interaction is resistant to and even enhanced by acidic pH, which allows attachment to host cells. Our results suggest a switch in receptor usage of enteric HAdVs after exposure to acidic pH and add to the understanding of the function of the short fibers. These results may also be useful for antiviral drug development and the utilization of enteric HAdVs for clinical applications such as vaccine development.

Keywords: enteric adenovirus; heparan sulfate; short fibers; capsid proteins; fiber knobs

1. Introduction

The *Adenoviridae* family contains more than 100 types of human adenoviruses (HAdVs), which are classified into seven species (A to G) based on serum neutralization and genome sequence analyses [1,2]. HAdVs cause different types of infections such as ocular (species B, D and E), respiratory (species A, B, C and E) and enteric infections (species F) [3]. HAdV-F40 and -F41 are the only members of species F and are referred to as enteric human adenoviruses. They cause gastrointestinal infections primarily in children below five years of age [4,5]. With developments in molecular diagnostics, enteric HAdV-F40 and -F41 are now recognized as a leading cause of gastroenteritis and diarrhea-associated mortality in young children [6,7]. The HAdV particle consists of three major capsid proteins: fiber, penton base and hexon. The protruding fiber is needed for initial attachment to host cell receptors [8,9], followed by secondary binding of the penton base to integrins leading to internalization of the virus into the cells [10–12]. Several attachment receptors have been identified for HAdVs, including the Coxsackievirus and adenovirus receptor (CAR) [8], CD46 [13–15], desmoglein 2 [16] and sialic acid-containing glycans [9,17,18]. HAdV-C2 and -C5 [19,20] and, to some extent, HAdV-B3 and -B35 [21] bind heparan sulfate as receptors/coreceptors to facilitate infection. Strikingly, sulfated glycosaminoglycans (GAGs) can also act as decoy receptors for sialic acid-binding HAdV-D37, where binding of the virus to GAGs prevent efficient infection [22]. Thus, GAGs play an important role for several HAdV types, but have not been associated with enteric adenoviruses until now. HS and other GAGs such as chondroitin sulfate (CS), dermatan sulfate (DS), keratan sulfate (KS) and hyaluronic acid (HA) are long, unbranched polysaccharides consisting of repeating disaccharide units of an amino sugar (N-acetylglucosamine or N-acetylgalactosamine) and an uronic sugar (iduronic or glucuronic acid) or galactose [23].

Enteric HAdV-F40 and -F41 have two different types of fibers, long fibers (LF) and short fibers (SF), a characteristic shared only with HAdV-G52 and no other human adenovirus [9,24,25]. HAdV-G52 was isolated from a small outbreak of gastroenteritis [26], and, although morphologically similar to HAdV-F40 and -F41, this virus has been classified into a new species—species G. In HAdV-F40 and -F41, the SF is more abundant in the capsid than the LF with a ratio of 6:1 [25]. The LF binds CAR [8], but no cellular interaction partners have been identified for the SFs. The SF of HAdV-F41 stimulates enteroendocrine cells of the small intestine, called enterochromaffin cells, to secrete serotonin, which activates enteric glial cells causing diarrhea and vomiting [27].

Unlike most other HAdVs, which can infect cells of multiple organs, enteric HAdVs exclusively infect the intestinal tract and do not cause any infections at other sites. We recently showed that this restricted tropism can be attributed to differences in the external structure of the virion as compared to respiratory and ocular HAdVs [28]. The restricted tropism has also been suggested to depend on the ability of the SF of HAdV-F40 and HAdV-F41 to protect the virus against acidic gastric conditions [29,30]. Here, we identify HS as a cellular attachment factor for the SF. Our findings also suggest a switch in receptor usage after exposure to acidic pH from being LF:CAR dependent to SF:HS dependent, further supporting the theory that SFs confer the enteric tropism of these viruses.

2. Materials and Methods

2.1. Cells, Viruses and Antibodies

Cells: A549 cells (a kind gift from Dr. Alistair Kidd) were cultured in Dulbecco's Modified Eagle Medium (DMEM; Sigma-Aldrich, St. Louis, MO, USA) with 5% fetal bovine serum (FBS; Thermo Fisher Scientific, Waltham, MA, USA), 20 mM HEPES (Sigma-Aldrich, St. Louis, MO, USA) and 20 U/mL penicillin +20 μg/mL streptomycin (PEST; Thermo Fisher Scientific, Waltham, MA, USA); HAP1-WT, HAP1-ΔCAR (Horizon Discovery, Waterbeach, UK) and HAP1-ΔGAG (beta-1,3-galactosyltransferase deficient; a kind gift from Dr. Frank Kuppeveld) were cultured in Iscove's Modified Dulbecco's Medium (IMDM; Gibco, Thermo Fischer Scientific, Waltham, MA, USA) with 10% FBS and PEST; CHO-K1, and CHO-ΔHS (pgsD-677) [31] were cultured in Ham's F12 (Sigma-Aldrich, St. Louis, MO, USA) with 10% FBS and PEST. HuTu80 (ATCC, Manassas, VA, USA) cells were grown in DMEM with PEST and 10% FBS; Pro-5 (LGC Promochem, Teddington, UK) and Lec2 cells (LGC Promochem, Teddington, UK) were grown as described previously [32,33]. Human neuroblastoma SK-N-SH cells (LGC Promochem, Teddington, UK) were grown in DMEM supplemented with 10% FBS, 20 mM HEPES and PEST. SH-SY5Y cells (LGC Promochem, Teddington, UK) were grown in DMEM:Ham's-F12 (Sigma-Aldrich, St. Louis, MO, USA) at 1:1, with the same supplements as the parental SK-N-SH cell line.

Viruses/Vector: HAdV-F40 (strain Hovix), HAdV-F41 (strain Tak), HAdV-D36 (strain 275), HAdV-D37 (strain 1477), HAdV-C5 (strain adenoid 75) were grown in A549 cells with or without ^{35}S-labeling as described previously in [34] except that the elution buffer after NAP column (GE Healthcare, Chicago, IL, USA) purification and storage buffer was phosphate buffered saline (PBS). Virions were stored in PBS with 10% glycerol at −80 °C. HAdV-F41 GFP vector was produced as described in [35].

Antibodies: anti-RGS His mouse monoclonal antibody (Qiagen, Hilden, Germany) and Alexa Fluor 488 donkey antimouse IgG (Life technologies, Carlsbad, CA, USA) were used in flow cytometry to detect recombinant fiber knobs. Serotype-specific rabbit polyclonal antisera against HAdV-C5, -D36, -D37, -F40 and -F41 were a kind gift from Dr. Göran Wadell [36]. Anti-HS (clone F58-10E4, Amsbio, Abingdon, UK) and anti-CAR (clone RmcB, Sigma-Aldrich, St. Louis, MO, USA) monoclonal antibodies were used to check expression levels of HS and CAR, respectively.

2.2. Recombinant Fiber Knobs, Enzymes, Metabolic Inhibitors, Soluble GAGs and Cholera Toxin

Recombinant FKs: Cloning, expression and purification of 40SFK (amino acids 215 to 387), 40LFK (amino acids 348 to 547), 41LFK (amino acids 363 to 562), 5FK (amino acids 387 to 581) and 37FK (amino acids 172 to 365) were performed in a similar manner as described in [9] and 52SFK was produced as described in [9].

Enzymes: *Vibrio cholera* neuraminidase (Sigma-Aldrich, St. Louis, MO, USA), fig tree latex ficin (Sigma-Aldrich, St. Louis, MO, USA), *Engyodontium album* proteinase K (Sigma-Aldrich, St. Louis, MO, USA) proteases and *Flavobacterium heparinum* heparinase III (Sigma-Aldrich, St. Louis, MO, USA).

Metabolic inhibitors: P4 [(1R,2R)-1-phenyl-2-hexadecanoylamino-3-pyrrolidino-1-propanol] and inactive enantiomer of P4 (1S,2S, both kindly provided by Dr. Ronald L. Schnaar).

Soluble GAGs: Heparin (from porcine intestinal mucosa; Sigma-Aldrich, St. Louis, MO, USA), chondroitin sulfate A (from bovine trachea; Sigma-Aldrich, St. Louis, MO, USA), dermatan sulfate (from porcine intestinal mucosa; Sigma-Aldrich, St. Louis, MO, USA), chondroitin sulfate mix A and C (from shark cartilage; Sigma-Aldrich, St. Louis, MO, USA) and hyaluronic acid (from *Streptococcus equi*; Sigma-Aldrich, St. Louis, MO, USA). Alexa Fluor 488 conjugated cholera toxin subunit B, which binds the ganglioside GM1 (Molecular probes, Invitrogen, Carlsbad, CA, USA).

Recombinant human CAR (CXADR Fc chimera full length extracellular D1D2 domain; R&D systems, Minneapolis, MN, USA).

2.3. Fiber Knob Binding Assay

FK binding to cells was evaluated by flow cytometry. Cells (A549, CHO-K1, CHO-ΔHS, Pro-5, Lec2, SK-N-SH, SH-SY5Y or HuTu80) were detached from culture flasks by PBS-EDTA (0.05% EDTA), counted and then reactivated in growth media for one hour at 37 °C under shaking conditions. After this incubation, the cells (2×10^5 cells/mL) were added to a V-bottom 96-well plate and washed with serum-free media. We added 10 μg/mL of fiber knobs in serum-free media to the cells and incubated for one hour at 4 °C. The cell and fiber knob mix was washed with FACS buffer (PBS with 2% FBS) to remove unbound fiber knobs. To detect bound fiber knobs, cells were first incubated with anti-RGS His antibody (diluted 1:200 in FACS buffer) for 30 min at 4 °C, washed and then incubated with Alexa Fluor 488 donkey antimouse IgG antibody for another 30 min at 4 °C. After a final wash, samples were analyzed using FACSLSRII flow cytometer (Becton Dickinson, Franklin Lakes, NJ, USA) and results were analyzed by FACSDiva software (Becton Dickinson, Franklin Lakes, NJ, USA). The results were expressed as geometrical mean of Alexa Fluor 488 fluorescence. FK binding experiment was performed with the following variations: Before addition of the FKs, cells were incubated with or without (a) different concentrations of ficin and proteinase K for 30 min at 37 °C to degrade proteins on the cell surface (b) 2.5 μM each of P4 or inactive P4 for 5 days at 37 °C to inhibit glycolipid biosynthesis via the glycosylceramide synthase enzyme as described in [37], (c) 10 mU/mL of *Vibrio cholera* neuraminidase for one hour at 37 °C to remove cell surface sialic acids, or (d) 1 U/mL of *Flavobacterium heparinum* heparinase III for one hour at 37 °C to degrade HS. Another variation to this experiment was (e) the preincubation of fiber knobs with a wide range of concentrations of soluble GAGs for one hour at 4 °C before addition to cells.

2.4. Infection Assay

A549, CHO-WT, CHO-ΔHS or HuTu80 cells were grown as monolayers overnight at 37 °C in 96-well clear bottom plates (Greiner bio-one, Kremsmünster, Austria). For experiments blocking infection with soluble GAGs, the cell monolayers were washed three times with serum-free medium and incubated with heparin-treated virions for one hour on ice. Unbound virions were removed by washing the cells three times with serum-free medium. Medium with 1% FBS was then added and the cells were incubated for 44 h at 37 °C. Next, the cells were washed with PBS and fixed with ice-cold methanol for 10 min at −20 °C. Viral capsid proteins were stained with polyclonal rabbit antibodies diluted in PBS for 30 min at room temperature. The cells were washed twice for 15 min with PBS and then incubated with Alexa Fluor 647 goat antirabbit IgG (H+L) secondary antibody for detection of infected cells, and Hoechst 33342 (Thermo Fisher Scientific, Waltham, MA, USA) for visualization of cellular nuclei. After two more washes with PBS, number of transfected cells were analyzed using the Tina program of the TROPHOS (Luminy Biotech Enterprises, Marseille, France). For transduction with HAdV-F41 vector, the vector was incubated with CHO cells for two hours on ice before adding medium with 1% FBS and incubating the cells at 37 °C for 72 h. The cells were given a final PBS wash before analyzing them on the TROPHOS.

2.5. ^{35}S-Labeled Virion Binding Assay

Cells were detached with PBS-EDTA, reactivated in growth medium for one hour at 37 °C, pelleted in V-bottom 96-well plates (2×10^5 cells/mL) and then washed with serum-free media. We preincubated 2×10^9 ^{35}S-labeled virions with or without different concentration of heparin in serum-free media for one hour at 4 °C and then added to the cells in the plate and incubated for another hour at 4 °C (all steps were done in suspension). Unbound virions were washed away with PBS and cell associated radioactivity was measured in a Wallac 1450 Microbeta liquid scintillation counter (Perkin Elmer, Waltham, MA, USA). In the experiment with HAP1 cells, cells were first incubated with 2 μg/mL of 41LFKs for one hour on ice. Meanwhile, the virions were treated with simulated gastric fluid without enzyme (pH 1.1–1.3; 2.0 g/l NaCl and 3.0 g/l HCl; based on synthetic gastric fluid formulation by Ricca Chemical Company catalog #7108-16) or incubated with plain DMEM for five minutes at 37 °C, and then neutralized with 20× volume of serum-free media. Virions were subsequently incubated with or without 100 μM soluble heparin for one hour on ice. The fiber knobs were washed away from the cells with serum-free media and the virus-heparin mixture was added to the cells for one hour on ice.

2.6. HAdV Uptake Assay

Virus labeling: Purified HAdV-F40, -F41 and -C5 virions were labeled on free amine groups with Alexa Fluor 488 NHS ester (Thermo Fisher Scientific, Waltham, MA, USA) by incubating with a 10-fold molar excess of the dye in PBS, pH 7.4, for one hour under rotation. Virions were then separated from free dye by CsCl gradient centrifugation as in the initial purification described under Viruses/Vectors, with subsequent desalting on NAP columns and elution in PBS, pH 7.4. Virions were supplemented with 4% glycerol and frozen at −80 °C.

Imaging-based HAdV entry assay: A549 cells were seeded in 8-well μ-slides (Ibidi, Gräfelfing, Germany; 45,000 cells/well) 24 h prior to experimentation. Alexa Fluor 488-labeled HAdVs were preincubated with 0 μM, 100 μM or 1000 μM soluble heparin in medium without FBS on ice for one hour. The virus-heparin mixtures were then added to the cells and incubated for one hour on ice. Unbound virions were removed, and cells were supplemented with growth medium to allow for virus entry for five hours at 37 °C. The cells were then transferred back on ice, washed with cold PBS, blocked with cold

PBS containing 3% BSA for 15 min and stained for remaining extracellular virus particles using anti-Alexa Fluor 488 antibody (Thermo Fisher Scientific, Waltham, MA, USA) and antirabbit Alexa Fluor 568 antibody (Thermo Fisher Scientific, Waltham, MA, USA) in PBS with 3% BSA for 45 min each. Cells were washed with cold PBS and subsequently fixed with 4% PFA at room temperature for 20 min. Cells were then stained with Hoechst 33342 (Thermo Fisher Scientific, Waltham, MA, USA) and Alexa Fluor 647-labeled wheat germ agglutinin (WGA; Thermo Fisher Scientific, Waltham, MA, USA) for visualization of cellular nuclei and plasma membrane, respectively. Images were acquired on a Leica SP8 confocal microscope with a 63× oil objective at the BICU imaging unit of Umeå University. Sufficient fields of view were acquired for each condition in independent experiments to reach a total of more than 100 cells. For HAdV-C5 and HAdV-F40, single medial focal planes from random fields of view were acquired. For HAdV-F41, cell spanning stacks of 10 slices with 0.5 µm step size from random fields of view were imaged and converted by maximum intensity projection (Fiji; [38]). Internalization ratio was determined by primary object identification in CellProfiler (version 3.1.9; [39]) as follows: cell outlines were determined using the cell counter stain WGA. Cell mask was transferred onto the channels for all virus (Alexa Fluor 488) and extracellular virus (Alexa Fluor 568). Within the masked images, virus particles per cell area were identified by primary object detection.

2.7. Surface Plasmon Resonance

SPR experiments were performed using BIAcore T200 optical biosensors (GE Healthcare, Chicago, IL, USA). CAR was covalently immobilized to CM5 sensor chips by standard EDC/NHS coupling for 420 s at a flow rate of 10 µL/minute using a 50 µg/mL concentration in 10 mM sodium acetate, pH 4.0. For heparin interaction a heparin chip (Xantec, Düsseldorf, Germany) was used. For each BIAcore kinetic experiment, HAdV-F40, -F41 or 40SFK was pretreated with or without acidic pH as described under ^{35}S-labeled virion binding assay, followed by a buffer exchange to running buffer (20 mM sodium phosphate, 150 mM NaCl, with 0.05% (v/v) surfactant P20, pH 7.4) and injected for 120 s at 30 µL/minute followed by 120 s of dissociation. All covalent surfaces were regenerated with one 30 s pulse of 10 mM glycine–HCl (pH 1.5, GE Healthcare, Chicago, IL, USA). All experiments were conducted at 25 °C. Virus binding to CAR was corrected for binding to an empty channel. Sensorgrams were calculated and processed using BIAcore T200 evaluation software (version 2.0, GE Healthcare, Chicago, IL, USA).

2.8. Statistical Analysis

The results are expressed as mean ± standard error of mean (SEM) and either t-test or one-way ANOVA with Dunnett's multiple comparison's test was performed using GraphPad Prism version 8.2.1. p-values < 0.05 were considered statistically significant. All experiments were performed at least three times with duplicate samples.

3. Results

3.1. HAdV-F40 Short Fiber Knob Binding to Cells Requires Heparan Sulfate-Containing Proteins

The knob domain of the fiber protein (FK) mediates the attachment of HAdVs to their cellular receptors on host cells. Assuming that the short fiber knob (SFK) of enteric HAdVs also contributes to cell attachment, we first characterized the nature of HAdV-F40 SFK interaction partners by flow cytometry using A549 cells, a cell line that supports productive infection of enteric HAdVs [40]. To test for dependence of proteins and glycolipids, we first treated cells with either ficin, a cysteine endopeptidase or proteinase K, a broad-spectrum serine protease to remove cell surface proteins, or with P4, an inhibitor of *de novo* glycolipid biosynthesis. Treatment with ficin and proteinase K reduced binding of 40SFK by more than 80% (Figure 1A), while the glycolipid biosynthesis inhibitor P4 had no effect on binding of any of the FKs used (Figure 1B). We used

CAR-binding HAdV-C5 fiber knob (5FK) and HAdV-F40 long fiber knob (40LFK) as controls for the protease treatment and the glycolipid-binding cholera toxin subunit B as a control for P4 treatment. As expected, protease treatment reduced 5FK and 40LFK binding and P4 treatment diminished binding of cholera toxin. Both 5FK and 40LFK bind CAR, but we observed a remarkable difference in sensitivity to protease treatment between the two, with 40LFK being substantially more sensitive than 5FK. In addition, we observed an increase in 5FK binding at low proteinase K concentrations that could potentially result from the partial removal of the protein factors from the cell surface revealing novel binding sites for 5FK. We did not evaluate these results further as it did not fall within the scope of this study. The strong reduction observed for 40SFK binding after protease treatment indicated that a protein component is required for its efficient binding to A549 cells. Since P4 did not inhibit binding of 5FK, 40SFK or 40LFK, we concluded that glycolipids are not a target for these FKs. Next, we investigated whether the 40SFK:cell interaction was a pure protein-protein interaction or if glycans could also be involved. Along with being receptors for other HAdV types, sialic acid-containing glycans [41] and HS proteoglycans [20] are abundantly expressed in the intestinal glycocalyx [42,43], which makes them ideal attachment factors for viruses attacking from the lumen. We treated A549 cells with *V. cholerae* neuraminidase, which cleaves all sialic acids, and evaluated the binding of FKs. Neuraminidase pretreatment reduced the binding of 40SFK by 30% and 40LFK by 10% (Figure 1C), while the sialic acid-binding HAdV-G52 short fiber knob (52SFK) control [9] was reduced by more than 85%. Since neuraminidase reduced both 40SFK and 40LFK binding to A549 cells, we also analyzed FK binding to sialic acid-expressing Pro-5 cells and Pro-5-derived, sialic acid-lacking Lec2 cells [32,33]. We observed no significant reduction in 40SFK binding to neuraminidase-treated Pro-5 cells as compared to untreated Pro-5 cells. However, a slightly reduced binding was detected to sialic acid-deficient Lec2 cells as compared to sialic acid-expressing Pro-5 cells (Figure 1D). As expected, a more significant reduction was seen for the sialic acid-binding 52SFK control, while the CAR-binding 5FK and 40LFK displayed no or low binding, respectively. Since 52SFK, the only other SFK found in HAdVs, preferentially binds α2,8-linked polysialic acid [18], we also evaluated the binding of 40SFK to polysialic acid-expressing SHSY-5Y cells and the polysialic acid-deficient parental SK-N-SH cell line. We found that 40SFK bound to both cell types to a similar extent (Figure 1E). Taken together, these results suggested that sialic acid-containing glycans are not important attachment factors for 40SFK.Next, we treated A549 cells with heparinase III from *Flavobacterium heparinum*, which cleaves cellular HS. The removal of HS drastically reduced 40SFK binding, whereas there was little effect on the other FKs (Figure 1F). An anti-HS antibody bound substantially less efficiently to heparinase III-treated cells as compared to nontreated cells, demonstrating that heparinase III treatment was efficient in cleaving HS from the cell surface. The reduced binding after both protease and heparinase III treatment of target cells suggested that HS-containing proteins could function as attachment factors for 40SFK on the cell surface.

3.2. HAdV-F40 SFK Binding to A549 Cells Requires Sulfated GAGs

To validate the effect of heparinase III treatment, and to determine the potential specificity of 40SFK for specific GAG types, we next investigated binding of 40SFK to A549 cells in the presence of different, soluble GAGs potentially blocking cell binding. Flow cytometry analysis of 40SFK binding after preincubation with soluble GAGs demonstrated that all GAGs except HA reduced 40SFK binding but with varying efficiency (Figure 2A). Heparin (a soluble analogue of HS) most efficiently inhibited binding with an inhibitory concentration (IC50) as low as 1 nM. CS and DS also inhibited 40SFK binding but were less efficient, being 50 and 100 times higher IC50, respectively, as compared to heparin. As expected, we did not observe any significant inhibition of GAGs on 5FK binding to cells (data not shown) since the domain of HAdV-5 that interacts with GAG is suggested to be present on the fiber shaft and not the knob domain [20]. The inability of HA—

a GAG structurally similar to HS lacking sulfate groups—to decrease 40SFK binding indicated that sulfation of GAGs was needed for efficient interaction. To further validate the interaction with HS, we studied knob binding to CHO cells lacking or expressing HS. We observed that 40SFK binding to CHO-ΔHS cells (expressing all GAGs except HS) was completely depleted as compared to the parental control cell line CHO-K1 (Figure 2B). Neither 5FK nor 40LFK bound efficiently to these cells, which was expected since these cells do not express CAR [44]. An anti-HS antibody was used to verify the absence of HS on CHO-ΔHS. In summary, these findings show that 40SFK interact with sulfated GAGs, preferably with HS.

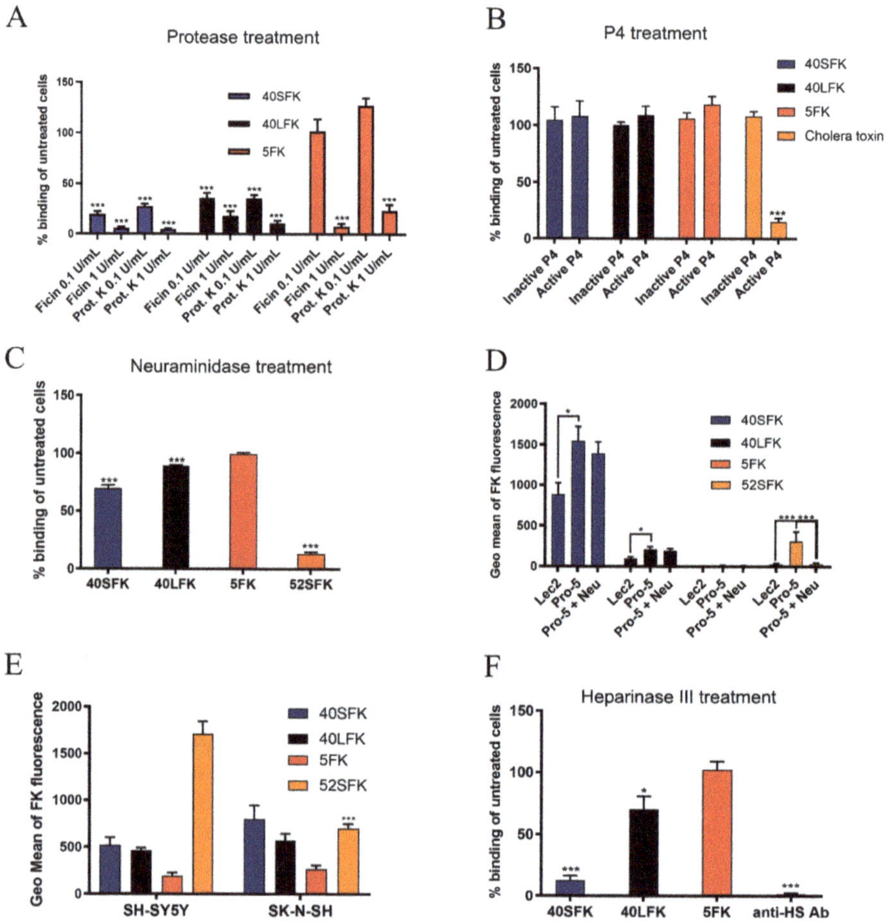

Figure 1. Receptor characterization for the short fiber of HAdV-F40: recombinant fiber knob binding to (**A**) A549 cells pretreated with the proteases—ficin and proteinase K—which degrade cell surface proteins; (**B**) A549 cells pretreated with P4—which inhibits glycolipid synthesis; (**C**) A549 cells pretreated with *V. cholerae* neuraminidase—which cleaves sialic acid. (**D**) CHO cells expressing or lacking sialic acid pretreated with *V. cholerae* neuraminidase (Neu). Pro-5 cells express sialic acid and Lec2 cells lack sialic acid expression. (**E**) SHSY-5Y cells expressing polysialic acid and SK-N-SH lacking polysialic acid, (**F**) A549 cells pretreated with heparinase III—which cleaves heparan sulfate. Binding was determined by geometric mean of fluorescence on a FACSLSRII flow cytometer and shown as % binding of fiber knobs on untreated cells in (**A–C**,**F**). All experiments were performed three times with duplicate samples in each experiment. Error bars represent mean ± SEM. * $P < 0.05$ and *** $P < 0.001$ versus control.

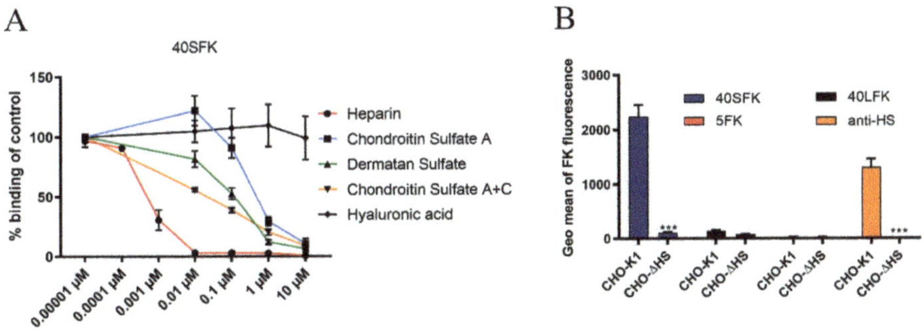

Figure 2. Effect of sulfation on FK binding: (**A**) 40SFK binding to A549 cells. Fiber knobs were pretreated with a broad range of concentrations of different glycosaminoglycans (GAGs). (**B**) Fiber knob binding to CHO cells expressing or lacking heparan sulfate (HS). HS expression levels are shown with an anti-HS antibody. The parental cell line CHO-K1 expresses GAGs and CHO-ΔHS specifically lacks HS but expresses other GAGs. Binding was determined by geometric mean of fluorescence on a FACSLSRII flow cytometer and shown as % binding of fiber knobs on untreated cells in (**A**). All experiments were performed three times with duplicate samples in each experiment. Error bars represent mean ± SEM. Error bars represent mean ± SEM. *** $P < 0.001$ versus control.

3.3. Sulfated GAGs Play an Important Role in HAdV-F40 and -F41 Infection of A549 Cells

Since soluble, sulfated GAGs reduced 40SFK binding to A549 cells, we tested their effect on HAdV infection of A549 cells. Due to the close relationship between HAdV-F40 and -F41 SFKs, we hypothesized that GAGs may be of importance for both viruses and therefore included HAdV-F41 in the study. To investigate the importance of HS for these viruses, we first analyzed the effect of preincubating wild type (WT) viruses with different concentrations of heparin in infection experiments. A dose-dependent decrease in HAdV-F40 and -F41 infection was observed, with about 80% reduction at the highest concentration (Figure 3A). Heparin did not affect HAdV-C5 or HAdV-D36 infection but inhibited HAdV-D37, -F40 and -F41 infection with comparable efficiency. HS-binding HAdV-D37 was included as a positive control [22,36]. The reduction of HAdV-F40 and -F41 infection after preincubation with heparin indicated that the enteric HAdVs required an interaction with HS for infection of A549 cells. To further examine the SFK:HS interaction without interference from the LFK:CAR interaction, we used CHO-cells since they do not express human CAR. We transduced CHO-K1 and CHO-ΔHS cells with a GFP-expressing HAdV-F41 vector since CHO cells do not support viral hexon production used to quantify infection by WT viruses. Here, we observed an 80% decrease in transduction of the cells lacking HS (Figure 3B). No difference in transduction levels of CHO-ΔHS compared to CHO-K1 was observed with a GFP-expressing HAdV-C5 vector (data not shown). Additionally, upon pretreatment of HAdV-F41 vector with heparin, transduction was lowered to the same amount as in the CHO-ΔHS cells (Figure 3C). Taken together these results show that HS is important for the infection of cells by both HAdV-F40 and -F41.

3.4. HAdV-F40 and -F41 Binding to and Uptake in A549 Cells Is Not Affected by Heparin

As cell surface HS serves as an attachment factor for the SFKs of enteric HAdVs, we next analyzed the role of HS during virion binding to A549 cells by preincubating ^{35}S-labeled HAdV-F40 or -F41 virions with soluble heparin before attachment to cells. Surprisingly, we did not see a decrease in HAdV-F40 or -F41 virion binding in the presence of heparin, not even at a concentration of 1 mM (Figure 4A). This was peculiar since 40SFK binding to A549 cells was completely blocked with as little as 10 nM of heparin (Figure 2A), and infection was reduced with 100 µM heparin (Figure 3A). As expected, the binding of the control virions to A549 cells was reduced in a dose-dependent manner (HAdV-D37) or was unaffected (HAdV-C5) (Figure 4A).

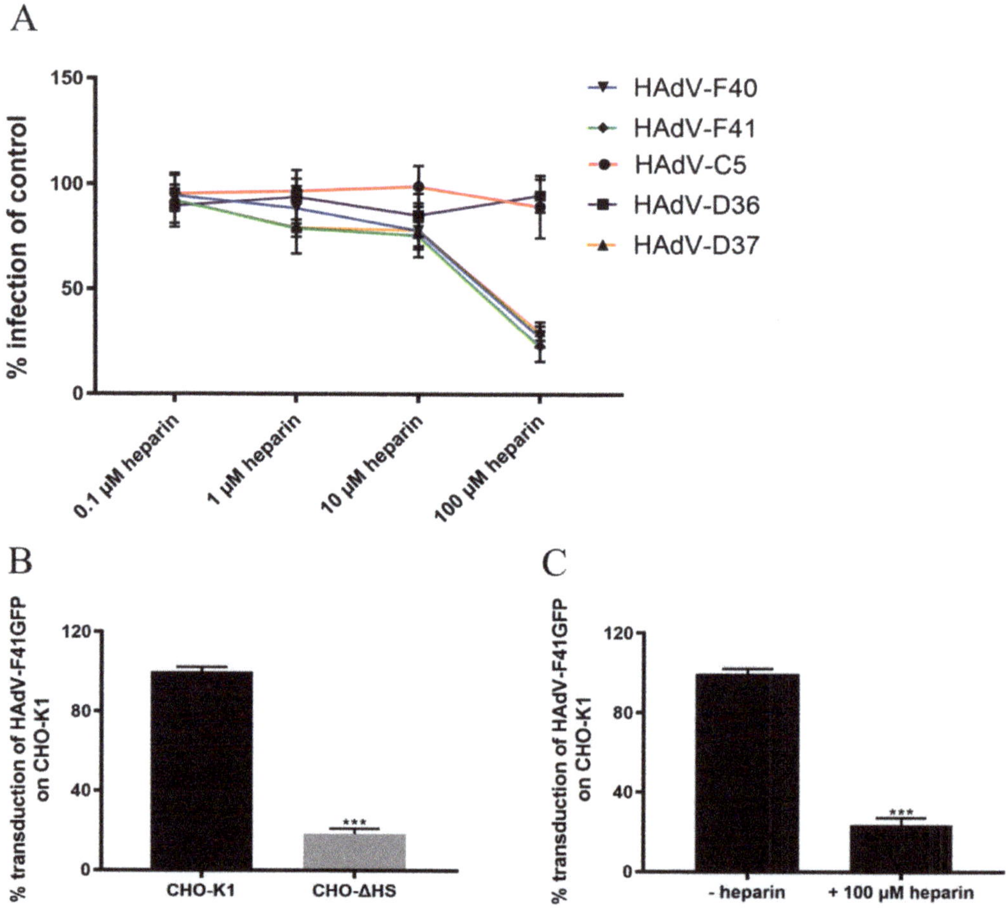

Figure 3. Importance of heparan sulfate on infection and transduction. (**A**) Infection of heparin pretreated HAdVs on A549 cells. The number of infected cells were quantified by immunofluorescence and shown as % infection of untreated HAdVs. (**B**) Transduction of HAdV-F41 GFP vector on CHO cells expressing or lacking heparan sulfate (HS) and (**C**) transduction of heparin pretreated HAdV-F41 GFP vector on CHO-K1 cells. The parental cell line CHO-K1 expresses glycosaminoglycans and CHO-ΔHS specifically lacks heparan sulfate but expresses other glycosaminoglycans. The number of infected/transduced cells were quantified by immunofluorescence and shown as % infection of untreated A549 cells. All experiments were performed three times with duplicate samples in each experiment. Error bars represent mean ± SEM. *** $P < 0.001$ versus control.

To investigate if HS was involved in virion internalization, Alexa Flour 488-labeled HAdV-F40 and -F41 virions were preincubated with heparin followed by a synchronized uptake into A549 cells for five hours at 37 °C, after which the number of extracellular and intracellular virions were quantified by confocal microscopy. After five hours at 37 °C, approximately 50% of HAdV-F40 and 20% of HAdV-F41 virions were internalized into the cells (Figure 4B). Preincubation with heparin did not reduce internalization even at high concentrations. Almost all HAdV-C5 virions were internalized regardless of heparin treatment (Figure 4B). Since heparin did not affect the number of bound or internalized HAdV-F40 and -F41 virions, these data suggest that the SFK:HS interactions do not contribute to virion attachment or internalization at neutral pH.

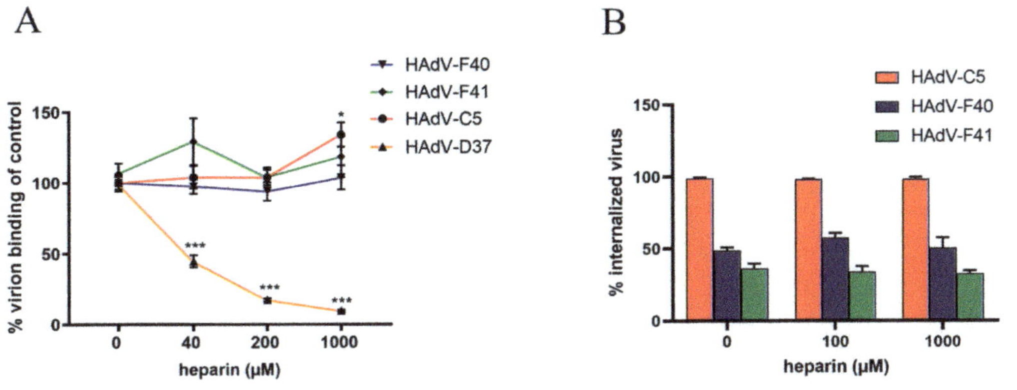

Figure 4. Effect of soluble heparin on virion binding and uptake in A549 cells. (**A**) ^{35}S-labeled HAdV-F40 and -F41 virion binding to A549 cells. Virions were pretreated with different concentrations of soluble heparin. Binding to cells was quantified as counts per minute (CPM) by a liquid scintillation counter and represented as % binding of untreated virions. (**B**) HAdV-C5, -F40 and –F41 uptake in A549 cells. The number of intracellular virions were quantified by immunofluorescence and shown as a relative % of internalized virus in the untreated sample. The experiments were performed three times with duplicate samples. Error bars represent mean ± SEM. * $P < 0.05$ and *** $P < 0.001$ versus control.

3.5. LFK:CAR-Dependent Cell Attachment of Enteric HAdVs Is Switched to SFK:HS-Dependent Cell Attachment after Virion Exposure to Acidic pH

Previous studies on SFs of enteric HAdVs suggest that the SF is important for the enteric tropism of these viruses since it imparts resistance to acidic pH [29,30]. To further understand the role of the SFK:HS interaction during enteric HAdV infection, we addressed the potential impact of acidic pH on virion attachment to host cells. We analyzed the binding of ^{35}S-labeled virions to WT haploid HAP1 cells, derived from a human chronic myelogenous leukemia cell line and to HAP1 cells devoid of CAR or GAGs. The absence of CAR and HS from these cells was verified by flow cytometry using anti-CAR and anti-HS monoclonal antibodies (data not shown). At neutral pH (Figure 5A), HAdV-F40 bound with similar efficiency to HAP1 WT cells and HAP1-ΔGAG cells, but to a substantially lower degree to HAP1-ΔCAR cells, suggesting that at this pH, HAdV-F40 virions bound to HAP1 cells mainly in a LF:CAR dependent manner, and did not rely on SF:GAG interactions. This was further supported by the observation that the preincubation of cells with soluble, CAR-binding 41LFK reduced binding of HAdV-F40 virions to CAR-expressing cells, while the preincubation of virions with heparin did not affect binding to these cells (Figure 5A). After the exposure of virions to acidic pH (pH 1.1 to 1.3) (Figure 5B) the binding of HAdV-F40 to WT HAP1 cells remained high, while binding to HAP1-ΔGAG cells decreased and binding to HAP1-ΔCAR cells increased, as compared to binding at neutral pH. Interestingly, at acidic pH the preincubation of virions with heparin reduced HAdV-F40 binding to both HAP1 WT cells and HAP1-ΔCAR cells efficiently (Figure 5B).

The binding to HAP1-ΔGAG cells in the presence of heparin was also reduced, but to a lesser extent. On the other hand, the preincubation of cells with 41LFK did not affect acidic pH-exposed virion binding. Similar results were observed when these experiments were performed with HAdV-F41 (Figure 5C,D). Taken together, these results suggested that virion exposure to acidic pH changes the mechanism by which enteric HAdVs attach to host cells. The LFK:CAR interaction dominates in a neutral environment, but exposure to acidic pH switches the mechanism of virion binding to be SFK:HS dependent.

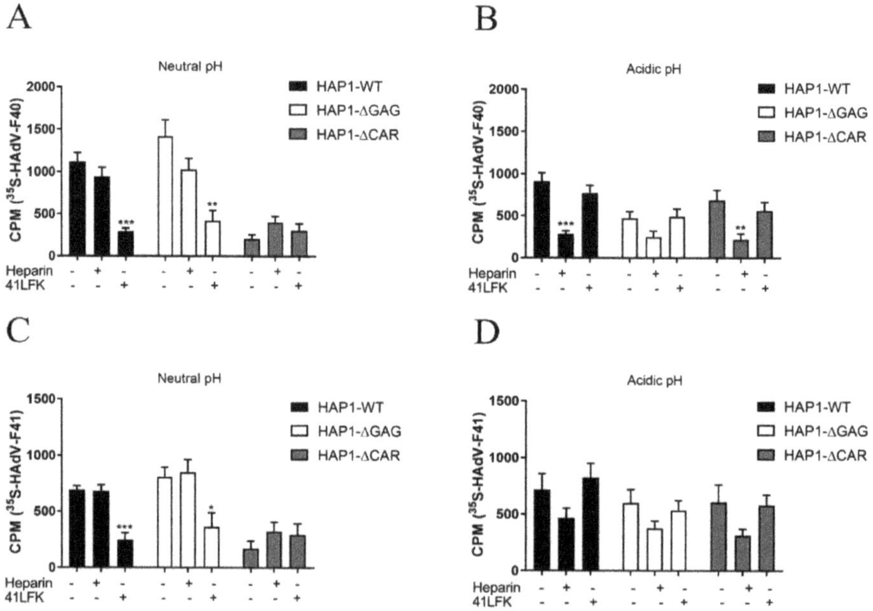

Figure 5. Effect of pH treatment on receptor usage: ^{35}S-labeled HAdV-F40 and HAdV-F41 virion binding to HAP1 WT cells, HAP1-ΔGAG cells and HAP1-ΔCAR cells. The parental cell line HAP1 WT expresses glycosaminoglycans (GAGs) and CAR. HAP1-ΔGAG lacks expression of GAGs, while HAP1-ΔCAR lacks expression of CAR. (**A**) Untreated (neutral pH) HAdV-F40 or (**B**) acidic pH treated HAdV-F40 or (**C**) untreated (neutral pH) HAdV-F41 or (**D**) acidic pH treated HAdV-F41 were preincubated with heparin and cells were preincubated with 41LFKs before binding. Binding to cells was quantified as counts per minute (CPM) by a liquid scintillation counter. All experiments were performed three times with duplicate samples in each experiment. Error bars represent mean ± SEM. * $P < 0.05$, ** $P < 0.01$ and *** $P < 0.001$ versus control.

3.6. Acidic pH-Exposed HAdV-F40 and -F41 Virions Lose CAR-Binding Capacity But Retain HS-Binding Capacity

To validate the HAdV-F40 and -F41 preference for HS binding after exposure to acidic pH, we performed surface plasmon resonance (SPR) studies with CAR or heparin attached on the sensor chip and virions as the analyte. At neutral pH, both HAdV-F40 and -F41 virions displayed a very strong binding to CAR, which was almost completely abrogated after exposure to acidic pH (Figure 6A–D). On the other hand, virion binding to heparin was maintained at similar levels before and after acidic pH treatment (Figure 6E–H). When the binding between 40SFK and heparin was tested (Figure 6I), we observed a substantially increased binding after exposure of 40SFK to acidic pH (Figure 6J). These results support our theory that acidic pH exposed HAdV-F40 and -F41 lose their ability to bind CAR, but instead bind HS on target cells through an interaction with the SF.

3.7. Heparin Specifically Reduces Infection and Short Fiber Knob Binding on a Small Intestinal Cell Line

HAdV-F40 and -F41 cause infection in the small intestine [45]. To validate the role of HS in a more physiologically relevant cell line, we studied the importance of the SFK:HS interaction using a small intestinal cell line, HuTu80. The results obtained with HuTu80 are largely consistent with those obtained using A549 cells. Preincubation with soluble heparin reduced 40SFK binding to HuTu80 cells very efficiently with more than 50% of the binding reduced at 1 nM heparin (Figure 7A). As expected, heparin also inhibited 37FK binding but not that of 40LFK or 5FK. We also observed that 40SFK binding was inhibited by CS, but only at a much higher concentrations (Figure 7B), and not at all by HA (Figure 7C) suggesting that 40SFK interacts mainly with HS on these cells. Finally, the preincubation of virions with heparin also reduced HAdV-F40 and -F41 (and positive control HAdV-D37) infection of HuTu80 cells but did not influence infection of the negative control HAdV-C5 (Figure 7D). Neither CS (Figure 7E) nor HA (Figure 7F) reduced HAdV-F40 and -F41 infection as efficiently as heparin. Taken together, these results demonstrate that enteric HAdVs interact with HS through their short fibers on multiple cell lines, including small intestinal cells.

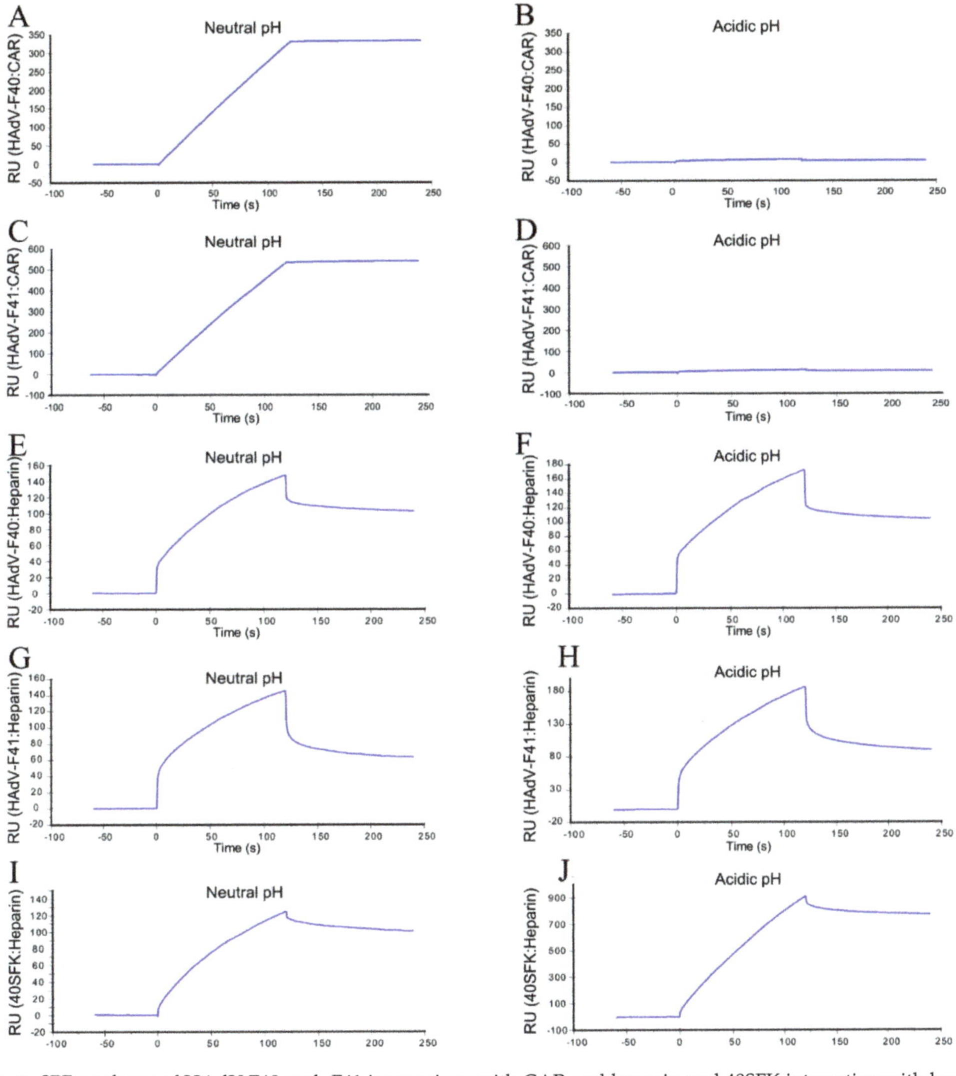

Figure 6. SPR analyses of HAdV-F40 and -F41 interactions with CAR and heparin and 40SFK interaction with heparin: (**A**) Untreated (neutral pH) HAdV-F40 or (**B**) acidic pH-treated HAdV-F40 or (**C**) untreated (neutral pH) HAdV-F41 or (**D**) acidic pH treated HAdV-F41 binding to immobilized CAR. (**E**) Untreated (neutral pH) HAdV-F40 or (**F**) acidic pH-treated HAdV-F40 or (**G**) untreated (neutral pH) HAdV-F41 or (**H**) acidic pH treated HAdV-F41 or (**I**) untreated (neutral pH) 40SFK or (**J**) acidic pH treated 40SFK binding to immobilized heparin. Interactions are displayed in response units (RU). All experiments were performed three times. Displayed is one representative set from each experiment.

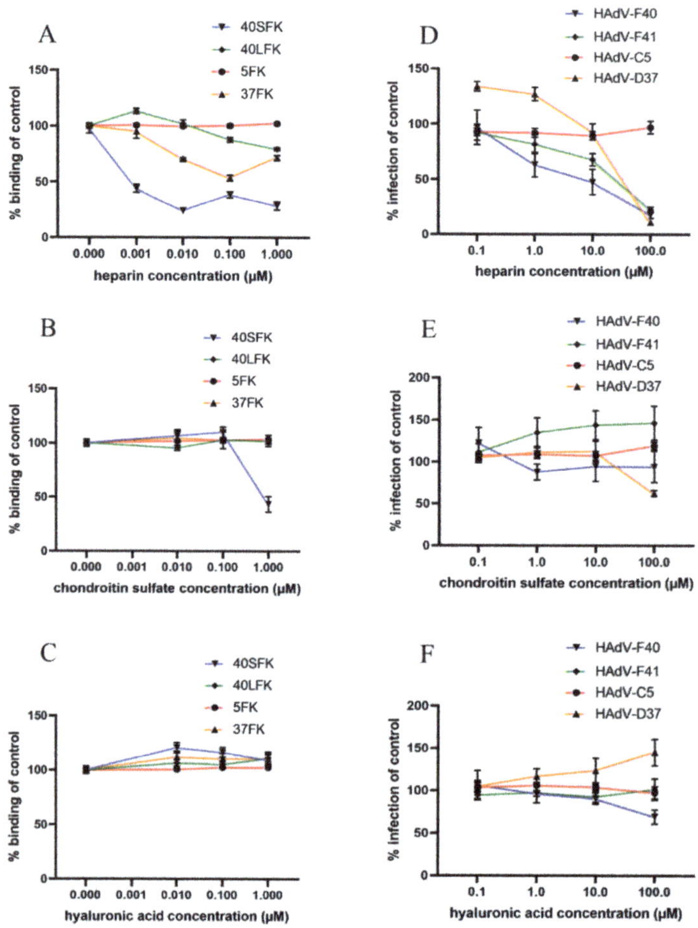

Figure 7. Effect of GAGs on fiber knob binding to and infection of HuTu80 cells: Binding of (**A**) heparin or (**B**) chondroitin sulfate mix or (**C**) hyaluronic acid pretreated HAdVs on HuTu80 cells. Binding was determined by geometric mean of fluorescence on a FACSLSRII flow cytometer and shown as % binding of untreated fiber knobs. Infection of (**D**) heparin or (**E**) chondroitin sulfate mix or (**F**) hyaluronic acid pretreated virions to HuTu80 cells. The number of infected cells were quantified by immunofluorescence and shown as % infection of untreated HAdVs. All experiments were performed three times with duplicate samples in each experiment.

4. Discussion

A distinguishing feature of enteric HAdVs is the presence of SFs in the viral capsid. The relatively high ratio of SFs to LFs suggests that SFs play an important role in the infectious cycle of these viruses. Despite being a leading cause of diarrhea and diarrhea-associated mortality in young children, no cellular attachment factors have been identified for the SFs of enteric HAdVs. Here, we demonstrate for the first time that HS is an important host factor for infection by enteric HAdVs. We pinpoint this interaction to occur through the SF by the observations that SF does not bind—(i) when cells are treated with heparinase III, (ii) in the presence of heparin and (iii) when cells are devoid of HS. We further demonstrate that enteric HAdVs undergo a switch in receptor usage when exposed to pH mimicking the acidic environment in the stomach. Exposure of enteric HAdVs to acidic pH resulted in a loss of LFK-dependent CAR-binding capacity but maintained SFK-dependent HS-binding to cells. Since enteric HAdVs are transmitted through the fecal–oral route by the ingestion of contaminated food or water, we propose a novel function of the short fiber, where it compensates for the loss of CAR-binding occurring during passage through the harsh acidic environment of the digestive tract and facilitates a first round of infection when reaching the permissive, HS-expressing cells in the small intestine. Subsequent rounds of infection result in production and release of progeny virions to an environment with neutral pH, where the LFK:CAR interaction can be of importance for further spread of the infection. It has previously been shown that cells infected with CAR-binding and CD46 binding HAdVs produce an excess of soluble fibers, which upon release can disrupt intercellular CAR and CD46 homodimers and destroy the integrity of tight junctions, thereby facilitating spread of progeny virions [46,47]. It is possible that the LF of HAdV-F40 and -F41 could exert a similar function and enable the transmission of the released progeny virions within a tissue, or between tissues. The released virions will not be subjected to acidic pH and will therefore maintain their ability to bind CAR in order to further propagate the infection. Exposure to gastrointestinal fluids is very likely to cause structural alterations in virus particles, making them noninfectious or, as for rotavirus, activating them to enable infection. Rotavirus requires proteolytic cleavage of proteins in the outer capsid layer by enzymes present in digestive fluids to render it more infectious [48]. For enteric HAdVs, the harsh environment of the stomach may destroy the LF:CAR interaction that would make the virions less infectious without the rescuing effect of the SF:HS interaction. Further studies are needed to address the exact molecular mechanism behind the effect of acidic gastric juice on SFs.

The ability of 40SFK to bind cellular HS was efficiently blocked with nanomolar concentrations of heparin. A significantly higher concentration was needed to reduce, but not completely block, HAdV-F40 and -F41 infection. This can partly be explained by the presence of the LF in the virion and its ability to interact with CAR as a receptor for attachment and entry, independent of the SF. To our surprise neither virion binding nor entry was affected by preincubation with heparin at neutral pH, indicating that the SF:HS interaction is not of importance for these steps in the viral life cycle under physiological conditions. Nevertheless, we observed a specific reduction of infection by heparin that could not be seen with CS or HA: nor could it be explained by a general negative effect on the cells since no effect was observed on the control viruses HAdV-C5 and HAdV-D36. We speculate that under neutral pH, the intact LF:CAR interaction is mainly responsible for virus entry and that heparin bound to the SF may block a subsequent interaction of the SF with an intracellular protein needed for efficient infection. This theory is supported by the suggestion that 41SF can interact with a novel intracellular protein, ParAd41, which may facilitate attachment to the nuclear membrane prior to injection of DNA to the nucleus [49].

The only other HAdV to have a SF is HAdV-G52 [26]. The SF of HAdV-52 binds to long negatively charged glycan chains of polymerized sialic acid residues known as polysialic acid [18]. We show here that 40SFK does not bind polysialic acid as indicated by similar binding levels to cells expressing and lacking polysialic acid. Instead 40SFK cellular binding shows a dependency of sulfated GAGs, preferably HS. Polysialic acid is a posttranslational modification present on proteins such as NCAM [50], SynCAM-1 [51], neuropilin-2 [52] and CCR7 [53]. Since these proteins are absent or only expressed in low amounts on target cells of the small intestine, they are unlikely receptor candidates for the SF of enteric HAdVs. Although HAdV-G52 was isolated from a case of gastroenteritis, it was classified in a separate species and is not considered one of the enteric HAdVs [26], which is the opposite of HAdV-F40 and -F41, which are common in the human population [54,55]. HAdV-G52 is rarely isolated [56] and even though it is closely related to HAdV-F40 and -F41 it is actually more similar to simian AdVs [26]. An interesting observation on the divergence of species F and G, both containing short fibers, from the other HAdVs species, is the evolved ability of short fiber binding to negatively charged glycans. While 52SFK has evolved to bind polysialic acid, 40SFK and 41SFK binds heparan sulfate.

It still needs to be explored whether exposure to acidic pH causes a receptor switch for HAdV-G52 as well.

The precise function of the short fibers of enteric HAdVs has been under investigation for a long time. The 41SFK is suggested to function as a viral enterotoxin by stimulating release of serotonin from enterochromaffin cells which could result in symptoms such as vomiting and diarrhea. Furthermore, the SF of enteric HAdVs have been suggested to be responsible for the restricted intestinal tropism since these viruses were shown to be more resistant to acidic pH than other HAdV types [29,30]. This claim was refuted by another study describing that a recombinant HAdV-C5 vector carrying the 41SFs showed reduced affinity for enterocytes [57]. The latter study, however, did not subject the virions to acidic pH, which could have affected their results and interpretations. Here, we identified HS as an interaction partner to the SF and explained its role in cellular attachment after exposure to acidic pH. In our opinion, the SF contributes to infection and pathogenesis through at least two distinct mechanisms. Firstly, the SF preserves the infectivity of the virion through an acidic gastric environment and enables the virions to attach to their host cell through an interaction with HS. Secondly, as shown in another study, the SF stimulates the production of serotonin, causing the symptoms of diarrhea and vomiting. This points out the short fibers as key proteins, contributing not only to pathogenesis when serving as viral enterotoxins, but also to infection of host cells. The ability of enteric HAdVs to interact with HS adds to the knowledge about their distinct tropism. These results may also be useful for the design of novel antiviral drugs and to explore enteric HAdVs as tools for gene therapy and vaccine development.

Author Contributions: Conceptualization, A.R. and A.L.; methodology, A.R., E.P., F.T., M.B., B.D.P., L.F. and A.L.; formal analysis, A.R., M.B., L.F. and A.L.; resources, A.R., M.B. and A.L.; writing—original draft preparation, A.R. and A.L.; writing—review and editing, A.R., E.P., F.T., S.M., M.B., B.D.P., L.F. and A.L.; supervision, A.R. and A.L.; funding acquisition, A.L. All authors have read and agreed to the published version of the manuscript.

Funding: This research was funded by the Swedish Society for Medical Research.

Institutional Review Board Statement: Not applicable.

Informed Consent Statement: Not applicable.

Data Availability Statement: A549 cell line was provided by Alistair Kidd. HAP1-WT and HAP1-ΔCAR cell lines were purchased from Horizon Discovery (Waterbeach, UK). HAP1-ΔGAG was provided by Frank Kuppeveld. HuTu80 cell line was purchased from ATCC (Manassas, VA, USA). Pro-5, Lec2, SK-N-SH and SH-SY5Y cells were purchased from LGC Promochem (Teddington, UK).

Acknowledgments: We thank Kristina Lindman for technical assistance provided in maintaining cells, producing virions and assistance in performing experiments. Purification of recombinant fiber knob proteins was performed by the Umeå Protein Expertise Platform. We acknowledge the Biochemical Imaging Center at Umeå University and the National Microscopy Infrastructure, NMI (VR-RFI 2016-00968) for providing assistance in microscopy.

Conflicts of Interest: The authors declare no conflict of interest.

References

1. Lefkowitz, E.J.; Dempsey, D.M.; Hendrickson, R.C.; Orton, R.J.; Siddell, S.G.; Smith, D.B. Virus taxonomy: The database of the International Committee on Taxonomy of Viruses (ICTV). *Nucleic Acids Res.* **2018**, *46*, D708–D717. [CrossRef] [PubMed]
2. Brister, J.R.; Chodosh, J.; Curiel, D.T.; Heim, A.; Jones, M.S.; Kajon, A.; Lion, T.; Seto, D.; Zhang, Q. HAdV Working Group. Available online: http://hadvwg.gmu.edu/ (accessed on 31 December 2019).
3. Wold, W.S.M.; Ison, M.G. Adenoviruses. In *Fields Virology*, 6th ed.; Knipe, D.M., Howley, P.M., Eds.; Lippincott Williams & Wilkins: Philadelphia, PA, USA, 2013; Volume 2, pp. 1732–1767.
4. Kidd, A.H.; Banatvala, J.E.; de Jong, J.C. Antibodies to fastidious faecal adenoviruses (species 40 and 41) in sera from children. *J. Med. Virol.* **1983**, *11*, 333–341. [CrossRef]
5. Tiemessen, C.T.; Kidd, A.H. The subgroup F adenoviruses. *J. Gen. Virol.* **1995**, *76*, 481–497. [CrossRef]

6. Liu, L.; Oza, S.; Hogan, D.; Chu, Y.; Perin, J.; Zhu, J.; Lawn, J.E.; Cousens, S.; Mathers, C.; Black, R.E. Global, regional, and national causes of under-5 mortality in 2000-15: An updated systematic analysis with implications for the Sustainable Development Goals. *Lancet* **2016**, *388*, 3027–3035. [CrossRef]
7. GBD 2016 Diarrhoeal Disease Collaborators. Estimates of the global, regional, and national morbidity, mortality, and aetiologies of diarrhoea in 195 countries: A systematic analysis for the Global Burden of Disease Study 2016. *Lancet Infect Dis.* **2018**, *18*, 1211–1228. [CrossRef]
8. Roelvink, P.W.; Lizonova, A.; Lee, J.G.; Li, Y.; Bergelson, J.M.; Finberg, R.W.; Brough, D.E.; Kovesdi, I.; Wickham, T.J. The coxsackievirus-adenovirus receptor protein can function as a cellular attachment protein for adenovirus serotypes from subgroups A, C, D, E, and F. *J. Virol.* **1998**, *72*, 7909–7915. [CrossRef] [PubMed]
9. Lenman, A.; Liaci, A.M.; Liu, Y.; Ardahl, C.; Rajan, A.; Nilsson, E.; Bradford, W.; Kaeshammer, L.; Jones, M.S.; Frangsmyr, L.; et al. Human adenovirus 52 uses sialic acid-containing glycoproteins and the coxsackie and adenovirus receptor for binding to target cells. *PLoS Pathog.* **2015**, *11*, e1004657. [CrossRef] [PubMed]
10. Wickham, T.J.; Filardo, E.J.; Cheresh, D.A.; Nemerow, G.R. Integrin alpha v beta 5 selectively promotes adenovirus mediated cell membrane permeabilization. *J. Cell. Biol.* **1994**, *127*, 257–264. [CrossRef] [PubMed]
11. Wickham, T.J.; Mathias, P.; Cheresh, D.A.; Nemerow, G.R. Integrins alpha v beta 3 and alpha v beta 5 promote adenovirus internalization but not virus attachment. *Cell* **1993**, *73*, 309–319. [CrossRef]
12. Rajan, A.; Persson, B.D.; Frangsmyr, L.; Olofsson, A.; Sandblad, L.; Heino, J.; Takada, Y.; Mould, A.P.; Schnapp, L.M.; Gall, J.; et al. Enteric Species F Human Adenoviruses use Laminin-Binding Integrins as Co-Receptors for Infection of Ht-29 Cells. *Sci. Rep.* **2018**, *8*, 10019. [CrossRef]
13. Gaggar, A.; Shayakhmetov, D.M.; Lieber, A. CD46 is a cellular receptor for group B adenoviruses. *Nat. Med.* **2003**, *9*, 1408–1412. [CrossRef]
14. Persson, B.D.; Reiter, D.M.; Marttila, M.; Mei, Y.F.; Casasnovas, J.M.; Arnberg, N.; Stehle, T. Adenovirus type 11 binding alters the conformation of its receptor CD46. *Nat. Struct. Mol. Biol.* **2007**, *14*, 164–166. [CrossRef] [PubMed]
15. Marttila, M.; Persson, D.; Gustafsson, D.; Liszewski, M.K.; Atkinson, J.P.; Wadell, G.; Arnberg, N. CD46 is a cellular receptor for all species B adenoviruses except types 3 and 7. *J. Virol.* **2005**, *79*, 14429–14436. [CrossRef]
16. Wang, H.; Li, Z.Y.; Liu, Y.; Persson, J.; Beyer, I.; Moller, T.; Koyuncu, D.; Drescher, M.R.; Strauss, R.; Zhang, X.B.; et al. Desmoglein 2 is a receptor for adenovirus serotypes 3, 7, 11 and 14. *Nat. Med.* **2011**, *17*, 96–104. [CrossRef] [PubMed]
17. Nilsson, E.C.; Storm, R.J.; Bauer, J.; Johansson, S.M.; Lookene, A.; Angstrom, J.; Hedenstrom, M.; Eriksson, T.L.; Frangsmyr, L.; Rinaldi, S.; et al. The GD1a glycan is a cellular receptor for adenoviruses causing epidemic keratoconjunctivitis. *Nat. Med.* **2011**, *17*, 105–109. [CrossRef]
18. Lenman, A.; Liaci, A.M.; Liu, Y.; Frangsmyr, L.; Frank, M.; Blaum, B.S.; Chai, W.; Podgorski, I.I.; Harrach, B.; Benko, M.; et al. Polysialic acid is a cellular receptor for human adenovirus 52. *Proc. Natl. Acad. Sci. USA* **2018**, *115*, E4264–E4273. [CrossRef] [PubMed]
19. Dechecchi, M.C.; Melotti, P.; Bonizzato, A.; Santacatterina, M.; Chilosi, M.; Cabrini, G. Heparan sulfate glycosaminoglycans are receptors sufficient to mediate the initial binding of adenovirus types 2 and 5. *J. Virol.* **2001**, *75*, 8772–8780. [CrossRef]
20. Dechecchi, M.C.; Tamanini, A.; Bonizzato, A.; Cabrini, G. Heparan sulfate glycosaminoglycans are involved in adenovirus type 5 and 2-host cell interactions. *Virology* **2000**, *268*, 382–390. [CrossRef]
21. Tuve, S.; Wang, H.; Jacobs, J.D.; Yumul, R.C.; Smith, D.F.; Lieber, A. Role of cellular heparan sulfate proteoglycans in infection of human adenovirus serotype 3 and 35. *PLoS Pathog.* **2008**, *4*, e1000189. [CrossRef]
22. Chandra, N.; Liu, Y.; Liu, J.X.; Frangsmyr, L.; Wu, N.; Silva, L.M.; Lindstrom, M.; Chai, W.; Pedrosa Domellof, F.; Feizi, T.; et al. Sulfated Glycosaminoglycans as Viral Decoy Receptors for Human Adenovirus Type 37. *Viruses* **2019**, *11*, 247. [CrossRef]
23. Esko, J.D.; Kimata, K.; Lindahl, U. Proteoglycans and Sulfated Glycosaminoglycans. In *Essentials of Glycobiology*; Varki, A., Cummings, R.D., Esko, J.D., Freeze, H.H., Stanley, P., Bertozzi, C.R., Hart, G.W., Etzler, M.E., Eds.; The Consortium of Glycobiology Editors: La Jolla, CA, USA, 2009.
24. Kidd, A.H.; Chroboczek, J.; Cusack, S.; Ruigrok, R.W. Adenovirus type 40 virions contain two distinct fibers. *Virology* **1993**, *192*, 73–84. [CrossRef]
25. Song, J.D.; Liu, X.L.; Chen, D.L.; Zou, X.H.; Wang, M.; Qu, J.G.; Lu, Z.Z.; Hung, T. Human adenovirus type 41 possesses different amount of short and long fibers in the virion. *Virology* **2012**, *432*, 336–342. [CrossRef]
26. Jones, M.S., 2nd; Harrach, B.; Ganac, R.D.; Gozum, M.M.; Dela Cruz, W.P.; Riedel, B.; Pan, C.; Delwart, E.L.; Schnurr, D.P. New adenovirus species found in a patient presenting with gastroenteritis. *J. Virol.* **2007**, *81*, 5978–5984. [CrossRef]
27. Westerberg, S.; Hagbom, M.; Rajan, A.; Loitto, V.; Persson, B.D.; Allard, A.; Nordgren, J.; Sharma, S.; Magnusson, K.E.; Arnberg, N.; et al. Interaction of Human Enterochromaffin Cells with Human Enteric Adenovirus 41 Leads to Serotonin Release and Subsequent Activation of Enteric Glia Cells. *J. Virol.* **2018**, *92*, e00026-18. [CrossRef] [PubMed]
28. Rafie, K.; Lenman, A.; Fuchs, J.; Rajan, A.; Arnberg, N.; Carlson, L.-A. The structure of enteric human adenovirus 41—A leading cause of diarrhea in children. *Sci. Adv.* **2021**, *7*, eabe0974. [CrossRef] [PubMed]
29. Rodriguez, E.; Romero, C.; Rio, A.; Miralles, M.; Raventos, A.; Planells, L.; Burgueno, J.F.; Hamada, H.; Perales, J.C.; Bosch, A.; et al. Short-fiber protein of ad40 confers enteric tropism and protection against acidic gastrointestinal conditions. *Hum. Gene Ther. Methods* **2013**, *24*, 195–204. [CrossRef] [PubMed]
30. Favier, A.L.; Burmeister, W.P.; Chroboczek, J. Unique physicochemical properties of human enteric Ad41 responsible for its survival and replication in the gastrointestinal tract. *Virology* **2004**, *322*, 93–104. [CrossRef]

31. Lidholt, K.; Weinke, J.L.; Kiser, C.S.; Lugemwa, F.N.; Bame, K.J.; Cheifetz, S.; Massague, J.; Lindahl, U.; Esko, J.D. A single mutation affects both N-acetylglucosaminyltransferase and glucuronosyltransferase activities in a Chinese hamster ovary cell mutant defective in heparan sulfate biosynthesis. *Proc. Natl. Acad. Sci. USA* **1992**, *89*, 2267–2271. [CrossRef] [PubMed]
32. Deutscher, S.L.; Nuwayhid, N.; Stanley, P.; Briles, E.I.; Hirschberg, C.B. Translocation across Golgi vesicle membranes: A CHO glycosylation mutant deficient in CMP-sialic acid transport. *Cell* **1984**, *39*, 295–299. [CrossRef]
33. Stanley, P.; Caillibot, V.; Siminovitch, L. Selection and characterization of eight phenotypically distinct lines of lectin-resistant Chinese hamster ovary cell. *Cell* **1975**, *6*, 121–128. [CrossRef]
34. Johansson, S.M.; Arnberg, N.; Elofsson, M.; Wadell, G.; Kihlberg, J. Multivalent HSA conjugates of 3′-sialyllactose are potent inhibitors of adenoviral cell attachment and infection. *Chembiochem* **2005**, *6*, 358–364. [CrossRef] [PubMed]
35. Lemiale, F.; Haddada, H.; Nabel, G.J.; Brough, D.E.; King, C.R.; Gall, J.G. Novel adenovirus vaccine vectors based on the enteric-tropic serotype 41. *Vaccine* **2007**, *25*, 2074–2084. [CrossRef]
36. Arnberg, N.; Kidd, A.H.; Edlund, K.; Nilsson, J.; Pring-Akerblom, P.; Wadell, G. Adenovirus type 37 binds to cell surface sialic acid through a charge-dependent interaction. *Virology* **2002**, *302*, 33–43. [CrossRef]
37. Mistry, N.; Inoue, H.; Jamshidi, F.; Storm, R.J.; Oberste, M.S.; Arnberg, N. Coxsackievirus A24 variant uses sialic acid-containing O-linked glycoconjugates as cellular receptors on human ocular cells. *J. Virol.* **2011**, *85*, 11283–11290. [CrossRef] [PubMed]
38. Schindelin, J.; Arganda-Carreras, I.; Frise, E.; Kaynig, V.; Longair, M.; Pietzsch, T.; Preibisch, S.; Rueden, C.; Saalfeld, S.; Schmid, B.; et al. Fiji: An open-source platform for biological-image analysis. *Nat. Methods* **2012**, *9*, 676–682. [CrossRef] [PubMed]
39. McQuin, C.; Goodman, A.; Chernyshev, V.; Kamentsky, L.; Cimini, B.A.; Karhohs, K.W.; Doan, M.; Ding, L.; Rafelski, S.M.; Thirstrup, D.; et al. CellProfiler 3.0: Next-generation image processing for biology. *PLoS Biol.* **2018**, *16*, e2005970. [CrossRef]
40. Witt, D.J.; Bousquet, E.B. Comparison of Enteric Adenovirus Infection in Various Human Cell-Lines. *J. Virol. Methods* **1988**, *20*, 295–308. [CrossRef]
41. Arnberg, N.; Kidd, A.H.; Edlund, K.; Olfat, F.; Wadell, G. Initial interactions of subgenus D adenoviruses with A549 cellular receptors: Sialic acid versus alpha(v) integrins. *J. Virol.* **2000**, *74*, 7691–7693. [CrossRef]
42. Varki, A. Sialic acids in human health and disease. *Trends Mol Med.* **2008**, *14*, 351–360. [CrossRef]
43. Sarrazin, S.; Lamanna, W.C.; Esko, J.D. Heparan sulfate proteoglycans. *Cold Spring Harb Perspect. Biol.* **2011**, *3*, a004952. [CrossRef]
44. Bergelson, J.M.; Cunningham, J.A.; Droguett, G.; Kurt-Jones, E.A.; Krithivas, A.; Hong, J.S.; Horwitz, M.S.; Crowell, R.L.; Finberg, R.W. Isolation of a common receptor for Coxsackie B viruses and adenoviruses 2 and 5. *Science* **1997**, *275*, 1320–1323. [CrossRef]
45. Parizhskaya, M.; Walpusk, J.; Mazariegos, G.; Jaffe, R. Enteric adenovirus infection in pediatric small bowel transplant recipients. *Pediatr. Dev. Pathol.* **2001**, *4*, 122–128. [CrossRef] [PubMed]
46. Walters, R.W.; Freimuth, P.; Moninger, T.O.; Ganske, I.; Zabner, J.; Welsh, M.J. Adenovirus fiber disrupts CAR-mediated intercellular adhesion allowing virus escape. *Cell* **2002**, *110*, 789–799. [CrossRef]
47. Rebetz, J.; Na, M.; Su, C.; Holmqvist, B.; Edqvist, A.; Nyberg, C.; Widegren, B.; Salford, L.G.; Sjogren, H.O.; Arnberg, N.; et al. Fiber mediated receptor masking in non-infected bystander cells restricts adenovirus cell killing effect but promotes adenovirus host co-existence. *PLoS ONE* **2009**, *4*, e8484. [CrossRef] [PubMed]
48. Estes, M.K.; Graham, D.Y.; Mason, B.B. Proteolytic enhancement of rotavirus infectivity: Molecular mechanisms. *J. Virol.* **1981**, *39*, 879–888. [CrossRef] [PubMed]
49. Favier, A.-L.; Grzela, R.; Tcherniuk, S.; Harsi, M.C.; Chroboczek, J. A Novel Human Membrane Protein Interacting with the Short Fiber of Enteric Adenovirus. *ISRN Virol.* **2013**, *2013*, 709734. [CrossRef]
50. Finne, J.; Finne, U.; Deagostini-Bazin, H.; Goridis, C. Occurrence of alpha 2-8 linked polysialosyl units in a neural cell adhesion molecule. *Biochem. Biophys. Res. Commun.* **1983**, *112*, 482–487. [CrossRef]
51. Galuska, S.P.; Rollenhagen, M.; Kaup, M.; Eggers, K.; Oltmann-Norden, I.; Schiff, M.; Hartmann, M.; Weinhold, B.; Hildebrandt, H.; Geyer, R.; et al. Synaptic cell adhesion molecule SynCAM 1 is a target for polysialylation in postnatal mouse brain. *Proc. Natl. Acad. Sci. USA* **2010**, *107*, 10250–10255. [CrossRef]
52. Werneburg, S.; Muhlenhoff, M.; Stangel, M.; Hildebrandt, H. Polysialic acid on SynCAM 1 in NG2 cells and on neuropilin-2 in microglia is confined to intracellular pools that are rapidly depleted upon stimulation. *Glia* **2015**, *63*, 1240–1255. [CrossRef]
53. Kiermaier, E.; Moussion, C.; Veldkamp, C.T.; Gerardy-Schahn, R.; de Vries, I.; Williams, L.G.; Chaffee, G.R.; Phillips, A.J.; Freiberger, F.; Imre, R.; et al. Polysialylation controls dendritic cell trafficking by regulating chemokine recognition. *Science* **2016**, *351*, 186–190. [CrossRef]
54. Yang, W.X.; Zou, X.H.; Jiang, S.Y.; Lu, N.N.; Han, M.; Zhao, J.H.; Guo, X.J.; Zhao, S.C.; Lu, Z.Z. Prevalence of serum neutralizing antibodies to adenovirus type 5 (Ad5) and 41 (Ad41) in children is associated with age and sanitary conditions. *Vaccine* **2016**, *34*, 5579–5586. [CrossRef]
55. Shinozaki, T.; Araki, K.; Ushijima, H.; Fujii, R. Antibody response to enteric adenovirus types 40 and 41 in sera from people in various age groups. *J. Clin. Microbiol.* **1987**, *25*, 1679–1682. [CrossRef]
56. Banyai, K.; Martella, V.; Meleg, E.; Kisfali, P.; Peterfi, Z.; Benko, M.; Melegh, B.; Szucs, G. Searching for HAdV-52, the putative gastroenteritis-associated human adenovirus serotype in Southern Hungary. *New Microbiol.* **2009**, *32*, 185–188.
57. Kesisoglou, F.; Chamberlain, J.R.; Schmiedlin-Ren, P.; Kaz, A.; Fleisher, D.; Roessler, B.; Zimmermann, E.M. Chimeric Ad5 vectors expressing the short fiber of Ad41 show reduced affinity for human intestinal epithelium. *Mol. Pharm.* **2005**, *2*, 500–508. [CrossRef]

viruses

MDPI

Article

Structure-Based Modeling of Complement C4 Mediated Neutralization of Adenovirus

Corey C. Emerson and Phoebe L. Stewart *

Department of Pharmacology and Cleveland Center for Membrane and Structural Biology, Case Western Reserve University, Cleveland, OH 44106, USA; cxe81@case.edu
* Correspondence: pls47@case.edu

Abstract: Adenovirus (AdV) infection elicits a strong immune response with the production of neutralizing antibodies and opsonization by complement and coagulation factors. One anti-hexon neutralizing antibody, called 9C12, is known to activate the complement cascade, resulting in the deposition of complement component C4b on the capsid, and the neutralization of the virus. The mechanism of AdV neutralization by C4b is independent of downstream complement proteins and involves the blockage of the release of protein VI, which is required for viral escape from the endosome. To investigate the structural basis underlying how C4b blocks the uncoating of AdV, we built a model for the complex of human adenovirus type-5 (HAdV5) with 9C12, together with complement components C1 and C4b. This model positions C4b near the Arg-Gly-Asp (RGD) loops of the penton base. There are multiple amino acids in the RGD loop that might serve as covalent binding sites for the reactive thioester of C4b. Molecular dynamics simulations with a multimeric penton base and C4b indicated that stabilizing interactions may form between C4b and multiple RGD loops. We propose that C4b deposition on one RGD loop leads to the entanglement of C4b with additional RGD loops on the same penton base multimer and that this entanglement blocks AdV uncoating.

Keywords: adenovirus; neutralization; neutralizing antibody; complement C1; complement C4; molecular dynamics

1. Introduction

There are multiple parallel pathways for neutralizing pathogens such as adenovirus (AdV). While neutralization pathways are beneficial in the case of natural infections, they represent roadblocks in the development of virus-based therapeutics, such as oncolytic viruses [1], and gene therapy vectors [2,3]. Both pre-clinical and clinical data showed that anti-AdV-specific neutralizing immunity reduce efficacy of AdV-based vaccines, including against HIV-1 [4], and SARS-CoV-2 [5]. Therefore, a better understanding of the molecular mechanisms underlying host neutralization pathways, specifically involving neutralizing antibodies and complement, would be beneficial for engineering AdV-based therapeutics with improved safety and efficacy.

Following AdV infection, both the innate and adaptive arms of the immune system are involved in the clearance of the virus. When human species C HAdV-C5 is injected into the bloodstream, the innate immune system responds with natural immunoglobulin M (IgM) antibodies [6–8], and coagulation factor X (FX) [9,10], to opsonize the virus and target it for clearance. For HAdV-C5, natural IgM binds to the hypervariable region 1 (HVR1) of hexon, the major capsid protein, which forms a repetitive, negatively charged pattern on the capsid surface [11]. IgM binding to AdV activates the complement cascade, leading to the covalent binding of first complement component C4b and then C3b to the virus [12]. The blood coagulation factor, FX, binds species C HAdV-C2 and HAdV-C5 with high affinity via the major capsid protein, hexon, and helps to target the virus to the liver for clearance [9,10]. Effectively, the FX-decorated surface of AdV becomes a pathogen-associated molecular pattern (PAMP), which, after internalization into a macrophage cell, serves to activate innate immunity via the TLF/NF-κB pathway [13]. The binding of IgM and FX to AdV represent parallel host–virus neutralization pathways, as FX binding to AdV protects the virus from complement-mediated inactivation [12].

During the initial exposure to a particular virus, innate immune responses activate and stimulate adaptive immune responses, which are ultimately responsible for complete viral clearance [14]. Adaptive immunity includes both a humoral immune response, involving B cells and CD4 helper T cells, and a cell-mediated immune response, involving CD8+ T cells. B cells produce virus-specific antibodies that can neutralize and inactivate virions. Virus-specific immunoglobulin G (IgG) antibodies, similar to IgM, can activate the complement system after binding to a virus particle [15]. Complement proteins serve to opsonize pathogens and induce inflammatory responses that help fight infection. The complement system is an integral effector part of both the innate and adaptive immune response to viral infections.

The classical pathway of complement activation begins with the binding of the C1 complex (C1q, C1r$_2$, C1s$_2$) to antigen-bound IgM or IgG [15]. IgM exists in circulation as planar pentameric and hexameric assemblies with its C1q binding site hidden [16]. After antigen binding, a conformational change occurs in IgM to convert it into a staple-like conformation with exposed C1q binding sites [17–19]. Only one antigen-bound IgM is needed to activate complement, whereas several IgG molecules bound to the antigen in close proximity are required for activation [16]. Several studies have shown that IgG antibodies oligomerize and form platforms with their F$_C$ domains to present appropriately spaced C1q binding sites [20–22]. C1q is a hexamer formed by heterotrimeric chains A, B and C, assembled into a bundle of six collagen helices and six globular recognition domains that bind immunoglobins [23]. C1r and C1s are both serine proteases. After the C1q globular domains interact with antigen-bound IgM or IgG, C1r is activated, which in turn activates C1s [16]. Activated C1s cleaves complement component C4 into C4a, which is released, and C4b, which has a highly reactive thioester that can react with hydroxyl or amino groups near the antibody binding site on the pathogen. The classical pathway continues with an enzyme cascade involving complement components C2 and C3. Like C4b, C3b has a reactive thioester that can opsonize the pathogen and C3b opsonization is involved in complement signal amplification [15].

Recently, Bottermann et al. have shown that, after an anti-hexon neutralizing IgG, called 9C12, binds HAdV-C5, complement components C1 and C4 mediate the potent neutralization of HAdV-C5 and that this antiviral activity is independent of downstream complement components [24]. The binding of C4b to HAdV-C5 does not block cell entry, but rather blocks the release of the virally encapsidated membrane lytic factor, protein VI, which is essential for escape from the endosome [25,26]. The antiviral mechanism of C4 is reminiscent of AdV neutralization by α-defensins, which are proteins of the innate immune system that suppress viral and bacterial infections [27]. Human α-defensin 5 (HD5) has potent antiviral activities against many but not all human AdV types [28]. HD5 binds at the interface between the AdV penton base and the fiber capsid proteins, stabilizes the vertex region of sensitive AdV types, and prevents the release of protein VI [28,29].

A cryo-electron microscopy (cryo-EM) study of antigen-bound IgG in complex with C1 revealed that bound IgG's adopt a bent conformation and form a hexameric F$_C$ platform, which binds four, five or six C1q globular domains [22]. The moderate resolution cryo-EM structure of the complex, combined with atomic resolution structures of the C1q globular domain [30], and IgG F$_C$ domains [31], facilitated the generation of a structural model for C1q-IgG interaction. A cryo-electron tomography (cryo-ET) study of IgM, C1, and C4b complexes formed on antigen-bearing lipid membranes revealed pentameric and hexameric IgM complexes in dome-shape conformations with bound C1 and C4 [19]. These IgG-C1 and IgM-C1 complex structures reveal similarly spaced C1q binding sites of the periphery of their respective F_C platforms. The structure of the IgM complex also revealed additional density for C4b oriented next to the IgM-Fab arms. The thioester-containing domain of C4b is positioned such that the reactive thioester points toward the antigenic surface.

In this study, we used structural information on the binding of the IgG 9C12 to HAdV-C5 [32,33], together with cryo-EM and cryo-ET structures of IgG-C1 and IgM-C1-C4 complexes [19,22], to build a composite model for HAdV-C5 with bound IgG, C1, and C4b. The goals of this study were to evaluate likely C4b binding sites on the HAdV-C5 capsid and investigate the structural mechanisms underlying the C4b neutralization of HAdV-C5. A prior cryo-EM structure of HAdV-C5 with 9C12 IgG molecules indicated strong density for 9C12 bound to the peripentonal hexons, which are the five hexons surrounding the penton base capsid protein at the vertices of the AdV capsid [33]. The pentameric penton base has five intrinsically disordered Arg-Gly-Asp (RGD)-containing loops that protrude from penton base and interact with αv integrins on host cells triggering internalization of the virus [29,34,35]. The composite model we built for HAdV-C5 with IgG 9C12, C1 and C4b indicated that C4b might bind to various solvent accessible hydroxyl and amino groups within the penton base RGD loop. We performed molecular dynamics simulations with C4b covalently bound to two possible sites within the RGD loop. The results of these simulations indicate that C4b binding to one RGD loop of a HAdV-C5 penton base will likely result in additional stabilizing interactions between C4b and another RGD loop of penton base. In addition, since the highly reactive thioester of C4b can react with a water molecule before reaching the pathogen [15], we also performed a molecular dynamics simulation with C4b positioned near, but not covalently bound to, the penton base RGD loops. This simulation indicates that, when C4b is positioned near one RGD loop, even without being covalently bound to the penton base, C4b may form additional stabilizing interactions with nearby RGD loops. This work revealed alternate mechanisms of how C4b might block AdV uncoating by entangling the RGD loops of the penton base with or without covalent binding to the virus and suggests strategies that might be used to modulate the interaction of AdV with the complement system. Our computational modeling-based analyses may prove useful in designing future biological experiments to evaluate complement-AdV interactions through the introduction of targeted mutations in the AdV capsid to reduce virus sensitivity to complement and, thus, aid in designing therapeutic vectors resistant to complement-mediated neutralization.

2. Methods

2.1. Model Building

For the penton base, Rosetta-based models for the RGD loop aa 297–376 [29] were added to the cryo-EM HAdV-C5 penton base coordinates (PDB: 6B1T) [36]. Five different RGD loop models were used, one for each subunit of the penton base pentamer (chains A–E). One icosahedral facet of hexons (12 trimers), plus two edge hexons from each of the three adjacent facets (6 additional trimers), were selected from the cryo-EM HAdV-C5 structure (PDB: 6B1T) with UCSF ChimeraX v1.1 [37]. Three penton base pentamers with modeled RGD loops were added to the adjacent vertex sites to form a model of the HAdV-C5 facet. Coordinates for 9C12 Fab fragments were positioned above hexon epitopes by aligning the crystal structure of the hexon in complex with the 9C12 Fab (PDB: 5LDN) with each hexon subunit in the HAdV-C5 facet [32]. The UCSF Chimera v1.15 MatchMaker tool was used for alignment [38]. The partially occupied facet/Fab model, with two thirds of possible hexon epitopes occupied with 9C12 Fab, was generated by removing Fab fragments from the fully occupied facet/Fab model with UCSF Chimera v1.15. Fab fragments were selected for removal to minimize steric clashes between Fab fragments and to approximate the Fab density observed in the cryo-EM structure of HAdV-C5 with 9C12 IgG [33].

To build a model of an HAdV-C5 facet with a hexameric IgG F_C platform, the F_C platform coordinates from a cryo-EM structure of an IgG-C1 complex (PDB: 6FCZ) were used [22]. The hinge region of one F_C was positioned near the exposed CL, CH_1 domains of a 9C12 Fab fragment positioned on a peripentonal hexon. The additional five hinge regions of the hexameric F_C platform were positioned more approximately over other Fab fragments in the partially occupied facet/Fab model. The hexameric F_C platform positioned over the HAdV-C5 facet was used as a guide to add in coordinates for six C1q globular domains (PDB: 6FCZ) and the cryo-EM density for IgG-C1 (EMD-4232) [22].

Given the strong similarity between the cryo-EM structure of an IgG-C1 complex and the cryo-ET structure of an IgM-C1-C4 complex [19,22], we used the position of C4b in the later structure to guide the positioning of C4b relative to the C1 complex modeled with HAdV-C5. Coordinates from the crystal structure of C4b (PDB: 4XAM) were used to complete the HAdV-C5/9C12/C1/C4b model [39]. All graphic figures were prepared with UCSF Chimera v1.15 [38].

2.2. Molecular Dynamics Simulations

A molecular dynamics simulation was performed to assess the solvent accessibility of possible C4b opsonization sites within the penton base RGD loop. A simulation for the penton base pentamer with Rosetta-based RGD loop models was performed with NAMD v2.12 on the Case Western Reserve University (CWRU) high-performance computing (HPC) cluster [40]. The molecular system was minimized for 50 ps, followed by slow heating to 300 K. A molecular dynamics simulation was run for 5 ns using the Chemistry at Harvard Molecular Mechanics (CHARMM) force field [41], with Generalized Born implicit solvent (GBIS). The solvent accessibility of the atoms in the hydroxyl groups of serines and threonines, and the amino groups of lysines and arginines, within the five RGD loops was assessed for the starting and ending coordinates with the UCSF ChimeraX v1.1 "measure sasa" command and a probe radius of 1.4 Å [37].

Molecular dynamics simulations were performed to assess the possibility that C4b would interact with multiple RGD loops on one penton base pentamer. The coordinates used for C4b were from the crystal structure (PDB: 4XAM) [39]. Six different starting models for C4b relative to a penton base pentamer were prepared for molecular dynamics simulations. Four of the starting models (models 1, 2, 4, and 5) were generated with a covalent bond between Cys1010 of C4b and a residue within one penton base RGD loop (Thr343, chain C; Arg347, chain C; Thr346, chain C; or Lys297, chain E, respectively). We used UCSF Chimera v1.13 to prepare the chosen RGD loop residue, position the sulfur atom of the reactive thioester on C4b near the appropriate atom of the RGD loop residue, form a covalent bond with the "bond sel" command, and change the chain IDs to be the same for the two covalently linked polypeptides [38]. Two starting models (models 3 and 6) were generated without a covalent bond between C4b and the penton base. For these models, the Cys1010 of C4b was positioned ~10 Å from one RGD loop of the penton base (chain C or chain E, respectively). Molecular dynamics simulations were performed with NAMD v2.12 on the Case Western Reserve University (CWRU) high-performance computing (HPC) cluster [40]. The molecular systems were minimized for 50 ps followed by slow heating to 300 K. Molecular dynamics simulations were run for 12 ns using the Chemistry at Harvard Molecular Mechanics (CHARMM) force field [41] with generalized born implicit solvent (GBIS).

2.3. Calculation of Non-Bonded Interaction Energies

Non-bonded interaction energies, including van der Waals and electrostatic components, were calculated between C4b and each chain of the penton base individually, as well as between C4b and the penton base multimer (chains A–E) as a whole. The energy calculations were performed for both starting and ending coordinates of the molecular dynamics simulations of models 1–6. NAMD v.2.14 [40] and the NAMD Energy plugin of VMD v1.9.3 [42], both running on a Windows 10 PC, were used to calculate the interaction energies.

3. Results

3.1. Modeling of HAdV-C5 with Antihexon Neutralizing Antibody

Previous work has shown that a particular anti-hexon neutralizing IgG, called 9C12, stimulates binding of complement component C4b to the capsid and mediates potent neutralization of HAdV-C5 [24]. Although there is both cryo-EM and crystallographic structural information on the binding of 9C12 to HAdV-C5 [32,33], there are still open questions about how this particular IgG interacts with the full virion. It has been shown that the minimum ratio of 9C12 to HAdV-C5 for neutralization is 240 antibody molecules per virus particle, which is equivalent to an average of two Fab fragments per hexon trimer [33]. In other words, assuming IgG binds bivalently, only two thirds of the available epitopes need to be occupied to achieve neutralization. The crystal structure of the isolated HAdV-C5 hexon with 9C12 Fab fragments revealed that the epitope includes hexon hypervariable regions (HVRs) 2 and 8, which form the outer corner of each of the three towers of a hexon trimer [32]. The cryo-EM structure of HAdV-C5 with intact 9C12 indicated bivalent binding for the IgG [33]. Strong density was observed for Fab fragments on the peripentonal hexons, hexons adjacent to penton base (Figure 1A), as well as a meshwork of Fab density covering the rest of the hexon capsid surface. The cryo-EM density was interpreted as indicating 100% occupancy of Fab at the peripentonal hexon sites and a spatial average of many alternate bivalent binding combinations for 9C12 on the rest of the capsid.

We built a model for the interaction of 9C12 Fabs with one facet of HAdV-C5 based on the available cryo-EM and crystallographic structural information. Initially we positioned three Fab fragments on each hexon trimer in the facet, as indicated by the crystal structure (Figure 1B). This resulted in a fully occupied model with multiple clashes between Fab fragments (Figure 1C). Clashes between Fab fragments, as well as the consideration of expected steric hindrance between IgG F_C fragments, led us to conclude that the fully occupied model is unrealistic. Therefore, we reduced the number of Fab fragments bound to one facet so that two thirds of the possible epitopes were occupied with a 9C12 Fab (Figure 1D). This partially occupied model displays fewer steric hindrances between neighboring IgG Fab and F_C fragments and better resembles the cryo-EM structure of the HAdV-C5/9C12 complex [33]. There is undoubtedly variation in how the 9C12 antibody binds to hexons in each icosahedral facet and in each virus particle. Therefore, the model shown in Figure 1D is meant only to be a representative approximation. In building the partially occupied model, we left all three epitopes on each peripentonal hexon occupied with Fabs in accordance with the strong cryo-EM density observed at these sites.

As a result of leaving all peripentonal hexon epitopes occupied with a Fab in the partially occupied model, major steric clashes are observed between the CL and CH_1 domains of two Fabs (Figure 2). However, it was noted in the cryo-EM structure of the HAdV-C5/9C12 complex that the IgG density at this site had one well-shaped Fab arm and one somewhat distorted Fab arm [33]. The cryo-EM structure indicated that the binding of 9C12 to these two peripentonal epitopes resulted in an unusually acute angle between the long axes of the Fab fragments. Varghese et al. concluded that bivalent binding of 9C12 to this site was likely facilitated by the inherent segmental flexibility of IgG molecules [33]. In addition, we note that the crystal structure of the isolated hexon with the 9C12 Fab reveals the epitope to be composed mainly of two HVR regions [32]. We suspect that the conformational flexibility of the epitope region might contribute to the bivalent binding of 9C12 to this apparently strained IgG binding site. Indeed, Bottermann et al. note that 9C12 does not display a particularly fast on-rate with hexon and that this observation might be explained by an entropic cost associated with engaging a structurally variable epitope [32].

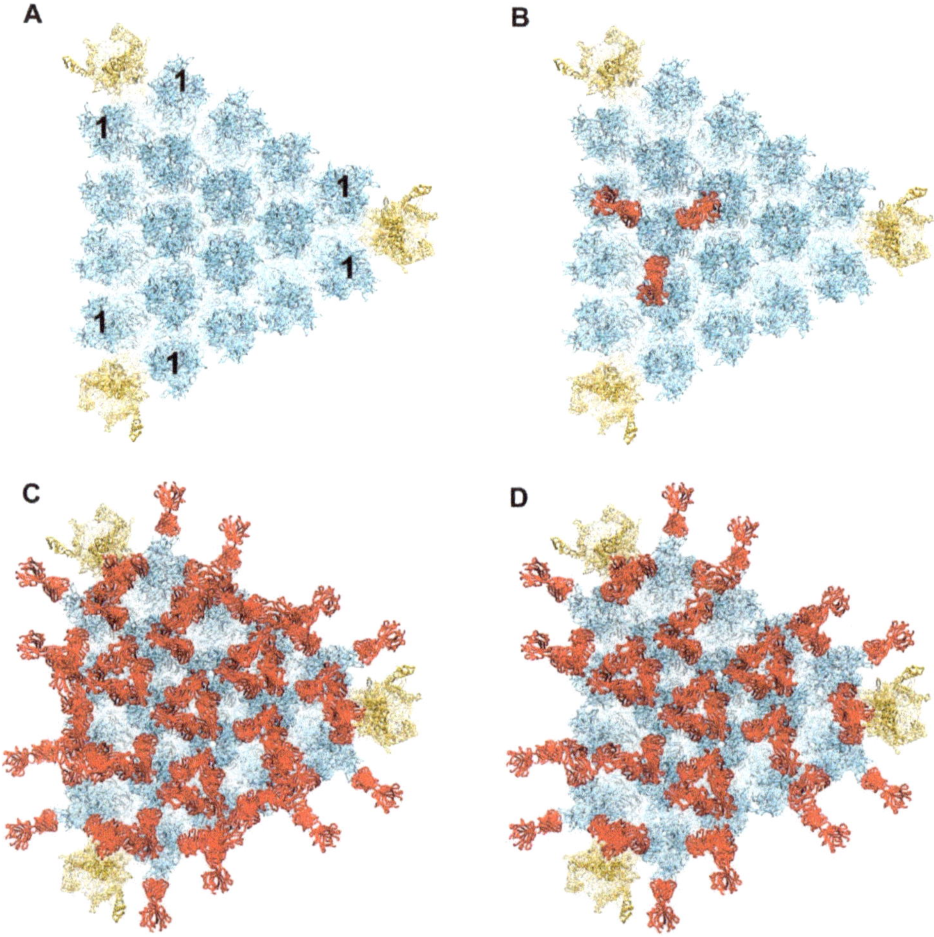

Figure 1. Model for binding of antihexon monoclonal antibody 9C12 to a facet of HAdV-C5. (**A**) Major capsid proteins in one facet of HAdV-C5 with hexons forming an array in the middle (light blue) and penton bases at the vertices (gold) (PDB: 6B1T) [36]. The peripentonal hexons are denoted with the number 1. Each penton base is shown with Rosetta-based models for the RGD loops [29]. (**B**) One hexon trimer shown with three 9C12 Fab fragments (red) positioned as in the crystal structure of isolated hexon with 9C12 Fab (PDB: 5LDN) [32]. (**C**) Fully occupied model of HAdV-C5 facet with all hexon epitopes occupied with a 9C12 Fab. (**D**) Partially occupied model with two thirds of the possible hexon epitopes occupied with a 9C12 Fab. The occupied Fab binding sites were selected so that the model would resemble the cryo-EM structure of HAdV-C5 with 9C12 IgG [33].

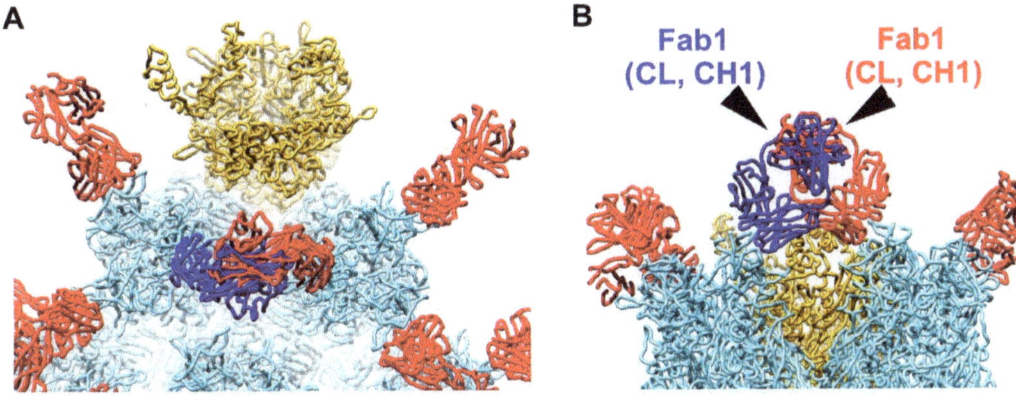

Figure 2. Steric clashes are modeled between two 9C12 Fab fragments bound to neighboring peripentonal hexons. (**A**) Enlarged view of vertex region with two hexon trimers and one penton base with selected 9C12 Fab fragments as in Figure 1D. A severe steric clash is modeled between Fab fragments (blue and red) bound to neighboring peripentonal hexons. (**B**) Perpendicular view of panel A. The clashing domains (CL, CH_1) of each Fab are indicated.

3.2. Modeling of HAdV-C5 with IgG and Complement Components C1 and C4b

The cryo-EM structure of the HAdV-C5/9C12 complex did not reveal defined density for the F_C regions, indicating variability in the F_C positions relative to the HAdV-C5 capsid [33]. The lack of observed F_C density in the cryo-EM structure of the HAdV-C5/IgG complex is not surprising given the known flexibility of IgG molecules [43]. The antihexon antibody 9C12 is of the IgG1 subclass of antibodies. Extensive structural flexibility has been observed for IgG1 molecules by individual-particle electron tomography 3D reconstruction [44]. The partially occupied Fab model shown in Figure 1D does not include modeled F_C regions for the bound 9C12 IgG molecules. However, modeling the locations of the 9C12 IgG F_C regions is important for adding complement components C1 and C4b to the HAdV-C5/9C12 model, since the F_C regions contain binding sites for the globular recognition domains of C1q [43]. The C1q binding site is near the IgG hinge region and is thought to be partially or completed shielded by the Fab arms when IgG is not bound to an antigen [43]. It has been suggested that the F_C regions of multiple IgG molecules form hexamers when opsonized on target surfaces [21]. Mutations can be introduced in IgG that drive the formation of IgG hexamers in solution [20,21,45]. The cryo-EM structure of a soluble C1-IgG complex was formed with hexamer-promoting IgG molecules [22]. The docking of the cryo-EM density for the soluble C1–IgG complex with known atomic resolution structures of the component domains resulted in the identification of the C1q binding residues within the two F_C CH2 domains of an IgG. These C1q binding residues were corroborated with mutagenesis studies. The cryo-EM and cryo-ET structures of IgG-C1 and IgM–C1–C4 complexes indicate that the Fab arms of an IgG hexamer and IgM fold so that they are nearly perpendicular to their respective F_C region when C1 is bound to the complex [19,22].

In the HAdV-C5/9C12 model, we added a hexamer of F_C domains with one IgG hinge region near the Fabs bound to the peripentonal hexons. A preference for 9C12 binding to the peripentonal hexons was noted in the cryo-EM structure of the HAdV-C5/9C12 complex [33]. The other IgG hinge regions of the F_C hexamer were positioned roughly near other Fabs bound to hexons in the facet (Figure 3A,B). It was not possible to align the additional five hinge regions of the F_C hexamer with particular Fab fragments without distorting the underlying hexon epitopes, bound Fab fragments, or the hexameric FC platform coordinates. Therefore, the FC portion of the HAdV-C5/9C12 IgG model shown in Figure 3 is likely more hexameric than can exist in reality. This is in accord with the lack of a clear hexameric pattern of Fab arms in the partially occupied model of

Fabs bound to one facet (Figure 1D). Additionally, we noted that the cryo-EM structure of the HAdV-C5/9C12 complex indicates heterogeneity of occupied Fab binding sites in the middle of the facet [33]. Therefore, we suspect that, in reality, perhaps only four or five F_C domains assemble over each HAdV-C5 facet to form imperfect F_C hexamers. Nevertheless, the imperfect F_C hexamers may still attract the binding of the complement C1 complex if the IgG molecules are bent and if the spacing of C1q binding sites is appropriate. Ugurlar found, in their cryo-EM analysis of soluble C1-IgG complexes, that classification results in separate classes with four, five or six globular C1q domains in contact with F_C platforms [22]. Their soluble C1-IgG complexes were formed with IgG molecules mutated to induce hexamer formation, which was undoubtedly useful for structural analysis but which may not represent all F_C assemblies that can bind C1. We propose that, with native IgG molecules, such as 9C12, perhaps F_C aggregation does not need to form perfect hexamers to induce C1 binding.

With an F_C hexamer in the HAdV-C5/9C12 model, it was possible to add in models for six C1q globular domains and cryo-EM density for the soluble IgG-C1 complex (Figure 3C,D). In reality, we expect that the assembly of HAdV-C5/9C12/C1 is more heterogeneous in nature with variations in the F_C aggregates. The key factors that the model shown in Figure 3 revealed are (1) that 9C12 IgG molecules bound bivalently to the peripentonal hexons may form F_C interactions with other 9C12 IgG molecules bound to the array of hexons in the middle of the HAdV-C5 facet, and (2) that the C1 complex may bind preferentially to the corners of the HAdV-C5 facets near the penton bases (Figure 3C), rather than to the middle of a HAdV-C5 facet. Once a model for HAdV-C5/9C12/C1 was built, it was possible to add in a molecule of complement C4b (Figure 4). We show C4b positioned over a peripentonal hexon, putting C4b in close proximity to a penton base. This position is based on the location of density for C4b in the cryo-ET structures of IgM–C1–C4b complexes [19]. In these structures, Sharp et al. detected density for one or two C4b molecules per complex and found C4b positioned next to Fab arms of IgM in a bent conformation. We admit that, in building the HAdV-C5/9C12/C1/C4b model shown in Figure 4B, we chose to position C4b close to the penton base, when in reality C4b might equally well be located over the middle of the facet. However, as C1 complexes preferentially bind near peripentonal hexons, our model predicts that at least some of the bound C1 complexes would present C4b molecules near a penton base.

3.3. *Possible Covalent Binding Sites for C4b on HAdV-C5 Penton Base*

After recognition of a pathogen by IgG or IgM and recruitment of the C1 complex, the activated C1s serine protease in the C1 complex cleaves complement protein C4 into a C4a fragment (9k Da), which is released into the solvent, and C4b (195 kDa), which acts as an opsonizing factor. C4b has an internal thioester bond that is exposed after conformational changes induced after its cleavage by C1s [39,46]. The thioester of C4b is highly reactive and rapidly forms a covalent bond with a nearby hydroxyl or amino group [15,16]. Often the C4b thioester forms a covalent bond with the surface of the pathogen by interacting with hydroxyl or amino groups, but it can also react with nearby water molecules. Our model for HAdV-C5/9C12/C1/C4b indicates that at least some molecules of C4b will be near the penton base of HAdV-C5 and that the reactive thioester will be oriented toward the penton base (Figure 4B).

The cryo-EM structure of HAdV-C5 and the crystal structure of HAdV-C2 penton base both indicate that the integrin-interacting, RGD-containing loops of the penton base are flexible [36,47]. In the HAdV-C5 penton base structure, over 80aa (aa297–376) are missing in the RGD loop due to flexibility and predicted intrinsic disorder [29]. Flatt et al. built Rosetta-based models for the HAdV-C5 RGD loops, which extend ~50 Å above the top of the ordered portion of penton base (Figure 5A) [29]. Intrinsic disorder within the RGD loops may provide a functional advantage for interaction with αv integrins, which serve as internalization receptors for HAdV [35]. The flexible and extended nature of the penton base RGD loops may also make them likely targets for C4b opsonization.

Examination of the HAdV-C5 penton base RGD loop sequence indicates eight residues with a hydroxyl group in their sidechain (serines and threonines) and eight residues an amino group (lysines and arginines), all of which might serve as binding sites for the reactive thioester of C4b (Figure 5). A molecular dynamics simulation of the HAdV-C5 penton base pentamer with modeled RGD loops, combined with a calculation of the solvent accessible surface area for the hydroxyl and amino groups, indicated that these possible reactive thioester binding sites are all solvent accessible. The maximum solvent accessible surface area was found for all of these groups in both the starting and ending coordinates (528 Å^2 for hydroxyl oxygens; 575 Å^2 for amino nitrogens).

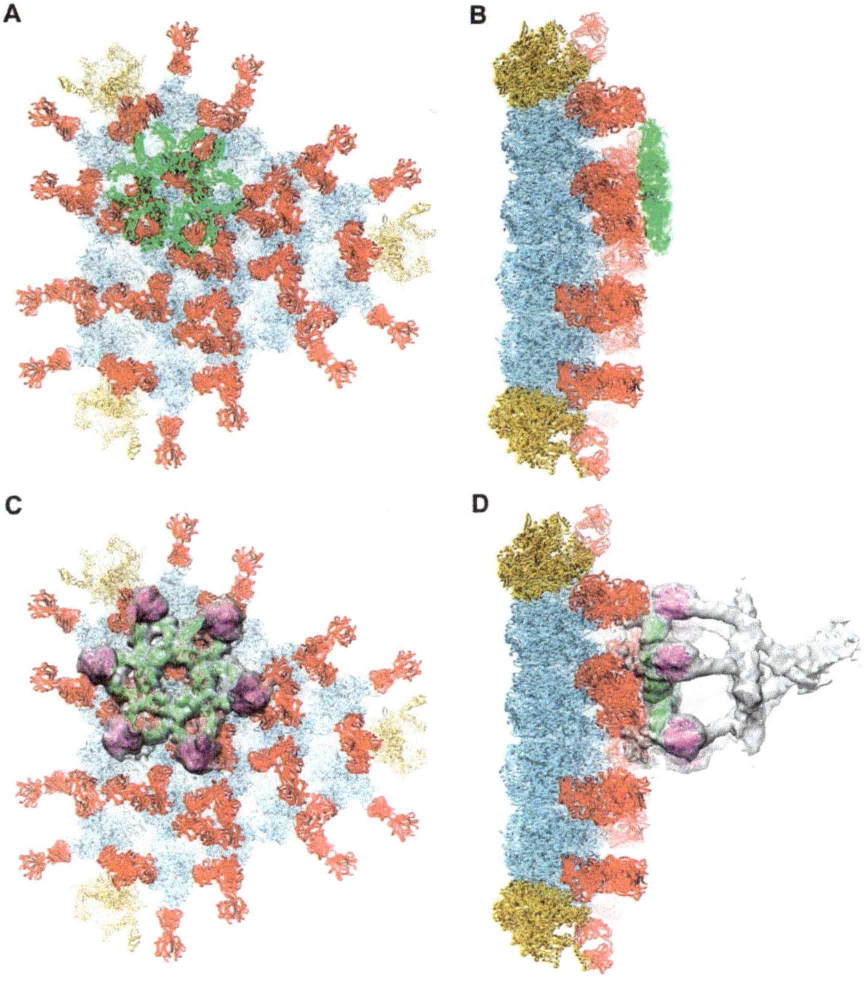

Figure 3. Model for binding of complement C1 complex to a facet of HAdV-C5. (**A**) One facet with selected 9C12 Fab fragments, as in Figure 1D, modeled with a hexameric F_C platform (green) from a cryo-EM structure of an IgG-C1 complex (PDB: 6FCZ) [22]. (**B**) Perpendicular view of panel A showing that the F_C platform is positioned just above the layer of Fab fragments (red) bound to the hexons. (**C**) One facet, as in panel A, modeled with six C1q globular domains (magenta) from the IgG-C1 structure, shown with a slab of cryo-EM density (transparent gray). (**D**) Perpendicular view of panel C showing the full cryo-EM density map for IgG-C1 (EMD-4232) [22].

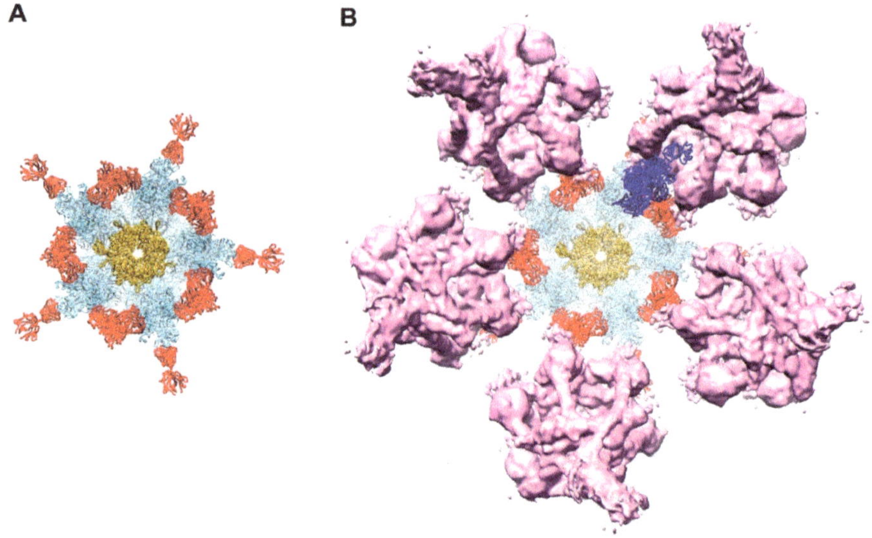

Figure 4. Model for complement C4b near penton base of HAdV-C5 vertex. (**A**) Capsid vertex with a central penton base (gold) surrounded by five peripentonal hexons (light blue) shown with modeled 9C12 Fab fragments (red). (**B**) Capsid vertex with 9C12 Fab fragments modeled with five copies of the IgG-C1 complex (purple) (EMD-4232) [22] near the peripentonal hexons and one copy of complement C4b (blue) (PDB: 4XAM) [39] positioned approximately based on the IgM-C1-C4 structure [19].

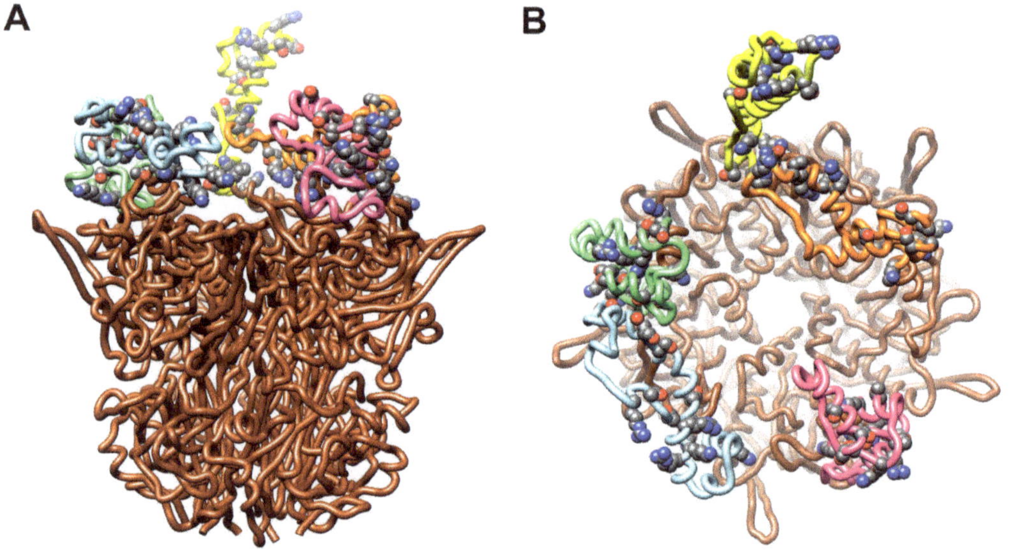

Figure 5. HAdV-C5 penton base with Rosetta-based models for the RGD loops. (**A**) Side view of pentameric penton base (PDB: 6B1T) [36] with different RGD loop models (aa297–376) for each of the five subunits (chain A, pink; chain B, light blue; chain C, green; chain D, yellow; chain E, orange) [29]. All of the sidechains in the RGD loop that contain hydroxyl or amino groups (serines, threonines, lysines and arginines) and that might serve as opsonization sites for the reactive thioester of C4b are shown in space filling representation. (**B**) Top view of panel A.

3.4. Molecular Dynamics Simulations with HAdV-C5 Penton Base and C4b

In order to test our hypothesis that C4b deposition on one RGD loop leads to the entanglement of C4b with additional RGD loops on the same penton base multimer, we built three starting models for molecular dynamics simulations. In model 1 the thioester of C4b is covalently bound to the hydroxyl group of Thr343 in one RGD loop (Figure 6A). In model 2, C4b is covalently bound to the amino group of Arg 347 in an RGD loop (Figure 6B). In model 3, we presume that the reactive thioester of C4b has reacted with a water molecule and position C4b near, but not covalently bound to, one RGD loop (Figure 6C). Using these three starting models we performed molecular dynamics simulations to observe whether nearby RGD loops would form favorable interactions with C4b. As noted by Flatt et al., the RGD loops move relatively quickly during molecular dynamics simulations, presumably because of their flexibility and intrinsic disorder [29]. We found that, within relatively short simulations (12 ns), stabilizing interactions formed between C4b and a nearby RGD loop (Figure 6).

Using the NAMD Energy plugin, we evaluated the stabilizing non-bonded interactions formed between C4b and each penton base RGD loop. In the model 1 simulation, C4b was covalently bound to a hydroxyl group in the RGD loop of the penton base chain C in the starting model. By the end of the simulation, favorable interactions had formed with RGD loops of two neighboring penton base subunits (chains C and D), with an overall strongly favorable interaction between C4b and penton base of −318 kcal/mol (Table 1). In the model 2 simulation, C4b was covalently bound to an amino group in the RGD loop of the penton base chain C in the starting model. Similar to the results of the model 1 simulation, by the end of the model 2 simulation, favorable interactions had formed with RGD loops of two neighboring penton base subunits (chains C and D). The calculated non-bonded interaction energy between C4b and penton base at the end of the model 2 simulation was even more favorable, −594 kcal/mol, (Table 2) than found for the model 1 simulation. In the model 3 simulation, C4b was positioned near the RGD loop of penton base chain C in the starting model. Similar to the results of the model 1 and model 2 simulations, by the end of the model 3 simulation, favorable interactions had formed with RGD loops of two neighboring penton base subunits (chains C and D). The non-bonded interaction energy between C4b and penton base at the end of the model 3 simulation, −315 kcal/mol (Table 3), was similar to that found for model 1. In each model simulation, the final non-bonded interaction energy was a combination of van der Waals (VdW) and electrostatic (Elec) components: model 1 (VdW: −132 kcal/mol; Elec: −186 kcal/mol), model 2 (VdW: −189 kcal/mol; Elec: −406 kcal/mol), and model 3 (VdW: −135 kcal/mol; Elec: −181 kcal/mol).

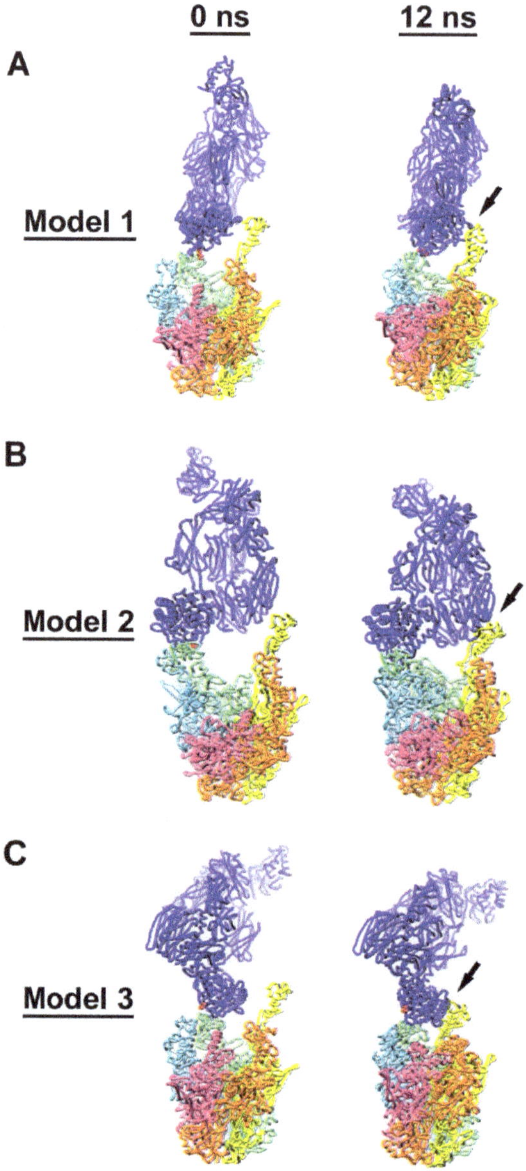

Figure 6. Interaction of C4b with multiple RGD loops of HAdV-C5 penton base. (**A**) Initial (0 ns) and final (12 ns) coordinates from the model 1 molecular dynamics simulation with a covalent bond between Thr343 of penton base chain C (green) and C4b (blue). The covalently linked residues, Thr343 of penton base and Cys1010 of C4b, are in red. (**B**) Coordinates from the model 2 simulation with a covalent bond between Arg347 of penton base chain C (green) and C4b. The covalently linked residues, Arg347 of penton base and Cys1010 of C4b, are in red. (**C**) Coordinates from the model 3 simulation with no covalent bond between penton base and C4b. Cys1010 of C4b is in red. Penton base subunits are colored as in Figure 5. Penton base is shown in a different orientation for each simulation. The final coordinates from all three simulations show additional favorable interactions indicated with arrows between C4b and penton base chain D (yellow).

Table 1. Total non-bonded interaction energy between C4b and penton base for model 1 with covalent linkage to penton base Thr343.

	0 ns (kcal/mol)	12 ns (kcal/mol)
Penton base Chain A (pink)	0	0
Penton base Chain B (blue)	0	0
Penton base Chain C (green)	−51	−226
Penton base Chain D (yellow)	6	−91
Penton base Chain E (orange)	0	0
Penton base Chains A–E	−45	−318

Table 2. Total non-bonded interaction energy between C4b and penton base for model 2 with covalent linkage to penton base Arg347.

	0 ns (kcal/mol)	12 ns (kcal/mol)
Penton base Chain A (pink)	0	0
Penton base Chain B (blue)	4	−10
Penton base Chain C (green)	−121	−517
Penton base Chain D (yellow)	1	−67
Penton base Chain E (orange)	0	0
Penton base Chains A–E	−116	−594

Table 3. Total non-bonded interaction energy between C4b and penton base for model 3 with no covalent linkage to penton base.

	0 ns (kcal/mol)	12 ns (kcal/mol)
Penton base Chain A (pink)	0	0
Penton base Chain B (blue)	0	29
Penton base Chain C (green)	−61	−158
Penton base Chain D (yellow)	0	−186
Penton base Chain E (orange)	0	0
Penton base Chains A–E	−61	−315

All three molecular dynamics simulations, presented in detail (models 1–3), support the idea that one molecule of C4b may form stabilizing interactions with multiple RGD loops of a HAdV-C5 penton base multimer. While a covalent bond between C4b and one RGD loop may promote the entanglement of multiple RGD loops (models 1 and 2), the model 3 simulation indicates that just positioning C4b near the penton base will also result in entanglement. We noted that, for all three models, secondary non-bonded interactions formed between C4b and the most extended RGD loop model of chain D (Figure 6, Tables 1–3). In contrast, preliminary models that did not have C4b positioned near the most extended RGD loop model (chain D) did not show the entanglement of RGD loops by the end of 12 ns simulations, and these models were rejected. Although the RGD loop is highly flexible, we did not observe the RGD loops of chains A, B, C, or E to extend as fully as that of chain D during the relatively short (12 ns) simulations. It is likely that, over longer simulations, all five RGD loops would extend and contract and that all five chains would be equally likely to interact with C4b. However, within the constraints of our analysis protocol, it seems that positioning C4b near an extended RGD loop is a critical factor for the acceptance of a C4b/penton base model. To confirm this idea, we built two additional starting models with covalent bonds between C4b and the penton base (models 4 and 5). These covalent linkages were made with residues within the RGD loops of chains C or E on either side of the most extended chain D RGD loop. Both models 4 and 5 showed entanglement with the chain D RGD loop by the end of a 12 ns simulation (Figures S1 and S2; Tables S1 and S2). One additional starting model without a covalent bond between C4b and penton base was generated (model 6), with C4b positioned over the chain E RGD loop. By the end of a 12 ns simulation, model 6 showed entanglement with the chain D RGD loop (Figure S3, Table S3). These additional C4b/penton base models (models 4–6) support the idea that positioning C4b near an extended RGD loop is a key factor for model acceptance with our simulation protocol. It seems likely that an abundant number of acceptable starting models could be generated that would result in the entanglement of RGD loops. The spacing of RGD loops at the top of the penton base (35 Å), the dimensions of the C4b thioester domain (~50 Å in diameter), and the multi-domain nature of C4b, all contribute to the likelihood of RGD loop interactions with C4b. In addition, longer molecular dynamics simulations would likely result in the observation of more RGD loop movement and an increase in C4b/RGD loop entanglement.

It has been proposed that integrin binding to the RGD loops of the penton base may induce a conformational change, or untwisting, of the penton base multimer that initiates AdV uncoating [34]. Together with the results presented in this study, it seems reasonable that the structural mechanism underlying the C4b neutralization of HAdV-C5 is the entanglement of C4b with multiple RGD loops of penton base multimers at each capsid vertex. This entanglement may lead to the stabilization of penton base multimers, which, in turn, may block capsid uncoating and prevent the release of the virally encapsidated endosomal membrane lytic factor, protein VI.

4. Discussion

The complement system has been described as keeping a constant vigil against viruses [48]. This system has an ancient origin, existing in a primitive form in a "living fossil", the horseshoe crab (*Carcinoscorpius rotundicauda*) [49]. In humans, a proteolytic cascade of multiple complement proteins serves to detect and mark viruses and other pathogens for destruction. The interaction of either multiple IgG molecules or a single IgM molecule with an AdV virion can initiate the classical complement activation pathway. Bottermann et al. have shown that neutralizing antibodies act with complement components C1 and C4 to effect AdV neutralization by blocking the release of AdV/C4b complexes from the endosome [24]. They also showed that this complement-based antiviral pathway works in parallel with the tripartite motif-containing protein 21 (TRIM21) antiviral activity [50]. TRIM21 is an intracellular antibody receptor that triggers the proteosome-dependent degradation of antibody-virus complexes that enter the cytoplasm [51].

In this computational modeling study, we investigated the possibility that C4b might neutralize HAdV-C5 by binding and entangling the flexible penton base RGD loops at the capsid vertices. We reasoned that the entanglement of multiple RGD loops at the same vertex might stabilize the penton base and block the conformational changes needed for the release of the penton base and the membrane lytic factor protein VI. We used available structural information for HAdV-C5 [36], HAdV-C5 anti-hexon antibody 9C12 complexes [32,33], a cryo-EM structure of an IgG-C1 complex [22], and a cryo-ET structure of an IgM-C1-C4 complex [19], to build a composite HAdV-C5/9C12/C1/C4b model (Figure 4B). This model positions C4b over the penton base capsomers of the HAdV-C5 capsid with the C4b reactive thioester positioned near the intrinsically disordered RGD loops of the penton base. Our molecular dynamics simulations with C4b and penton base indicate that it is possible for C4b to interact with multiple RGD loops at the same vertex (Figure 6) and that favorable non-bonded interactions may be formed that could stabilize the penton base (Tables 1–3) and thus block capsid uncoating. The molecular dynamics results support our hypothesis that C4b neutralizes HAdV-C5 by stabilizing penton base capsomers via RGD loop entanglement. Thus, the intrinsically disordered RGD loops of the HAdV-C5 penton base may provide a functional advantage for interacting with αv integrins on host cells, while at the same time serving as an Achilles heel of the virus, which can be exploited by the complement system.

In building a model for 9C12 Fab fragments interacting with the HAdV-C5 capsid (Figure 1D), we observed a steric clash between Fab arms bound at neighboring peripentonal hexons (Figure 2). We reasoned that this steric clash might be resolved with conformational changes of the hexon epitopes or within the IgG molecule. In fact, the observation of a steric clash at this position is consistent with past observations. The cryo-EM structure of the HAdV-C5/9C12 complex indicated that 9C12 binds bivalently to neighboring peripentonal hexons with a distorted conformation for one of the two Fab arms [33]. Bottermann et al. found that 9C12 has a slow on-rate and that binding occurs with a concurrent cost in entropy [32]. In light of our current structural modeling study, it seems reasonable to propose that distorted, bivalently bound 9C12 IgGs at the peripentonal hexons might initiate C1 binding in the vertex region of HAdV-C5. It is known that the C1q binding sites on IgGs are normally shielded by the Fab arms and are generally only exposed after the binding of IgG to a pathogen [43]. The antihexon 9C12 IgG may be able to efficiently neutralize HAdV-C5 via C4b opsonization by virtue of its favorable epitope position and its distorted binding mode, which likely exposes C1q binding sites near the IgG hinge region. The C4b mediated neutralization of HAdV-C5 may be enhanced by the preferential binding of 9C12 to peripentonal hexons versus hexons in the middle of the facet [33]. Our model of the HAdV-C5/9C12/C1/C4b complex indicates that the preferential binding of 9C12 to peripentonal hexons would ensure that a good percentage of C4b would opsonize the virus in the vicinity of the penton base.

Interestingly, our structural modeling study may offer a possible explanation for why the minimum ratio of 9C12 to HAdV-C5 for neutralization is 240 antibody molecules per virus particle [33]. We propose that the complement-mediated neutralization mechanism of 9C12 is the recruitment of C1 and C4b to the vertex regions, the entanglement of the penton base RGD loops by C4b, and the stabilization of the penton base, which leads to the blockage of capsid uncoating steps, including the release of protein VI. If we are correct, then it would be important to stabilize all twelve of the penton base capsomers on a particular virus particle to achieve neutralization by the complement-based pathway. In other words, the binding of C4b to only a subset of the penton base capsomers would not be expected to completely block protein VI release and HAdV-C5 would not be neutralized by this antiviral pathway. Our model of the HAdV-C5/9C12 complex suggests that a ratio of 240 antibody molecules per virus particle would lead to the bivalent binding of 9C12 at all peripentonal hexons, as well as a sufficient number of IgG molecules bound in the middle of each facet to promote the formation of F_C platforms near the peripentonal hexons. These well-positioned F_C platforms would effectively prime the system for the recruitment of C1 and C4b to the vicinity of the capsid vertices and heighten the chances for stabilization of all twelve of the penton base capsomers.

A limitation of our study is that the HAdV-C5 fiber protein was not included in our model of the HAdV-C5/9C12/C1/C4b complex. Multiple studies indicate that fiber is released during early AdV cell entry steps, which occur at the plasma membrane [52]. Therefore, we reasoned that the neutralization mechanism of C4b was likely not dependent on the presence of fiber. If, however, the fiber is still present when C4b is opsonizing the HAdV-C5/9C12/C1 complex, then we would anticipate that the fiber would present additional possible opsonization sites and further opportunities for C4b to form stabilizing non-bonded interactions with HAdV-C5 capsid proteins. Indeed, Bottermann et al. observed that C4b deposition on the viral capsid interferes with fiber and penton base shedding during in vitro heat treatment assays [24]. The entanglement of both the fiber and penton base RGD loops by C4b is an alternative and plausible neutralization mechanism. Nevertheless, the molecular dynamics simulations presented in this study indicate that if C4b is positioned near a HAdV-C5 vertex, then the entanglement of penton base

RGD loops and stabilization of the multimeric penton base capsomer are likely outcomes (Figure 6). Previous cryo-EM studies of AdV/integrin complexes suggest that symmetry mismatched interactions between integrins and the penton base trigger the untwisting of the penton base pentamers and the release of the penton base from the capsid [34]. We envision that the C4b entanglement of the RGD loops would have the opposite effect and serve to lock penton the base capsomers firmly in the AdV capsid.

This work suggests that introducing mutations into the penton base RGD loops might be a feasible strategy to modulate the interaction of HAdV-C5 with the complement system. However, simply mutating the serine, threonine, lysine and arginine residues in the RGD loops to other residue types might not be sufficient, as our study indicates that C4b can entangle multiple RGD loops even without being covalently bound to the penton base. Analysis of the non-bonded interaction energies formed between RGD loops and C4b during molecular dynamics simulations indicates that both sizable van der Waals and electrostatic interactions are formed (Tables 1–3). Strategies to minimize the C4b entanglement of the penton base RGD loops might include shortening the RGD loops to diminish van der Waals interactions, and reducing the number of charged residues in the RGD loops to minimize possible electrostatic interactions with C4b. These proposed modification strategies would require experimental testing and verification. It is of interest to note that the vast majority of HAdV species have very short penton base RGD loops that may point to an evolutionary complement attack evasion mechanism [48,53]. Whether or not HAdV species with short penton base RGD loops are more resistant to C4 complement-mediated neutralization, compared to HAdv species C, also requires experimental verification. In addition, RGD loop modifications would have to be designed so that either they do not impair interactions with αv integrins on host cells or so that they provide targeting to alternative internalization receptors. Atasheva et al. have demonstrated that it is possible to replace HAdV-C5 RGD loops with sequences derived from human laminin-α1 to retarget the virus to use $\alpha3\beta1$, $\alpha6\beta1$, and $\alpha6\beta4$ integrins present on human epithelial tumor cells [11].

A potential limitation of our study is that our analysis and conclusions are based exclusively on computational modeling of AdV interactions with antibodies and complement components C1q and C4. As the detrimental effect of neutralizing antibodies and complement on the safety and efficacy of AdV-based vectors has been extensively reported, numerous approaches to shield AdV particles from blood factors have been proposed. Many of these approaches have been tested in pre-clinical models and in human clinical trials, including shielding AdV with polymers [54] and plasma proteins, such as albumin [55]. Our study may serve as a foundation for engineering novel AdV vectors that resist complement-mediated inactivation based on specific targeted mutations in the adenovirus hexon and penton base. While our study is purely theoretical, the analyses we have done point to the conceptual feasibility of designing AdV vectors resistant to C4 complement deposition on penton base capsomers. It is certain that the direct visualization of AdV in complex with complement components, using cryo-EM or cryo-ET approaches, will be required to provide exhaustive information on the mode of complement-AdV interaction. It is anticipated that these structural studies would lead to the design and experimental validation of mutant AdV variants with improved resistance to complement-mediated neutralization.

Together, our modeling and molecular dynamics results provide a structural hypothesis for complement C4 mediated neutralization of AdV. An enhanced understanding of the molecular mechanisms underlying the interaction of AdV with host factors, including complement proteins, should promote the development of AdV-based oncolytic viruses and gene therapy vectors.

Supplementary Materials: The following are available online at https://www.mdpi.com/1999-4915/13/1/111/s1, Figure S1: Interaction of C4b with multiple RGD loops of HAdV-C5 penton base for Model 4; Figure S2: Interaction of C4b with multiple RGD loops of HAdV-C5 penton base for Model 5; Figure S3: Interaction of C4b with multiple RGD loops of HAdV-C5 penton base for

Model 6; Table S1: Total Non-bonded Interaction Energy between C4b and Penton Base for Model 4; Table S2: Total Non-bonded Interaction Energy between C4b and Penton Base for Model 5; Table S3: Total Non-bonded Interaction Energy between C4b and Penton Base for Model 6.

Author Contributions: Conceptualization, P.L.S.; methodology, C.C.E. and P.L.S.; formal analysis, C.C.E. and P.L.S.; investigation, C.C.E. and P.L.S.; writing—original draft preparation, C.C.E.; writing—review and editing, C.C.E. and P.L.S.; visualization, C.C.E. and P.L.S.; supervision, P.L.S.; project administration, P.L.S.; funding acquisition, C.C.E. and P.L.S. All authors have read and agreed to the published version of the manuscript.

Funding: This work was supported by U.S. NIH grant AI107960 to P.L.S. and Dmitry M. Shayakhmetov. C.C.E. acknowledges support from the U.S. NIH T32 GM008803 training grant.

Institutional Review Board Statement: Not applicable.

Informed Consent Statement: Not applicable.

Data Availability Statement: The models presented in this study are available on request from the corresponding author.

Acknowledgments: We are thankful to D. Shayakhmetov for helpful discussions and critical reading of the manuscript. We thank the Case Western Reserve University High Performance Computing staff for assistance with campus computing resources. NAMD was developed by the Theoretical and Computational Biophysics Group in the Beckman Institute for Advanced Science and Technology at the University of Illinois at Urbana-Champaign. Molecular graphics and analyses performed with UCSF Chimera, developed by the Resource for Biocomputing, Visualization, and Informatics at the University of California, San Francisco, with support from NIH P41-GM103311. Molecular graphics and analyses performed with UCSF ChimeraX, developed by the Resource for Biocomputing, Visualization, and Informatics at the University of California, San Francisco, with support from National Institutes of Health R01-GM129325 and the Office of Cyber Infrastructure and Computational Biology, National Institute of Allergy and Infectious Diseases.

Conflicts of Interest: The authors declare no conflict of interest.

References

1. Niemann, J.; Woller, N.; Brooks, J.; Fleischmann-Mundt, B.; Martin, N.T.; Kloos, A.; Knocke, S.; Ernst, A.M.; Manns, M.P.; Kubicka, S.; et al. Molecular retargeting of antibodies converts immune defense against oncolytic viruses into cancer immunotherapy. *Nat. Commun.* **2019**, *10*, 3236. [CrossRef] [PubMed]
2. Allen, R.J.; Byrnes, A.P. Interaction of adenovirus with antibodies, complement, and coagulation factors. *FEBS Lett.* **2019**, *593*, 3449–3460. [CrossRef] [PubMed]
3. Shirley, J.L.; de Jong, Y.P.; Terhorst, C.; Herzog, R.W. Immune Responses to Viral Gene Therapy Vectors. *Mol. Ther.* **2020**, *28*, 709–722. [CrossRef] [PubMed]
4. Roberts, D.M.; Nanda, A.; Havenga, M.J.; Abbink, P.; Lynch, D.M.; Ewald, B.A.; Liu, J.; Thorner, A.R.; Swanson, P.E.; Gorgone, D.A.; et al. Hexon-chimaeric adenovirus serotype 5 vectors circumvent pre-existing anti-vector immunity. *Nature* **2006**, *441*, 239–243. [CrossRef]
5. Zhu, F.C.; Li, Y.H.; Guan, X.H.; Hou, L.H.; Wang, W.J.; Li, J.X.; Wu, S.P.; Wang, B.S.; Wang, Z.; Wang, L.; et al. Safety, tolerability, and immunogenicity of a recombinant adenovirus type-5 vectored COVID-19 vaccine: A dose-escalation, open-label, non-randomised, first-in-human trial. *Lancet* **2020**, *395*, 1845–1854. [CrossRef]
6. Ferguson, M.S.; Lemoine, N.R.; Wang, Y. Systemic delivery of oncolytic viruses: Hopes and hurdles. *Adv. Virol.* **2012**, *2012*, 805629. [CrossRef]
7. Khare, R.; Hillestad, M.L.; Xu, Z.; Byrnes, A.P.; Barry, M.A. Circulating antibodies and macrophages as modulators of adenovirus pharmacology. *J. Virol.* **2013**, *87*, 3678–3686. [CrossRef]
8. Xu, Z.; Tian, J.; Smith, J.S.; Byrnes, A.P. Clearance of adenovirus by Kupffer cells is mediated by scavenger receptors, natural antibodies, and complement. *J. Virol.* **2008**, *82*, 11705–11713. [CrossRef]
9. Kalyuzhniy, O.; Di Paolo, N.C.; Silvestry, M.; Hofherr, S.E.; Barry, M.A.; Stewart, P.L.; Shayakhmetov, D.M. Adenovirus serotype 5 hexon is critical for virus infection of hepatocytes in vivo. *Proc. Natl. Acad. Sci. USA* **2008**, *105*, 5483–5488. [CrossRef]
10. Waddington, S.N.; McVey, J.H.; Bhella, D.; Parker, A.L.; Barker, K.; Atoda, H.; Pink, R.; Buckley, S.M.; Greig, J.A.; Denby, L.; et al. Adenovirus serotype 5 hexon mediates liver gene transfer. *Cell* **2008**, *132*, 397–409. [CrossRef]
11. Atasheva, S.; Emerson, C.C.; Yao, J.; Young, C.; Stewart, P.L.; Shayakhmetov, D.M. Systemic cancer therapy with engineered adenovirus that evades innate immunity. *Sci. Transl. Med.* **2020**, *12*. [CrossRef]
12. Xu, Z.; Qiu, Q.; Tian, J.; Smith, J.S.; Conenello, G.M.; Morita, T.; Byrnes, A.P. Coagulation factor X shields adenovirus type 5 from attack by natural antibodies and complement. *Nat. Med.* **2013**, *19*, 452–457. [CrossRef] [PubMed]

13. Doronin, K.; Flatt, J.W.; Di Paolo, N.C.; Khare, R.; Kalyuzhniy, O.; Acchione, M.; Sumida, J.P.; Ohto, U.; Shimizu, T.; Akashi-Takamura, S.; et al. Coagulation factor X activates innate immunity to human species C adenovirus. *Science* **2012**, *338*, 795–798. [CrossRef] [PubMed]
14. Newton, A.H.; Cardani, A.; Braciale, T.J. The host immune response in respiratory virus infection: Balancing virus clearance and immunopathology. *Semin. Immunopathol.* **2016**, *38*, 471–482. [CrossRef]
15. Barnum, S.R. Complement: A primer for the coming therapeutic revolution. *Pharm. Ther.* **2017**, *172*, 63–72. [CrossRef] [PubMed]
16. Merle, N.S.; Church, S.E.; Fremeaux-Bacchi, V.; Roumenina, L.T. Complement System Part I—Molecular Mechanisms of Activation and Regulation. *Front. Immunol.* **2015**, *6*, 262. [CrossRef] [PubMed]
17. Burton, D.R. Is IgM-like dislocation a common feature of antibody function? *Immunol. Today* **1986**, *7*, 165–167. [CrossRef]
18. Feinstein, A.; Munn, E.A. Conformation of the free and antigen-bound IgM antibody molecules. *Nature* **1969**, *224*, 1307–1309. [CrossRef]
19. Sharp, T.H.; Boyle, A.L.; Diebolder, C.A.; Kros, A.; Koster, A.J.; Gros, P. Insights into IgM-mediated complement activation based on in situ structures of IgM-C1-C4b. *Proc. Natl. Acad. Sci. USA* **2019**, *116*, 11900–11905. [CrossRef]
20. de Jong, R.N.; Beurskens, F.J.; Verploegen, S.; Strumane, K.; van Kampen, M.D.; Voorhorst, M.; Horstman, W.; Engelberts, P.J.; Oostindie, S.C.; Wang, G.; et al. A Novel Platform for the Potentiation of Therapeutic Antibodies Based on Antigen-Dependent Formation of IgG Hexamers at the Cell Surface. *PLoS Biol.* **2016**, *14*, e1002344. [CrossRef]
21. Diebolder, C.A.; Beurskens, F.J.; de Jong, R.N.; Koning, R.I.; Strumane, K.; Lindorfer, M.A.; Voorhorst, M.; Ugurlar, D.; Rosati, S.; Heck, A.J.; et al. Complement is activated by IgG hexamers assembled at the cell surface. *Science* **2014**, *343*, 1260–1263. [CrossRef] [PubMed]
22. Ugurlar, D.; Howes, S.C.; de Kreuk, B.J.; Koning, R.I.; de Jong, R.N.; Beurskens, F.J.; Schuurman, J.; Koster, A.J.; Sharp, T.H.; Parren, P.; et al. Structures of C1-IgG1 provide insights into how danger pattern recognition activates complement. *Science* **2018**, *359*, 794–797. [CrossRef] [PubMed]
23. Reid, K.B.; Porter, R.R. Subunit composition and structure of subcomponent C1q of the first component of human complement. *Biochem. J.* **1976**, *155*, 19–23. [CrossRef] [PubMed]
24. Bottermann, M.; Foss, S.; Caddy, S.L.; Clift, D.; van Tienen, L.M.; Vaysburd, M.; Cruickshank, J.; O'Connell, K.; Clark, J.; Mayes, K.; et al. Complement C4 Prevents Viral Infection through Capsid Inactivation. *Cell Host Microbe* **2019**, *25*, 617–629.e7. [CrossRef] [PubMed]
25. Maier, O.; Marvin, S.A.; Wodrich, H.; Campbell, E.M.; Wiethoff, C.M. Spatiotemporal dynamics of adenovirus membrane rupture and endosomal escape. *J. Virol.* **2012**, *86*, 10821–10828. [CrossRef] [PubMed]
26. Wiethoff, C.M.; Wodrich, H.; Gerace, L.; Nemerow, G.R. Adenovirus protein VI mediates membrane disruption following capsid disassembly. *J. Virol.* **2005**, *79*, 1992–2000. [CrossRef] [PubMed]
27. Ganz, T. Defensins: Antimicrobial peptides of innate immunity. *Nat. Rev. Immunol.* **2003**, *3*, 710–720. [CrossRef]
28. Smith, J.G.; Nemerow, G.R. Mechanism of adenovirus neutralization by Human alpha-defensins. *Cell Host Microbe* **2008**, *3*, 11–19. [CrossRef]
29. Flatt, J.W.; Kim, R.; Smith, J.G.; Nemerow, G.R.; Stewart, P.L. An intrinsically disordered region of the adenovirus capsid is implicated in neutralization by human alpha defensin 5. *PLoS ONE* **2013**, *8*, e61571. [CrossRef]
30. Gaboriaud, C.; Juanhuix, J.; Gruez, A.; Lacroix, M.; Darnault, C.; Pignol, D.; Verger, D.; Fontecilla-Camps, J.C.; Arlaud, G.J. The crystal structure of the globular head of complement protein C1q provides a basis for its versatile recognition properties. *J. Biol. Chem.* **2003**, *278*, 46974–46982. [CrossRef]
31. Saphire, E.O.; Parren, P.W.; Pantophlet, R.; Zwick, M.B.; Morris, G.M.; Rudd, P.M.; Dwek, R.A.; Stanfield, R.L.; Burton, D.R.; Wilson, I.A. Crystal structure of a neutralizing human IGG against HIV-1: A template for vaccine design. *Science* **2001**, *293*, 1155–1159. [CrossRef] [PubMed]
32. Bottermann, M.; Lode, H.E.; Watkinson, R.E.; Foss, S.; Sandlie, I.; Andersen, J.T.; James, L.C. Antibody-antigen kinetics constrain intracellular humoral immunity. *Sci. Rep.* **2016**, *6*, 37457. [CrossRef]
33. Varghese, R.; Mikyas, Y.; Stewart, P.L.; Ralston, R. Postentry neutralization of adenovirus type 5 by an antihexon antibody. *J. Virol.* **2004**, *78*, 12320–12332. [CrossRef] [PubMed]
34. Lindert, S.; Silvestry, M.; Mullen, T.M.; Nemerow, G.R.; Stewart, P.L. Cryo-electron microscopy structure of an adenovirus-integrin complex indicates conformational changes in both penton base and integrin. *J. Virol.* **2009**, *83*, 11491–11501. [CrossRef] [PubMed]
35. Wickham, T.J.; Mathias, P.; Cheresh, D.A.; Nemerow, G.R. Integrins alpha v beta 3 and alpha v beta 5 promote adenovirus internalization but not virus attachment. *Cell* **1993**, *73*, 309–319. [CrossRef]
36. Dai, X.; Wu, L.; Sun, R.; Zhou, Z.H. Atomic Structures of Minor Proteins VI and VII in Human Adenovirus. *J. Virol.* **2017**, *91*. [CrossRef]
37. Pettersen, E.F.; Goddard, T.D.; Huang, C.C.; Meng, E.C.; Couch, G.S.; Croll, T.I.; Morris, J.H.; Ferrin, T.E. UCSF ChimeraX: Structure visualization for researchers, educators, and developers. *Protein Sci.* **2021**, *30*, 70–82. [CrossRef]
38. Pettersen, E.F.; Goddard, T.D.; Huang, C.C.; Couch, G.S.; Greenblatt, D.M.; Meng, E.C.; Ferrin, T.E. UCSF Chimera—A visualization system for exploratory research and analysis. *J. Comput. Chem.* **2004**, *25*, 1605–1612. [CrossRef]
39. Mortensen, S.; Kidmose, R.T.; Petersen, S.V.; Szilagyi, A.; Prohaszka, Z.; Andersen, G.R. Structural Basis for the Function of Complement Component C4 within the Classical and Lectin Pathways of Complement. *J. Immunol.* **2015**, *194*, 5488–5496. [CrossRef]

40. Phillips, J.C.; Hardy, D.J.; Maia, J.D.C.; Stone, J.E.; Ribeiro, J.V.; Bernardi, R.C.; Buch, R.; Fiorin, G.; Henin, J.; Jiang, W.; et al. Scalable molecular dynamics on CPU and GPU architectures with NAMD. *J. Chem. Phys.* **2020**, *153*, 044130. [CrossRef]
41. Best, R.B.; Zhu, X.; Shim, J.; Lopes, P.E.; Mittal, J.; Feig, M.; Mackerell, A.D., Jr. Optimization of the additive CHARMM all-atom protein force field targeting improved sampling of the backbone phi, psi and side-chain chi (1) and chi (2) dihedral angles. *J. Chem. Theory Comput.* **2012**, *8*, 3257–3273. [CrossRef] [PubMed]
42. Humphrey, W.; Dalke, A.; Schulten, K. VMD: Visual molecular dynamics. *J. Mol. Graph.* **1996**, *14*, 33–38. [CrossRef]
43. Vidarsson, G.; Dekkers, G.; Rispens, T. IgG subclasses and allotypes: From structure to effector functions. *Front. Immunol.* **2014**, *5*, 520. [CrossRef] [PubMed]
44. Zhang, X.; Zhang, L.; Tong, H.; Peng, B.; Rames, M.J.; Zhang, S.; Ren, G. 3D Structural Fluctuation of IgG1 Antibody Revealed by Individual Particle Electron Tomography. *Sci. Rep.* **2015**, *5*, 9803. [CrossRef] [PubMed]
45. Wang, G.; de Jong, R.N.; van den Bremer, E.T.; Beurskens, F.J.; Labrijn, A.F.; Ugurlar, D.; Gros, P.; Schuurman, J.; Parren, P.W.; Heck, A.J. Molecular Basis of Assembly and Activation of Complement Component C1 in Complex with Immunoglobulin G1 and Antigen. *Mol. Cell* **2016**, *63*, 135–145. [CrossRef]
46. Law, S.K. The covalent binding reaction of C3 and C4. *Ann. N. Y. Acad. Sci* **1983**, *421*, 246–258. [CrossRef]
47. Zubieta, C.; Schoehn, G.; Chroboczek, J.; Cusack, S. The structure of the human adenovirus 2 penton. *Mol. Cell* **2005**, *17*, 121–135. [CrossRef]
48. Agrawal, P.; Nawadkar, R.; Ojha, H.; Kumar, J.; Sahu, A. Complement Evasion Strategies of Viruses: An Overview. *Front. Microbiol.* **2017**, *8*, 1117. [CrossRef]
49. Zhu, Y.; Thangamani, S.; Ho, B.; Ding, J.L. The ancient origin of the complement system. *EMBO J.* **2005**, *24*, 382–394. [CrossRef]
50. Mallery, D.L.; McEwan, W.A.; Bidgood, S.R.; Towers, G.J.; Johnson, C.M.; James, L.C. Antibodies mediate intracellular immunity through tripartite motif-containing 21 (TRIM21). *Proc. Natl. Acad. Sci. USA* **2010**, *107*, 19985–19990. [CrossRef]
51. Foss, S.; Watkinson, R.; Sandlie, I.; James, L.C.; Andersen, J.T. TRIM21: A cytosolic Fc receptor with broad antibody isotype specificity. *Immunol. Rev.* **2015**, *268*, 328–339. [CrossRef] [PubMed]
52. Greber, U.F.; Flatt, J.W. Adenovirus Entry: From Infection to Immunity. *Annu. Rev. Virol.* **2019**, *6*, 177–197. [CrossRef] [PubMed]
53. Mellors, J.; Tipton, T.; Longet, S.; Carroll, M. Viral Evasion of the Complement System and Its Importance for Vaccines and Therapeutics. *Front. Immunol.* **2020**, *11*, 1450. [CrossRef] [PubMed]
54. Wu, Y.; Li, L.; Frank, L.; Wagner, J.; Andreozzi, P.; Hammer, B.; D'Alicarnasso, M.; Pelliccia, M.; Liu, W.; Chakrabortty, S.; et al. Patchy Amphiphilic Dendrimers Bind Adenovirus and Control Its Host Interactions and in Vivo Distribution. *ACS Nano* **2019**, *13*, 8749–8759. [CrossRef]
55. Rojas, L.A.; Condezo, G.N.; Moreno, R.; Fajardo, C.A.; Arias-Badia, M.; San Martin, C.; Alemany, R. Albumin-binding adenoviruses circumvent pre-existing neutralizing antibodies upon systemic delivery. *J. Control. Release* **2016**, *237*, 78–88. [CrossRef]

viruses MDPI

Review

Human Adenovirus Species D Interactions with Corneal Stromal Cells

Jaya Rajaiya *, Amrita Saha, Xiaohong Zhou and James Chodosh *

Massachusetts Eye and Ear, Harvard Medical School, Boston, MA 02114, USA;
Amrita_Saha@meei.harvard.edu (A.S.); Xiaohong_Zhou@meei.harvard.edu (X.Z.)
* Correspondence: jaya_rajaiya@meei.harvard.edu (J.R.); james_chodosh@meei.harvard.edu (J.C.)

Abstract: Notable among the many communicable agents known to infect the human cornea is the human adenovirus, with less than ten adenoviruses having corneal tropism out of more than 100 known types. The syndrome of epidemic keratoconjunctivitis (EKC), caused principally by human adenovirus, presents acutely with epithelial keratitis, and later with stromal keratitis that can be chronic and recurrent. In this review, we discuss the current state of knowledge regarding the molecular biology of adenovirus infection of corneal stromal cells, among which the fibroblast-like keratocyte is the most predominant, in order to elucidate basic pathophysiologic mechanisms of stromal keratitis in the human patient with EKC.

Keywords: adenovirus; human adenovirus species D; epidemic keratoconjunctivitis; adenovirus keratitis; intracellular signaling; keratocyte

1. Introduction

The cornea is the transparent window of the eye. The cornea is also a mucosal surface and frequently encounters external infectious agents, whether airborne, through hand-to-eye contact, on fomites or contact lenses, or due to trauma. Corneal infection can lead to scarring, reduced vision, and when particularly severe, perforation and loss of the eye. An important and often overlooked cause of corneal infection is the highly transmissible adenovirus, more often recognized as the most common cause of infectious conjunctivitis [1]. The conjunctiva is the mucous membrane that abuts the cornea, surfaces the rest of the external globe, and lines the internal surface of the eyelids. Infections of the conjunctiva are medically important; about 6 million infectious conjunctivitis cases present to clinicians annually in the United States, at an estimated cost of hundreds of millions of US dollars for diagnosis and treatment annually [2]. Over half of all conjunctivitis cases are caused by adenovirus [1,3,4], and in an undetermined but significant proportion of these cases, the cornea is also infected. In such cases, the presentation is referred to as epidemic keratoconjunctivitis, or EKC [5]. EKC is highly contagious, outbreaks occur world-wide [6–19], and to date, there is no effective therapy to safely mitigate morbidity from the disorder [20].

The adenovirus is non-enveloped, icosahedral shaped, and contains a double stranded DNA genome of ≈36,000 base pairs. The virion capsid contains 240 hexon and 12 penton capsomers, and assorted minor capsid proteins. Each penton capsomer forms the base (hence, "penton base") for a trimeric fiber protein, the principal ligand for the various host cell receptors [21–27]. Receptor binding leads to secondary interaction between an RGD motif in each penton base protein with cellular integrins [28–30], e.g., $\alpha_v\beta_3$, that then aggregate, phosphorylate, and activate Src [31,32], to induce endocytosis of the virus. Human adenoviruses (HAdV) divide phylogenetically into 7 species (A-G), with now over 100 types [33,34]. HAdV infections are an important source of morbidity and mortality world-wide, through readily transmittable infections at mucosal sites [35]. Infection may be especially lethal in infants [36–39] and the immune compromised [40–43], but can also lead to fatal acute respiratory distress syndrome in healthy adults [44,45]. HAdV infections of the eye present as either simple follicular conjunctivitis, pharyngoconjunctival fever, or EKC. The first two are self-limited and do not disturb the cornea. In contrast, in EKC, the cornea is directly infected. The major EKC pathogens fall within species D: types 8, 37, 53, 54, 56, 64 (previously 19a), and 85 (recently emerged) [20,46–54]. In a recent comprehensive phylogenomics study of archived adenovirus sequences, the members of HAdV-D were uniquely distinct from all other adenoviruses found in human and non-human primates [55].

2. Cornea Organogenesis, Structure, and Response to Injury/Infection

The anterior eye tissue is derived from the surface ectoderm, mesoderm, and neural crest, while the posterior eye tissue develops from the neural tube ectoderm [56]. The cornea's function is to focus light onto the (posterior) neural retina; visual information is then sent to the brain via the optic nerve. The cornea performs ≈80% of light bending/focusing for vision; the remainder is performed by the crystalline lens, which also serves to adjust the eye's focus from far to near (accommodation). The cornea is the most anterior eye tissue and, from front to back (Figure 1A), is composed of corneal epithelium, epithelial basement membrane, Bowman's layer, stroma, Descemet's membrane, and the single-cell layer-thick corneal endothelium. Descemet's membrane is the basement membrane of the corneal endothelium. The corneal epithelium measures about 50 μm in thickness and is composed of four to six layers of nonkeratinized, stratified squamous cells. There is constant repopulation of epithelial cells through differentiation and maturation of transient amplifying cells at the basal layer, which in turn have migrated through the deep corneal epithelial layer from the peripheral corneal "limbus", the site of corneal epithelial stem cells [57]. Superficial epithelial cells have a life span of 7–10 days, after which they undergo apoptosis and desquamation [58]. The corneal epithelium serves as a relative barrier to pathogens and other foreign particles through surface mucins and intercellular tight junctions, protecting the underlying corneal stroma. The corneal epithelium is the most highly innervated mucosal surface with ≈7000 nociceptors per square mm [59], rendering it more sensitive than skin by several orders of magnitude [60]. The corneal epithelium, like other epithelial surfaces, also expresses membrane-associated mucins (MAMs), and viruses must first negotiate a MAM-rich glycocalyx in order to infect corneal epithelial cells. One specific MAM, MUC16, has been shown to be particularly important in ocular surface defense [61]. MUC16 expressed by corneal epithelial cells has been shown to directly impact adenovirus tropism for the eye. Menon and coworkers demonstrated that the EKC-causing human adenovirus species D type 37 (HAdV-D37), but not the highly similar but non-EKC-associated HAdV-D19, induced ectodomain release of corneal epithelial MUC16, reducing barrier function to infection [62]. Corneal epithelial cells also express Toll-like receptor (TLR) 2 and TLR4, but their expression under normal conditions is intracellular, i.e., not on the cell surface [63]. Therefore, corneal epithelial TLRs are not directly activated by commensal bacteria on the ocular surface, thus rendering a relatively "immune-silent" environment. However, adenovirus infection of the conjunctiva induces significant inflammation, and the associated conjunctival discharge, replete with proinflammatory cytokines [64,65], very likely alters corneal epithelial cell TLR expression. Dendritic cells (DCs) are also present in normal human corneal epithelium, mostly in the corneal periphery. $CD11c^{+}CD16^{-}$ DCs are the most predominant, with smaller numbers of $CD11c^{+}CD16^{+}$ and $CD11c^{+}CD1c^{+}$ cells [66,67]. Innate immune responses upon adenovirus infection of dendritic cells at other mucosal sites have been well studied [68–73].

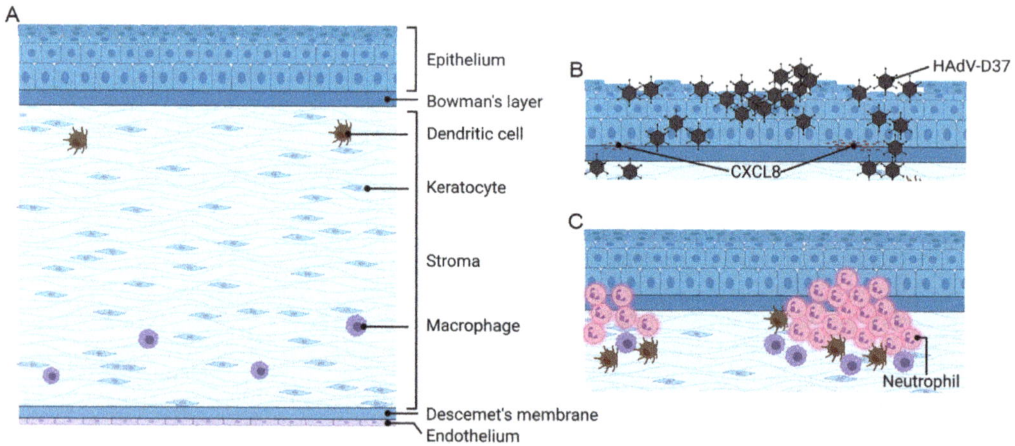

Figure 1. Schematic of the human cornea in cross-section. In the normal, uninfected eye (**A**), the central corneal epithelium is intact, and the corneal stroma is sparsely populated by fibroblast-like keratocytes and a smaller number of myeloid cells. In the cornea acutely infected by a cornea-tropic adenovirus (**B**), the corneal epithelium is disrupted by virus-induced cytopathic effect. Upon infection of superficial keratocytes, chemokines, e.g., CXCL8, accumulate in multifocal loci at the level of the corneal epithelial basement membrane. In a subset of infected eyes, after the epithelium has healed (**C**), neutrophils and other monocytes migrate from corneal limbal blood vessels to form multifocal, subepithelial infiltrates at foci of chemokine accumulation. Created with BioRender.com under a standard academic license.

Directly below the corneal epithelium is the epithelial basement membrane, laid down by basal epithelial cells. The corneal epithelial basement membrane is rich in heparan sulfate proteoglycans which promote epithelial cell migration, proliferation, and differentiation [74]. Corneal epithelial basement membrane very effectively binds positively charged chemokines, thus establishing relatively stable reservoirs that may serve as initiation sites for recurrent accumulations of leukocytes, characteristic of post-EKC keratitis, and manifest clinically by multifocal, corneal subepithelial infiltrates (SEI) [75]. The corneal stroma sits just beneath the epithelial basement membrane. The most superficial corneal stroma in humans, known as Bowman's layer, is distinct histologically as an acellular layer of 15–18 μm in thickness, consisting mostly of collagen type I. Its function is unknown, but it is permeable to macromolecules [76]. Below the Bowman's layer is a sparsely cellular stroma that contributes to the majority of the corneal thickness ($\approx$90%), approximately 550 μm centrally but closer to 1 mm in the cornea periphery. A highly organized network of lamellar collagen fibers in the stroma confers transparency and consists largely of heterodimeric complexes of collagen types I and V [77]. Other proteins in the stroma include lumican and keratocan, the major keratan sulfate proteoglycans. The principal cell type in the stroma (constituting $\approx$94% of corneal stromal cells) is the neural crest derived keratocyte. These cells produce collagen and glycosaminoglycans, and maintain corneal transparency through modulation of corneal stromal extracellular matrix [78–80] and expression of intracellular crystallins [81]. Keratocytes are relatively quiescent and rarely undergo cell division, but in injury or infection, they are transformed to fibroblasts and myofibroblasts [82]. Although keratocytes make up the great majority of cells in the corneal stroma, $\approx$6% of the total cell population in the corneal stroma are $CD45^+$ cells of monocytic lineage [83]. $CD45^+$ $CD11c^+$ cells are found mostly within the anterior stroma, and $CD11c^-$ $CD11b^+$ cells within the posterior stroma [67]. In a study utilizing the mouse model of adenovirus keratitis [84], macrophage Fas-Induced apoptosis (MaFIA) transgenic mice, in which the mouse colony stimulating factor 1 receptor promoter (*Csf1r*) was used to drive expression of a mutant human FK506 binding protein 1A in macrophages and dendritic cells to induce apoptosis, clinical keratitis was reduced and the recruitment of leukocytes was diminished as compared to controls [85].

In addition to monocytes in the cornea, very small numbers of plasmacytoid DCs are also present in the corneal stroma, as first shown by Sosnova and co-authors [86]. A role for these cells in adenovirus keratitis has not been explored. The very posterior cornea is bound by Descemet's membrane and the corneal endothelium.

Corneal epithelial and stromal cells communicate together to regulate and maintain corneal homeostasis [87–89]. Such crosstalk is critical to the corneal response to injury [90]. Upon disruption of the epithelial barrier, whether by injury or infection, even when isolated to the overlying corneal epithelium alone, those keratocytes that do not take on the phenotype of fibroblasts or myofibroblasts die due to apoptosis [91–94]. They are repopulated upon injury by a subpopulation of corneal stromal stem cells [95]. Myofibroblasts have reduced crystalline expression and therefore lack transparency and directly contribute to corneal opacity [82,96–98]. Keratocytes which have been activated to the fibroblast phenotype express TLR1-TLR7, TLR9, and TLR10 [99–104]. Both fibroblasts and myofibroblasts participate in the response to injury and in tissue remodeling. This is similar to the role of fibroblasts elsewhere, for example in the heart, in which crosstalk between fibroblasts and cardiomyocytes is critical to both cardiac development and repair after tissue injury [105,106]. Corneal fibroblasts ably detect and respond to pathogen-associated molecular patterns (PAMPs) from microbes through the expression of TLRs. They express cytokines, chemokines, and adhesion molecules that are responsible for the recruitment of inflammatory cells [102,107,108]. Our past studies of adenovirus infection of human corneal fibroblasts in vitro and in a mouse model of adenovirus keratitis in vivo showed early expression of CXCL8 and its murine homologue CXCL1, respectively; and CCL2, and ICAM-1, all within the first day post infection (pi) [32,75,84,109]. In addition to a role in the acute keratitis of EKC, and based on additional and compelling evidence for a critical role of tissue fibroblasts in the maintenance of chronic inflammation [110–113], we speculate that human corneal fibroblasts also play a role in the pathogenesis of the chronic, recurrent stromal keratitis that often follows EKC [5].

3. Epidemic Keratoconjunctivitis (EKC)

EKC is characterized clinically by follicular lymphoid hyperplasia of the conjunctiva, preauricular lymphadenopathy, and punctate or geographic epithelial keratitis (Figure 1B), with an explosive clinical course [114,115]. The contralateral eye is affected in ≈70% [116]. Inflammatory conjunctival membranes form in $\frac{1}{4}$–$\frac{1}{2}$ of infected eyes [5,115]; if untreated, they become incorporated into the host tissue and can form adhesions (symblephara) that may restrict ocular motility [117]. In EKC, rapid onset and severity are distinctive, but the hallmark is corneal involvement [115,118,119]. The epithelial keratitis induced by viral cytopathic effect resolves within days, but stromal keratitis in the form of SEI then ensues (in 60% of cases in a recent large study) [52], typically appearing at 14–21 days pi.

SEI can be recalcitrant to treatment. While full resolution can occur within a few weeks after onset, in a significant proportion of patients, the keratitis will persist or recur for months to years pi (in one study, 47% at 2 years pi) [5,120,121], causing scarring [122], irregular astigmatism, glare, foreign body sensation, and blurred vision [123]. Rajaiya and coworkers [75] studied the ontogeny of adenovirus-induced, corneal SEI, by modifying a previously published 3D human corneal "facsimile" of adenovirus keratitis [124]. The original facsimile was composed of human keratocytes and type I collagen plated on transwell plates; Rajaiya added an overlying layer of Matrigel® to the model to simulate an epithelial basement membrane. The facsimiles were plated on transwell plates with a 3 μm pore size and infected overnight with HAdV-D37. When freshly isolated human peripheral blood leukocytes were placed beneath the inserts, within 1 h, neutrophils had migrated upward (against gravity) to form focal infiltrates that mimicked SEI and co-localized with CXCL8 bound to heparan sulfate in the Matrigel.

HAdVs do not replicate in mouse cells [125–128]. However, a novel mouse model of adenovirus keratitis [46,84,129–131] has permitted study of the innate immune responses to the virus. Injection with $\geq 5 \times 10^4$ tissue culture infectious doses (TCID) of the virulent

EKC pathogen HAdV-D37 [50,132–135], directly into the mouse corneal stroma by a heat-pulled glass micropipette needle, bypasses the nonpermissive mouse ocular surface and induces a stromal keratitis that peaks ≈4 days pi [136], and then resolves, only to recur later in a subset of infected mice (Figure 1C) [84]. HAdV-D37 empty capsid (no viral DNA), is sufficient to induce clinically evident keratitis; viral replication is unnecessary (and does not occur) [130]. Neutrophil chemotaxis into the adenovirus infected cornea is dependent on CXCL1 and its receptor, CXCR2 [129]. Leukocytic infiltrates and CXCL1 expression in the mouse can be blocked by treatment with a monomer containing RGD but not by a control peptide, consistent with capsid driven inflammation [130]. Subsequent investigations into TLR activation in the mouse adenovirus keratitis model showed Src kinase associated activation of MyD88 [131]. Keratitis was reduced in MyD88 knockout mice, as well as in mice knocked out for both TLR2 and TLR9.

As viruses within HAdV-D do not replicate in murine cells, detailed in vitro studies of the cellular response to adenovirus infection of the cornea have relied extensively on human keratocytes cultured from deceased human corneal donors. These studies were initiated after publication of a seminal study showing that adenovirus type 5 (HAdV-C5) infection of HeLa cells induces interleukin-8 (CXCL8) expression through a Raf/mitogen-activated protein (MAP) kinase signaling pathway [137], and based on reasoning that CXCL8 expression by infected corneal cells might be responsible for the keratitis seen in EKC. The earliest studies utilized a virus isolated from the cornea of a patient with acute EKC that was initially characterized as HAdV-D19c [118]. This virus was later whole genome sequenced [51], and found to contain only the hexon hypervariable regions from type 19 (≈3% of the total genome), with the majority of the genome recombinant with HAdV-D37 [49], a highly virulent cause of EKC. HAdV-D19c was then classified as a novel type under criteria accepted by GenBank [138], and renamed as HAdV-D64 [49]. Primary cultures of keratocytes were chosen for the infection model, after it was found that corneal epithelial cells, although susceptible to adenovirus infection, expressed cytokines in very limited amounts in comparison to log-unit increases in expression by adenovirus-infected keratocytes [32].

As discussed above, primary keratocytes when cultured in the presence of serum take on the morphology of fibroblasts. Early studies using HAdV-D64 identified several tyrosine kinases as critical to the expression of CXCL8 (Figure 2). For example, cytosolic focal adhesion kinase (FAK) was shown to be activated (phosphorylated) within 15 min of infection, well before the onset of adenoviral gene expression, and its chemical inhibition reduced CXCL8 expression [139]. FAK activation is known to induce changes to the cytoskeleton [140], consistent with a role in the altered cellular morphology previously shown to occur in adenovirus infected cells [28]. In keratocytes infected with HAdV-D64, phosphorylated FAK accumulated in relatively greater abundance in the Triton X-insoluble cell pellet [139]. Further studies using ultraviolet (UV) light-inactivated virus, which enters cells like an unaltered wild-type virus but does not replicate, still induced CXCL8 expression [32]. Heat-inactivated viruses, which do not enter cells due to heat-induced damage to protein ligands on the external capsid surface, did not induce CXCL8. Furthermore, the general tyrosine kinase inhibitor, herbimycin, blocked CXCL8 expression in infected cells. Src was the first tyrosine kinase induced; phosphorylation was observed within 5 min of infection. Phosphorylation of the extracellular signal-regulated kinases (ERK) 1/2 was seen within 15 min of infection. Chemical inhibitors of Src and ERK1/2 blocked CXCL8 expression. It was subsequently shown that adenovirus infection of keratocytes also activated phosphoinositide 3-kinase (PI3K) and downstream protein kinase B (AKT) [31], in a pathway that appeared to also involve Src. Activation of the PI3K/AKT pathway was found to protect infected keratocytes from apoptotic cell death, whereas chemical or siRNA knock down led to rapid apoptotic cell death upon infection. Notably, early viral gene expression after infection occurred despite AKT knock down, but cells died prior to viral replication, suggesting that induction of this pathway by adenovirus acts to sustain cell viability and enable viral replication. Therefore, blockade of the PI3K/AKT pathway may represent a viable target for anti-adenoviral therapy.

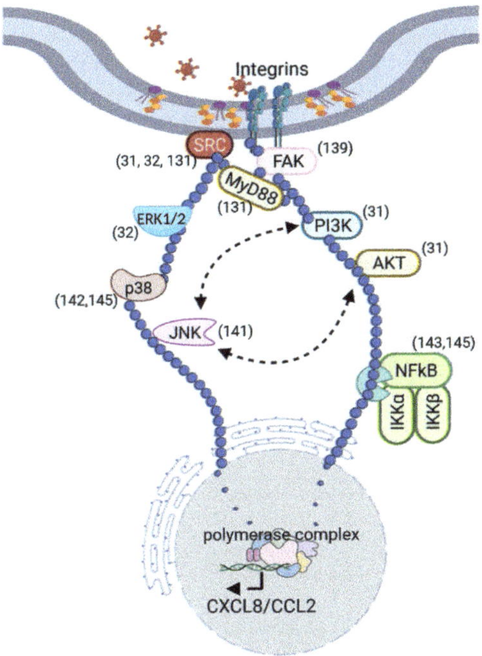

Figure 2. Schematic summarizes the known intracellular signaling events in human keratocytes upon adenovirus infection. Activation of intracellular signaling is initiated upon integrin binding and viral entry, prior to the earliest adenoviral gene expression. Signaling leads to chemokine expression with subsequent formation of subepithelial leukocyte infiltrates at chemokine sites of binding to epithelial basement membrane, and to maintenance of cell viability to enable viral replication. The dark blue circles reflect the relationships between intracellular signaling molecules that occur upon adenovirus infection, culminating in nuclear translocation of the transcription factors shown, and leading to host proinflammatory gene expression. Citations for each signaling molecule activated by adenovirus infection are shown in parentheses. Created with BioRender.com under a standard academic license.

In vitro studies showing activation of ERK1/2 downstream of Src led to investigation of other MAP kinases, including JNK [141] and p38 [142]; both were phosphorylated within 15 to 30 min pi. While ERK1/2 and p38 were each shown to activate expression of CXCL8, activation of JNK induced expression of CCL2 (monocyte chemoattractant protein-1, MCP-1) but not CXCL8, through activation of the transcription factor c-Jun. All three MAP kinases were induced by adenovirus infection through upstream activation of Src, and inhibitors of each pathway differentially inhibited translocation of different NFkappaB (NFκB) subunits [143]. Differential activation of NFκB was found to control the specific pattern of chemokine expression and was time dependent. Transactivation of the CXCL8 promotor occurred within one hour, upon binding of NFκB p65 and p50 subunits, while NFκB cREL binding to the CCL2 promotor occurred at ≈4 h pi. PP2, an inhibitor of Src, fully blocked NFκB translocation. These findings correlated well with those from the mouse model of adenovirus keratitis, in which CXCL1 expression and resultant neutrophil infiltration occurred rapidly, within 1 day pi, while CCL2 expression and monocyte infiltration were delayed [84]. Notably, injection of PP2 into the mouse cornea prior to infection reduced clinical signs of adenovirus keratitis and CXCL1 expression, as compared to controls [144]. PP2 also blocked CXCL8 expression and neutrophil infiltrates in the human corneal facsimile model of adenovirus keratitis [75].

Later studies showed a role for heat shock protein (HSP) 27 in infection by HAdV-D37 in a signalosome that included p38 and NFκB p65 [145]; siRNA knockdown of HSP27 reduced CXCL8 expression. Studies further upstream indicated that viral entry into keratocytes was dependent on cell membrane lipid rafts and caveolin-1 within minutes of infection [144]. HAdV-D37 DNA was found in caveolin-1 containing endosomal fractions and by immunoelectron microscopy; caveolin-1 and HAdV-37 virions were colocalized to intracellular vesicles. Importantly, Src phosphorylation and CXCL1 expression were both reduced in HAdV-D37 infection of corneas in caveolin-1 knockout mice, relative to wild-type mice. In contrast, entry of HAdV-D37 into human corneal epithelial cells was found to occur by a noncanonical clathrin-mediated pathway not involving caveolin-1 [146].

The cornea quite literally forms the window of the eye. The uniqueness of its design stems from its function as the major site of light bending and transparency for focusing of light on the retina. The cornea also serves as a barrier between the external world and the highly vulnerable internal eye. Its highly specific cellular and extracellular organization renders it uniquely susceptible to damage by infectious organisms that might otherwise be less impactful. For example, infections of the conjunctiva by the very same viruses that infect the cornea typically resolve without significant long-term sequelae. However, much remains unknown about adenovirus infection of the cornea, including, for example, the interactions between the virus and resident monocytes, and the potential interactions between keratocytes and resident monocytes in the initial response to infection. Also unknown is why subepithelial infiltrates can recur months to years after the acute infection. Future studies are needed to elucidate cellular interactions during acute infection, and how acute infection alters the cornea for the long term.

Author Contributions: Conceptualization, J.R. and J.C.; writing—original draft preparation, J.R.; writing—review and editing, J.R., A.S., X.Z. and J.C.; funding acquisition, J.R. and J.C. All authors have read and agreed to the published version of the manuscript.

Funding: Supported by National Institutes of Health grants EY013124, EY021558, and EY014104; an unrestricted grant to the Department of Ophthalmology, Harvard Medical School from Research to Prevent Blindness, Inc., New York, NY; and the Massachusetts Lions Eye Research Fund.

Institutional Review Board Statement: Not applicable.

Conflicts of Interest: The authors declare no conflict of interest.

References

1. Ishii, K.; Nakazono, N.; Fujinaga, K.; Fujii, S.; Kato, M.; Ohtsuka, H.; Aoki, K.; Chen, C.W.; Lin, C.C.; Sheu, M.M.; et al. Comparative studies on aetiology and epidemiology of viral conjunctivitis in three countries of East Asia—Japan, Taiwan and South Korea. *Int. J. Epidemiol.* **1987**, *16*, 98–103. [CrossRef]
2. Udeh, B.L.; Schneider, J.E.; Ohsfeldt, R.L. Cost effectiveness of a point-of-care test for adenoviral conjunctivitis. *Am. J. Med. Sci.* **2008**, *336*, 254–264. [CrossRef] [PubMed]
3. Stenson, S.; Newman, R.; Fedukowicz, H. Laboratory studies in acute conjunctivitis. *Arch. Ophthalmol.* **1982**, *100*, 1275–1277. [CrossRef] [PubMed]
4. Uchio, E.; Takeuchi, S.; Itoh, N.; Matsuura, N.; Ohno, S.; Aoki, K. Clinical and epidemiological features of acute follicular conjunctivitis with special reference to that caused by herpes simplex virus type 1. *Br. J. Ophthalmol.* **2000**, *84*, 968–972. [CrossRef]
5. Butt, A.L.; Chodosh, J. Adenoviral keratoconjunctivitis in a tertiary care eye clinic. *Cornea* **2006**, *25*, 199–202. [CrossRef] [PubMed]
6. Esparcia Rodriguez, O.; Gomez Martinez, A.; Martinez Nieto, M.J.; Salmeron Cifuentes, M.S.; Rodolfo Saavedra, R.; de la Cruz de Julian, I. Outbreak of epidemic keratoconjunctivitis caused by human adenovirus serotype 8 in a nursing home. *Rev. Esp. Salud Publica* **2020**, *94*, 32896840.
7. Li, D.; Zhou, J.N.; Li, H.; He, C.Y.; Dai, Q.S.; Li, X.L.; He, J.F.; He, H.; Li, M.B.; Jiang, L.I.; et al. An outbreak of epidemic keratoconjunctivitis caused by human adenovirus type 8 in primary school, southwest China. *BMC Infect. Dis.* **2019**, *19*, 624. [CrossRef]
8. Uemura, T.; Migita, H.; Ueno, T.; Tsukahara-Kawamura, T.; Saeki, Y.; Fujimoto, T.; Uchio, E. Clinical and virological analysis of epidemic keratoconjunctivitis caused by adenovirus type 54 in a regional ophthalmic clinic in Kyushu, Japan. *Clin. Ophthalmol.* **2018**, *12*, 511–517. [CrossRef] [PubMed]

9. Muller, M.P.; Siddiqui, N.; Ivancic, R.; Wong, D. Adenovirus-related epidemic keratoconjunctivitis outbreak at a hospital-affiliated ophthalmology clinic. *Am. J. Infect. Control* **2018**, *46*, 581–583. [CrossRef]
10. Lei, Z.; Zhu, Z.; Wang, B.M.C.; Mei, H.; Li, H.; Ga, D.Z.G.; Jie, G.; Chi, M.M.B.; Zhang, S.; Ma, C.; et al. Outbreaks of epidemic keratoconjunctivitis caused by human adenovirus type 8 in the Tibet Autonomous Region of China in 2016. *PLoS ONE* **2017**, *12*, e0185048. [CrossRef]
11. Killerby, M.E.; Stuckey, M.J.; Guendel, I.; Sakthivel, S.; Lu, X.; Erdman, D.D.; Schneider, E.; Fagan, R.; Davis, M.S.; Watson, J.T.; et al. Notes from the Field: Epidemic Keratoconjunctivitis Outbreak Associated with Human Adenovirus Type 8—U.S. Virgin Islands, June-November 2016. *MMWR Morb. Mortal. Wkly. Rep.* **2017**, *66*, 811–812. [CrossRef] [PubMed]
12. Hage, E.; Espelage, W.; Eckmanns, T.; Lamson, D.M.; Panto, L.; Ganzenmueller, T.; Heim, A. Molecular phylogeny of a novel human adenovirus type 8 strain causing a prolonged, multi-state keratoconjunctivitis epidemic in Germany. *Sci. Rep.* **2017**, *7*, 40680. [CrossRef]
13. Gopalkrishna, V.; Ganorkar, N.N.; Patil, P.R. Identification and molecular characterization of adenovirus types (HAdV-8, HAdV-37, HAdV-4, HAdV-3) in an epidemic of keratoconjunctivitis occurred in Pune, Maharashtra, Western India. *J. Med. Virol.* **2016**, *88*, 2100–2105. [CrossRef]
14. Huang, G.; Yao, W.; Yu, W.; Mao, L.; Sun, H.; Yao, W.; Tian, J.; Wang, L.; Bo, Z.; Zhu, Z.; et al. Outbreak of epidemic keratoconjunctivitis caused by human adenovirus type 56, China, 2012. *PLoS ONE* **2014**, *9*, e110781. [CrossRef]
15. Massey, J.; Henry, R.; Minnich, L.; Lamson, D.M.; St George, K. Notes from the Field: Health Care-Associated Outbreak of Epidemic Keratoconjunctivitis—West Virginia, 2015. *MMWR Morb. Mortal. Wkly. Rep.* **2016**, *65*, 382–383. [CrossRef]
16. Centers for Disease Control and Prevention. Adenovirus-associated epidemic keratoconjunctivitis outbreaks—four states, 2008–2010. *MMWR Morb. Mortal. Wkly. Rep.* **2013**, *62*, 637–641.
17. Calkavur, S.; Olukman, O.; Ozturk, A.T.; Kilic, F.K.; Gulfidan, G.; Devrim, I.; Malatyali, R.; Oruc, Y.; Atlihan, F. Epidemic adenoviral keratoconjunctivitis possibly related to ophthalmological procedures in a neonatal intensive care unit: Lessons from an outbreak. *Ophthalmic Epidemiol.* **2012**, *19*, 371–379. [CrossRef]
18. Doyle, T.J.; King, D.; Cobb, J.; Miller, D.; Johnson, B. An outbreak of epidemic keratoconjunctivitis at an outpatient ophthalmology clinic. *Infect. Dis. Rep.* **2010**, *2*, e17. [CrossRef]
19. Melendez, C.P.; Florentino, M.M.; Martinez, I.L.; Lopez, H.M. Outbreak of epidemic keratoconjunctivitis caused by adenovirus in medical residents. *Mol. Vis.* **2009**, *15*, 557–562. [PubMed]
20. Jonas, R.A.; Ung, L.; Rajaiya, J.; Chodosh, J. Mystery eye: Human adenovirus and the enigma of epidemic keratoconjunctivitis. *Prog. Retin. Eye Res.* **2020**, *76*, 100826. [CrossRef]
21. Goosney, D.L.; Nemerow, G.R. Adenovirus infection: Taking the back roads to viral entry. *Curr. Biol.* **2003**, *13*, R99–R100. [CrossRef]
22. Nemerow, G.R. Cell receptors involved in adenovirus entry. *Virology* **2000**, *274*, 1–4. [CrossRef]
23. Nemerow, G.R.; Pache, L.; Reddy, V.; Stewart, P.L. Insights into adenovirus host cell interactions from structural studies. *Virology* **2009**, *384*, 380–388. [CrossRef]
24. Nilsson, E.C.; Storm, R.J.; Bauer, J.; Johansson, S.M.; Lookene, A.; Angstrom, J.; Hedenstrom, M.; Eriksson, T.L.; Frangsmyr, L.; Rinaldi, S.; et al. The GD1a glycan is a cellular receptor for adenoviruses causing epidemic keratoconjunctivitis. *Nat. Med.* **2011**, *17*, 105–109. [CrossRef] [PubMed]
25. Wang, H.; Li, Z.Y.; Liu, Y.; Persson, J.; Beyer, I.; Moller, T.; Koyuncu, D.; Drescher, M.R.; Strauss, R.; Zhang, X.B.; et al. Desmoglein 2 is a receptor for adenovirus serotypes 3, 7, 11 and 14. *Nat. Med.* **2011**, *17*, 96–104. [CrossRef]
26. Huang, S.; Reddy, V.; Dasgupta, N.; Nemerow, G.R. A single amino acid in the adenovirus type 37 fiber confers binding to human conjunctival cells. *J. Virol.* **1999**, *73*, 2798–2802. [CrossRef] [PubMed]
27. Ismail, A.M.; Lee, J.S.; Dyer, D.W.; Seto, D.; Rajaiya, J.; Chodosh, J. Selection Pressure in the Human Adenovirus Fiber Knob Drives Cell Specificity in Epidemic Keratoconjunctivitis. *J. Virol.* **2016**, *90*, 9598–9607. [CrossRef] [PubMed]
28. Li, E.; Stupack, D.; Bokoch, G.M.; Nemerow, G.R. Adenovirus endocytosis requires actin cytoskeleton reorganization mediated by Rho family GTPases. *J. Virol.* **1998**, *72*, 8806–8812. [CrossRef]
29. Li, E.; Stupack, D.; Klemke, R.; Cheresh, D.A.; Nemerow, G.R. Adenovirus endocytosis via alpha(v) integrins requires phosphoinositide-3-OH kinase. *J. Virol.* **1998**, *72*, 2055–2061. [CrossRef]
30. Li, E.; Stupack, D.G.; Brown, S.L.; Klemke, R.; Schlaepfer, D.D.; Nemerow, G.R. Association of p130CAS with phosphatidylinositol-3-OH kinase mediates adenovirus cell entry. *J. Biol. Chem.* **2000**, *275*, 14729–14735. [CrossRef]
31. Rajala, M.S.; Rajala, R.V.; Astley, R.A.; Butt, A.L.; Chodosh, J. Corneal cell survival in adenovirus type 19 infection requires phosphoinositide 3-kinase/Akt activation. *J. Virol.* **2005**, *79*, 12332–12341. [CrossRef]
32. Natarajan, K.; Rajala, M.S.; Chodosh, J. Corneal IL-8 expression following adenovirus infection is mediated by c-Src activation in human corneal fibroblasts. *J. Immunol.* **2003**, *170*, 6234–6243. [CrossRef]
33. Ismail, A.M.; Zhou, X.; Dyer, D.W.; Seto, D.; Rajaiya, J.; Chodosh, J. Genomic foundations of evolution and ocular pathogenesis in human adenovirus species D. *FEBS Lett.* **2019**, *593*, 3583–3608. [CrossRef] [PubMed]
34. Ismail, A.M.; Cui, T.; Dommaraju, K.; Singh, G.; Dehghan, S.; Seto, J.; Shrivastava, S.; Fedorova, N.B.; Gupta, N.; Stockwell, T.B.; et al. Genomic analysis of a large set of currently-and historically-important human adenovirus pathogens. *Emerg. Microbes Infect.* **2018**, *7*, 10. [CrossRef] [PubMed]

35. Shenk, T. Adenoviridae: The Viruses and Their Replication. In *Fields Virology*; Fields, B.N., Knipe, D.M., Howley, P.M., Eds.; Lippincott-Raven: Philadelphia, PA, USA, 1996; Volume 2, pp. 2111–2148.
36. Bhat, A.M.; Meny, R.G.; Aranas, E.A.; Yehia, F. Fatal adenoviral (type 7) respiratory disease in neonates. *Clin. Ped.* **1984**, *23*, 409–411. [CrossRef]
37. Henquell, C.; Boeuf, B.; Mirand, A.; Bacher, C.; Traore, O.; Dechelotte, P.; Labbe, A.; Bailly, J.L.; Peigue-Lafeuille, H. Fatal adenovirus infection in a neonate and transmission to health-care workers. *J. Clin. Virol.* **2009**, *45*, 345–348. [CrossRef]
38. Chodosh, J. Neonatal Intensive Care Eye. *Ophthalmology* **2019**, *126*, 144–145. [CrossRef]
39. Sammons, J.S.; Graf, E.H.; Townsend, S.; Hoegg, C.L.; Smathers, S.A.; Coffin, S.E.; Williams, K.; Mitchell, S.L.; Nawab, U.; Munson, D.; et al. Outbreak of Adenovirus in a Neonatal Intensive Care Unit: Critical Importance of Equipment Cleaning During Inpatient Ophthalmologic Examinations. *Ophthalmology* **2019**, *126*, 137–143. [CrossRef]
40. Lion, T. Adenovirus persistence, reactivation, and clinical management. *FEBS Lett.* **2019**, *593*, 3571–3582. [CrossRef] [PubMed]
41. Lion, T. Adenovirus infections in immunocompetent and immunocompromised patients. *Clin. Microbiol. Rev.* **2014**, *27*, 441–462. [CrossRef]
42. Bhanthumkosol, D. Fatal adenovirus infections in infants probably infected with HIV. *J. Med. Assoc. Thai* **1998**, *81*, 214–222. [PubMed]
43. Wallot, M.A.; Dohna-Schwake, C.; Auth, M.; Nadalin, S.; Fiedler, M.; Malago, M.; Broelsch, C.; Voit, T. Disseminated adenovirus infection with respiratory failure in pediatric liver transplant recipients: Impact of intravenous cidofovir and inhaled nitric oxide. *Ped. Transplant.* **2006**, *10*, 121–127. [CrossRef] [PubMed]
44. Ryu, J.S.; Cho, J.H.; Han, H.S.; Jung, M.H.; Yoon, Y.H.; Song, E.S.; Lee, J.Y.; Kim, S.Y.; Lee, K.W.; Kwak, S.M.; et al. Acute respiratory distress syndrome induced by adenovirus in an otherwise healthy woman. *Yonsei Med. J.* **2003**, *44*, 732–735. [CrossRef] [PubMed]
45. Kujawski, S.A.; Lu, X.; Schneider, E.; Blythe, D.; Boktor, S.; Farrehi, J.; Haupt, T.; McBride, D.; Stephens, E.; Sakthivel, S.K.; et al. Outbreaks of Adenovirus-associated Respiratory Illness on 5 College Campuses in the United States, 2018–2019. *Clin. Infect. Dis.* **2021**, *72*, 1992–1999. [CrossRef] [PubMed]
46. Walsh, M.P.; Chintakuntlawar, A.; Robinson, C.M.; Madisch, I.; Harrach, B.; Hudson, N.R.; Schnurr, D.; Heim, A.; Chodosh, J.; Seto, D.; et al. Evidence of molecular evolution driven by recombination events influencing tropism in a novel human adenovirus that causes epidemic keratoconjunctivitis. *PLoS ONE* **2009**, *4*, e5635. [CrossRef]
47. Ishiko, H.; Aoki, K. Spread of epidemic keratoconjunctivitis due to a novel serotype of human adenovirus in Japan. *J. Clin. Microbiol.* **2009**, *47*, 2678–2679. [CrossRef]
48. Robinson, C.M.; Singh, G.; Henquell, C.; Walsh, M.P.; Peigue-Lafeuille, H.; Seto, D.; Jones, M.S.; Dyer, D.W.; Chodosh, J. Computational analysis and identification of an emergent human adenovirus pathogen implicated in a respiratory fatality. *Virology* **2011**, *409*, 141–147. [CrossRef]
49. Zhou, X.; Robinson, C.M.; Rajaiya, J.; Dehghan, S.; Seto, D.; Jones, M.S.; Dyer, D.W.; Chodosh, J. Analysis of human adenovirus type 19 associated with epidemic keratoconjunctivitis and its reclassification as adenovirus type 64. *Investig. Ophthalmol. Vis. Sci.* **2012**, *53*, 2804–2811. [CrossRef]
50. Robinson, C.M.; Shariati, F.; Gillaspy, A.F.; Dyer, D.W.; Chodosh, J. Genomic and bioinformatics analysis of human adenovirus type 37: New insights into corneal tropism. *BMC Genom.* **2008**, *9*, 213. [CrossRef]
51. Robinson, C.M.; Shariati, F.; Zaitshik, J.; Gillaspy, A.F.; Dyer, D.W.; Chodosh, J. Human adenovirus type 19: Genomic and bioinformatics analysis of a keratoconjunctivitis isolate. *Virus Res.* **2009**, *139*, 122–126. [CrossRef]
52. Lee, C.S.; Lee, A.Y.; Akileswaran, L.; Stroman, D.; Najafi-Tagol, K.; Kleiboeker, S.; Chodosh, J.; Magaret, A.; Wald, A.; Van Gelder, R.N.; et al. Determinants of Outcomes of Adenoviral Keratoconjunctivitis. *Ophthalmology* **2018**, *125*, 1344–1353. [CrossRef]
53. Hashimoto, S.; Gonzalez, G.; Harada, S.; Oosako, H.; Hanaoka, N.; Hinokuma, R.; Fujimoto, T. Recombinant type Human mastadenovirus D85 associated with epidemic keratoconjunctivitis since 2015 in Japan. *J. Med. Virol.* **2018**, *90*, 881–889. [CrossRef]
54. Kaneko, H.; Hanaoka, N.; Konagaya, M.; Kobayashi, M.; Nakagawa, H.; Hatano, H.; Ikuta, K.; Sekiryu, T.; Fujimoto, T. Five Cases of Epidemic Keratoconjunctivitis Due to Human Adenovirus Type 85 in Fukushima, Japan. *Jpn. J. Infect. Dis.* **2020**, *73*, 316–319. [CrossRef]
55. Kang, J.; Ismail, A.M.; Dehghan, S.; Rajaiya, J.; Allard, M.W.; Lim, H.C.; Dyer, D.W.; Chodosh, J.; Seto, D. Genomics-based re-examination of the taxonomy and phylogeny of human and simian Mastadenoviruses: An evolving whole genomes approach, revealing putative zoonosis, anthroponosis, and amphizoonosis. *Cladistics* **2020**, *36*, 358–373. [CrossRef] [PubMed]
56. Bales, T.R.; Lopez, M.J.; Clark, J. *Embryology, Eye*; StatPearls: Treasure Island, FL, USA, 2021.
57. Bonnet, C.; Gonzalez, S.; Roberts, J.S.; Robertson, S.Y.T.; Ruiz, M.; Zheng, J.; Deng, S.X. Human limbal epithelial stem cell regulation, bioengineering and function. *Prog. Retin. Eye Res.* **2021**, *85*, 100956. [CrossRef]
58. Lavker, R.M.; Kaplan, N.; Wang, J.; Peng, H. Corneal epithelial biology: Lessons stemming from old to new. *Exp. Eye Res.* **2020**, *198*, 108094. [CrossRef]
59. Muller, L.J.; Marfurt, C.F.; Kruse, F.; Tervo, T.M. Corneal nerves: Structure, contents and function. *Exp. Eye Res.* **2003**, *76*, 521–542. [CrossRef]
60. Zander, E.; Weddell, G. Observations on the innervation of the cornea. *J. Anat.* **1951**, *85*, 68–99.
61. Gipson, I.K.; Spurr-Michaud, S.; Tisdale, A.; Menon, B.B. Comparison of the transmembrane mucins MUC1 and MUC16 in epithelial barrier function. *PLoS ONE* **2014**, *9*, e100393. [CrossRef]

62. Menon, B.B.; Zhou, X.; Spurr-Michaud, S.; Rajaiya, J.; Chodosh, J.; Gipson, I.K. Epidemic Keratoconjunctivitis-Causing Adenoviruses Induce MUC16 Ectodomain Release To Infect Ocular Surface Epithelial Cells. *mSphere* **2016**, *1*. [CrossRef] [PubMed]
63. Ueta, M.; Nochi, T.; Jang, M.H.; Park, E.J.; Igarashi, O.; Hino, A.; Kawasaki, S.; Shikina, T.; Hiroi, T.; Kinoshita, S.; et al. Intracellularly expressed TLR2s and TLR4s contribution to an immunosilent environment at the ocular mucosal epithelium. *J. Immunol.* **2004**, *173*, 3337–3347. [CrossRef]
64. Muruve, D.A.; Barnes, M.J.; Stillman, I.E.; Libermann, T.A. Adenoviral gene therapy leads to rapid induction of multiple chemokines and acute neutrophil-dependent hepatic injury in vivo. *Hum. Gene Ther.* **1999**, *10*, 965–976. [CrossRef]
65. Yawata, N.; Arundhati, A.; Liu, Y.C.; Siak JJ, K.; Yawata, M.; Mehta, J.S. Human ocular surface immune profiling in situ in epidemic keratoconjunctivitis. In Proceedings of the Association for Research in Vision and Ophthalmology, Seattle, WA, USA, 1–5 May 2016; p. 5863.
66. Yamagami, S.; Yokoo, S.; Usui, T.; Yamagami, H.; Amano, S.; Ebihara, N. Distinct populations of dendritic cells in the normal human donor corneal epithelium. *Investig. Ophthalmol. Vis. Sci.* **2005**, *46*, 4489–4494. [CrossRef] [PubMed]
67. Mayer, W.J.; Irschick, U.M.; Moser, P.; Wurm, M.; Huemer, H.P.; Romani, N.; Irschick, E.U. Characterization of antigen-presenting cells in fresh and cultured human corneas using novel dendritic cell markers. *Investig. Ophthalmol. Vis. Sci.* **2007**, *48*, 4459–4467. [CrossRef] [PubMed]
68. Greber, U.F.; Flatt, J.W. Adenovirus Entry: From Infection to Immunity. *Annu. Rev. Virol.* **2019**, *6*, 177–197. [CrossRef]
69. Cheneau, C.; Kremer, E.J. Adenovirus-Extracellular Protein Interactions and Their Impact on Innate Immune Responses by Human Mononuclear Phagocytes. *Viruses* **2020**, *12*, 1351. [CrossRef]
70. Lore, K.; Adams, W.C.; Havenga, M.J.; Precopio, M.L.; Holterman, L.; Goudsmit, J.; Koup, R.A. Myeloid and plasmacytoid dendritic cells are susceptible to recombinant adenovirus vectors and stimulate polyfunctional memory T cell responses. *J. Immunol.* **2007**, *179*, 1721–1729. [CrossRef] [PubMed]
71. Miller, G.; Lahrs, S.; Pillarisetty, V.G.; Shah, A.B.; DeMatteo, R.P. Adenovirus infection enhances dendritic cell immunostimulatory properties and induces natural killer and T-cell-mediated tumor protection. *Cancer Res.* **2002**, *62*, 5260–5266.
72. Perreau, M.; Welles, H.C.; Pellaton, C.; Gjoksi, B.; Potin, L.; Martin, R.; Harari, A.; Bett, A.; Casimiro, D.; Gall, J.; et al. The number of Toll-like receptor 9-agonist motifs in the adenovirus genome correlates with induction of dendritic cell maturation by adenovirus immune complexes. *J. Virol.* **2012**, *86*, 6279–6285. [CrossRef]
73. Philpott, N.J.; Nociari, M.; Elkon, K.B.; Falck-Pedersen, E. Adenovirus-induced maturation of dendritic cells through a PI3 kinase-mediated TNF-alpha induction pathway. *Proc. Natl. Acad. Sci. USA* **2004**, *101*, 6200–6205. [CrossRef] [PubMed]
74. Wilson, S.E.; Torricelli, A.A.M.; Marino, G.K. Corneal epithelial basement membrane: Structure, function and regeneration. *Exp. Eye Res.* **2020**, *194*, 108002. [CrossRef]
75. Rajaiya, J.; Zhou, X.; Barequet, I.; Gilmore, M.S.; Chodosh, J. Novel model of innate immunity in corneal infection. *In Vitro Cell Dev. Biol. Anim.* **2015**, *51*, 827–834. [CrossRef]
76. Wilson, S.E. Bowman's layer in the cornea- structure and function and regeneration. *Exp. Eye Res.* **2020**, *195*, 108033. [CrossRef]
77. Espana, E.M.; Birk, D.E. Composition, structure and function of the corneal stroma. *Exp. Eye Res.* **2020**, *198*, 108137. [CrossRef] [PubMed]
78. Funderburgh, J.L.; Mann, M.M.; Funderburgh, M.L. Keratocyte phenotype mediates proteoglycan structure: A role for fibroblasts in corneal fibrosis. *J. Biol. Chem.* **2003**, *278*, 45629–45637. [CrossRef]
79. Hassell, J.R.; Birk, D.E. The molecular basis of corneal transparency. *Exp. Eye Res.* **2010**, *91*, 326–335. [CrossRef]
80. Yam, G.H.F.; Riau, A.K.; Funderburgh, M.L.; Mehta, J.S.; Jhanji, V. Keratocyte biology. *Exp. Eye Res.* **2020**, *196*, 108062. [CrossRef] [PubMed]
81. Jester, J.V.; Brown, D.; Pappa, A.; Vasiliou, V. Myofibroblast differentiation modulates keratocyte crystallin protein expression, concentration, and cellular light scattering. *Investig. Ophthalmol. Vis. Sci.* **2012**, *53*, 770–778. [CrossRef]
82. Wilson, S.E. Corneal myofibroblast biology and pathobiology: Generation, persistence, and transparency. *Exp. Eye Res.* **2012**, *99*, 78–88. [CrossRef] [PubMed]
83. Yamagami, S.; Ebihara, N.; Usui, T.; Yokoo, S.; Amano, S. Bone marrow-derived cells in normal human corneal stroma. *Arch. Ophthalmol.* **2006**, *124*, 62–69. [CrossRef] [PubMed]
84. Chintakuntlawar, A.V.; Astley, R.; Chodosh, J. Adenovirus type 37 keratitis in the C57BL/6J mouse. *Investig. Ophthalmol. Vis. Sci.* **2007**, *48*, 781–788. [CrossRef] [PubMed]
85. Ramke, M.; Zhou, X.; Materne, E.C.; Rajaiya, J.; Chodosh, J. Resident corneal c-fms(+) macrophages and dendritic cells mediate early cellular infiltration in adenovirus keratitis. *Exp. Eye Res.* **2016**, *147*, 144–147. [CrossRef] [PubMed]
86. Sosnova, M.; Bradl, M.; Forrester, J.V. CD34+ corneal stromal cells are bone marrow-derived and express hemopoietic stem cell markers. *Stem Cells* **2005**, *23*, 507–515. [CrossRef]
87. Ko, J.A.; Liu, Y.; Yanai, R.; Chikama, T.; Takezawa, T.; Nishida, T. Upregulation of tight-junctional proteins in corneal epithelial cells by corneal fibroblasts in collagen vitrigel cultures. *Investig. Ophthalmol. Vis. Sci.* **2008**, *49*, 113–119. [CrossRef]
88. Ko, J.A.; Yanai, R.; Morishige, N.; Takezawa, T.; Nishida, T. Upregulation of connexin43 expression in corneal fibroblasts by corneal epithelial cells. *Investig. Ophthalmol. Vis. Sci.* **2009**, *50*, 2054–2060. [CrossRef]
89. Ko, J.A.; Yanai, R.; Chikama, T.; Nishida, T. Downregulation of matrix metalloproteinase-2 in corneal fibroblasts by interleukin-1 receptor antagonist released from corneal epithelial cells. *Investig. Ophthalmol. Vis. Sci.* **2010**, *51*, 6286–6293. [CrossRef] [PubMed]
90. Wilson, S.E.; Liu, J.J.; Mohan, R.R. Stromal-epithelial interactions in the cornea. *Prog. Retin. Eye Res.* **1999**, *18*, 293–309. [CrossRef]

91. Kim, W.J.; Mohan, R.R.; Mohan, R.R.; Wilson, S.E. Caspase inhibitor z-VAD-FMK inhibits keratocyte apoptosis, but promotes keratocyte necrosis, after corneal epithelial scrape. *Exp. Eye Res.* **2000**, *71*, 225–232. [CrossRef] [PubMed]
92. Mohan, R.R.; Mohan, R.R.; Kim, W.J.; Stark, G.R.; Wilson, S.E. Defective keratocyte apoptosis in response to epithelial injury in stat 1 null mice. *Exp. Eye Res.* **2000**, *70*, 485–491. [CrossRef]
93. Mohan, R.R.; Mohan, R.R.; Kim, W.J.; Wilson, S.E. Modulation of TNF-alpha-induced apoptosis in corneal fibroblasts by transcription factor NF-kappaB. *Investig. Ophthalmol. Vis. Sci.* **2000**, *41*, 1327–1336.
94. Wilson, S.E.; He, Y.G.; Weng, J.; Li, Q.; McDowall, A.W.; Vital, M.; Chwang, E.L. Epithelial injury induces keratocyte apoptosis: Hypothesized role for the interleukin-1 system in the modulation of corneal tissue organization and wound healing. *Exp. Eye Res.* **1996**, *62*, 325–327. [CrossRef]
95. Pinnamaneni, N.; Funderburgh, J.L. Concise review: Stem cells in the corneal stroma. *Stem Cells* **2012**, *30*, 1059–1063. [CrossRef] [PubMed]
96. Fini, M.E. Keratocyte and fibroblast phenotypes in the repairing cornea. *Prog. Retin. Eye Res.* **1999**, *18*, 529–551. [CrossRef]
97. Jester, J.V.; Huang, J.; Barry-Lane, P.A.; Kao, W.W.; Petroll, W.M.; Cavanagh, H.D. Transforming growth factor(beta)-mediated corneal myofibroblast differentiation requires actin and fibronectin assembly. *Investig. Ophthalmol. Vis. Sci.* **1999**, *40*, 1959–1967.
98. Wilson, S.L.; El Haj, A.J.; Yang, Y. Control of scar tissue formation in the cornea: Strategies in clinical and corneal tissue engineering. *J. Funct. Biomater.* **2012**, *3*, 642–687. [CrossRef]
99. Cendra, M.D.M.; Christodoulides, M.; Hossain, P. Signaling Mediated by Toll-Like Receptor 5 Sensing of Pseudomonas aeruginosa Flagellin Influences IL-1beta and IL-18 Production by Primary Fibroblasts Derived from the Human Cornea. *Front Cell Infect. Microbiol.* **2017**, *7*, 130. [CrossRef]
100. Ebihara, N.; Yamagami, S.; Chen, L.; Tokura, T.; Iwatsu, M.; Ushio, H.; Murakami, A. Expression and function of toll-like receptor-3 and -9 in human corneal myofibroblasts. *Investig. Ophthalmol. Vis. Sci.* **2007**, *48*, 3069–3076. [CrossRef]
101. Jin, X.; Lin, Z.; Xie, X. The delayed response of Toll-like receptors may relate to Pseudomonas aeruginosa keratitis exacerbating rapidly at the early stages of infection. *Eur. J. Clin. Microbiol. Infect. Dis.* **2010**, *29*, 231–238. [CrossRef] [PubMed]
102. Kumagai, N.; Fukuda, K.; Fujitsu, Y.; Lu, Y.; Chikamoto, N.; Nishida, T. Lipopolysaccharide-induced expression of intercellular adhesion molecule-1 and chemokines in cultured human corneal fibroblasts. *Investig. Ophthalmol. Vis. Sci.* **2005**, *46*, 114–120. [CrossRef]
103. Redfern, R.L.; Reins, R.Y.; McDermott, A.M. Toll-like receptor activation modulates antimicrobial peptide expression by ocular surface cells. *Exp. Eye Res.* **2011**, *92*, 209–220. [CrossRef] [PubMed]
104. Rodriguez-Martinez, S.; Cancino-Diaz, M.E.; Cancino-Diaz, J.C. Expression of CRAMP via PGN-TLR-2 and of alpha-defensin-3 via CpG-ODN-TLR-9 in corneal fibroblasts. *Br. J. Ophthalmol.* **2006**, *90*, 378–382. [CrossRef] [PubMed]
105. Ottaviano, F.G.; Yee, K.O. Communication signals between cardiac fibroblasts and cardiac myocytes. *J. Cardiovasc. Pharmacol.* **2011**, *57*, 513–521. [CrossRef]
106. Martin, M.L.; Blaxall, B.C. Cardiac intercellular communication: Are myocytes and fibroblasts fair-weather friends? *J. Cardiovasc. Transl. Res.* **2012**, *5*, 768–782. [CrossRef] [PubMed]
107. Nomi, N.; Kimura, K.; Nishida, T. Release of interleukins 6 and 8 induced by zymosan and mediated by MAP kinase and NF-kappaB signaling pathways in human corneal fibroblasts. *Investig. Ophthalmol. Vis. Sci.* **2010**, *51*, 2955–2959. [CrossRef]
108. You, L.; Kruse, F.E.; Bacher, S.; Schmitz, M.L. Lipoteichoic acid selectively induces the ERK signaling pathway in the cornea. *Investig. Ophthalmol. Vis. Sci.* **2002**, *43*, 2272–2277.
109. Natarajan, K.; Shepard, L.A.; Chodosh, J. The use of DNA array technology in studies of ocular viral pathogenesis. *DNA Cell Biol.* **2002**, *21*, 483–490. [CrossRef]
110. Crowley, T.; Buckley, C.D.; Clark, A.R. Stroma: The forgotten cells of innate immune memory. *Clin. Exp. Immunol.* **2018**, *193*, 24–36. [CrossRef]
111. Owens, B.M.; Steevels, T.A.; Dudek, M.; Walcott, D.; Sun, M.Y.; Mayer, A.; Allan, P.; Simmons, A. CD90(+) Stromal Cells are Non-Professional Innate Immune Effectors of the Human Colonic Mucosa. *Front. Immunol.* **2013**, *4*, 307. [CrossRef]
112. Ara, T.; Kurata, K.; Hirai, K.; Uchihashi, T.; Uematsu, T.; Imamura, Y.; Furusawa, K.; Kurihara, S.; Wang, P.L. Human gingival fibroblasts are critical in sustaining inflammation in periodontal disease. *J. Periodontal. Res.* **2009**, *44*, 21–27. [CrossRef]
113. Tetlow, L.C.; Lees, M.; Ogata, Y.; Nagase, H.; Woolley, D.E. Differential expression of gelatinase B (MMP-9) and stromelysin-1 (MMP-3) by rheumatoid synovial cells in vitro and in vivo. *Rheumatol. Int.* **1993**, *13*, 53–59. [CrossRef]
114. Chodosh, J. Epidemic Keratoconjunctivitis. In *Atlas of Clinical Wisdom: Cornea, Refractive and External Disease*, 1st ed.; Melki, S., Ed.; Slack: Thorofare, NJ, USA, 2011; pp. 91–96.
115. Dawson, C.R.; Hanna, L.; Togni, B. Adenovirus type 8 infections in the United States. IV. Observations on the pathogenesis of lesions in severe eye disease. *Arch. Ophthalmol.* **1972**, *87*, 258–268. [CrossRef]
116. Kimura, R.; Migita, H.; Kadonosono, K.; Uchio, E. Is it possible to detect the presence of adenovirus in conjunctiva before the onset of conjunctivitis? *Acta Ophthalmol.* **2009**, *87*, 44–47. [CrossRef]
117. Chintakuntlawar, A.V.; Chodosh, J. Cellular and tissue architecture of conjunctival membranes in epidemic keratoconjunctivitis. *Ocul. Immunol. Inflamm.* **2010**, *18*, 341–345. [CrossRef] [PubMed]
118. Chodosh, J.; Miller, D.; Stroop, W.G.; Pflugfelder, S.C. Adenovirus epithelial keratitis. *Cornea* **1995**, *14*, 167–174. [CrossRef]

119. Hogan, M.J.; Crawford, J.W. Epidemic Keratoconjunctivitis: (Superficial Punctate Keratitis, Keratitis Subepithelialis, Keratitis Maculosa, Keratitis Nummularis) With a Review of the Literature and a Report of 125 Cases. *Am. J. Ophthalmol.* **1942**, *25*, 1059–1078. [CrossRef]
120. Pettit, T.H.; Holland, G.N. Chronic keratoconjunctivitis associated with ocular adenovirus infection. *Am. J. Ophthalmol.* **1979**, *88*, 748–751. [CrossRef]
121. Freyler, H.; Sehorst, W. The fate of corneal infiltrations in cases of epidemic keratoconjunctivitis. A follow-up study over two and a half years (author's transl). *Wien Klin. Wochenschr.* **1976**, *88*, 341–343.
122. Gokhale, N.S. Anterior segment optical coherence tomography to differentiate adenoviral subepithelial infiltrates and scars. *Indian J. Ophthalmol.* **2019**, *67*, 1725. [CrossRef]
123. Aydin Kurna, S.; Altun, A.; Oflaz, A.; Karatay Arsan, A. Evaluation of the impact of persistent subepithelial corneal infiltrations on the visual performance and corneal optical quality after epidemic keratoconjunctivitis. *Acta Ophthalmol.* **2015**, *93*, 377–382. [CrossRef]
124. Chodosh, J.; Astley, R.A.; Butler, M.G.; Kennedy, R.C. Adenovirus keratitis: A role for interleukin-8. *Investig. Ophthalmol. Vis. Sci.* **2000**, *41*, 783–789.
125. Duncan, S.J.; Gordon, F.C.; Gregory, D.W.; McPhie, J.L.; Postlethwaite, R.; White, R.; Willcox, H.N. Infection of mouse liver by human adenovirus type 5. *J. Gen. Virol.* **1978**, *40*, 45–61. [CrossRef] [PubMed]
126. Blair, G.E.; Dixon, S.C.; Griffiths, S.A.; Zajdel, M.E. Restricted replication of human adenovirus type 5 in mouse cell lines. *Virus Res.* **1989**, *14*, 339–346. [CrossRef]
127. Jogler, C.; Hoffmann, D.; Theegarten, D.; Grunwald, T.; Uberla, K.; Wildner, O. Replication properties of human adenovirus in vivo and in cultures of primary cells from different animal species. *J. Virol.* **2006**, *80*, 3549–3558. [CrossRef]
128. Younghusband, H.B.; Tyndall, C.; Bellett, A.J. Replication and interaction of virus DNA and cellular DNA in mouse cells infected by a human adenovirus. *J. Gen. Virol.* **1979**, *45*, 455–467. [CrossRef]
129. Chintakuntlawar, A.V.; Chodosh, J. Chemokine CXCL1/KC and its receptor CXCR2 are responsible for neutrophil chemotaxis in adenoviral keratitis. *J. Interferon. Cytokine Res.* **2009**, *29*, 657–666. [CrossRef]
130. Chintakuntlawar, A.V.; Zhou, X.; Rajaiya, J.; Chodosh, J. Viral capsid is a pathogen-associated molecular pattern in adenovirus keratitis. *PLoS Pathog.* **2010**, *6*, e1000841. [CrossRef]
131. Zhou, X.; Ramke, M.; Chintakuntlawar, A.V.; Lee, J.Y.; Rajaiya, J.; Chodosh, J. Role of MyD88 in adenovirus keratitis. *Immunol. Cell Biol.* **2017**, *95*, 108–116. [CrossRef] [PubMed]
132. Keenlyside, R.A.; Hierholzer, J.C.; D'Angelo, L.J. Keratoconjunctivitis associated with adenovirus type 37: An extended outbreak in an ophthalmologist's office. *J. Infect. Dis.* **1983**, *147*, 191–198. [CrossRef] [PubMed]
133. Harding, S.P.; Mutton, K.J.; van der Avoort, H.; Wermenbol, A.G. An epidemic of keratoconjunctivitis due to adenovirus type 37. *Eye* **1988**, *2 Pt 3*, 314–317. [CrossRef] [PubMed]
134. Takeuchi, R.; Nomura, Y.; Kojima, M.; Uchio, E.; Kobayashi, N.; Matumoto, M. A nosocomial outbreak of epidemic keratoconjunctivitis due to adenovirus type 37. *Microbiol. Immunol.* **1990**, *34*, 749–754. [CrossRef] [PubMed]
135. Curtis, S.; Wilkinson, G.W.; Westmoreland, D. An outbreak of epidemic keratoconjunctivitis caused by adenovirus type 37. *J. Med. Microbiol.* **1998**, *47*, 91–94. [CrossRef]
136. Mukherjee, S.; Zhou, X.; Rajaiya, J.; Chodosh, J. Ultrastructure of adenovirus keratitis. *Investig. Ophthalmol. Vis. Sci.* **2015**, *56*, 472–477. [CrossRef] [PubMed]
137. Bruder, J.T.; Kovesdi, I. Adenovirus infection stimulates the Raf/MAPK signaling pathway and induces interleukin-8 expression. *J. Virol.* **1997**, *71*, 398–404. [CrossRef] [PubMed]
138. Seto, D.; Chodosh, J.; Brister, J.R.; Jones, M.S.; Members of the Adenovirus Research Community. Using the whole-genome sequence to characterize and name human adenoviruses. *J. Virol.* **2011**, *85*, 5701–5702. [CrossRef]
139. Natarajan, K.; Ghalayini, A.J.; Sterling, R.S.; Holbrook, R.M.; Kennedy, R.C.; Chodosh, J. Activation of focal adhesion kinase in adenovirus-infected human corneal fibroblasts. *Investig. Ophthalmol. Vis. Sci.* **2002**, *43*, 2685–2690.
140. Tapial Martinez, P.; Lopez Navajas, P.; Lietha, D. FAK Structure and Regulation by Membrane Interactions and Force in Focal Adhesions. *Biomolecules* **2020**, *10*, 179. [CrossRef]
141. Xiao, J.; Chodosh, J. JNK regulates MCP-1 expression in adenovirus type 19-infected human corneal fibroblasts. *Investig. Ophthalmol. Vis. Sci.* **2005**, *46*, 3777–3782. [CrossRef]
142. Rajaiya, J.; Xiao, J.; Rajala, R.V.; Chodosh, J. Human adenovirus type 19 infection of corneal cells induces p38 MAPK-dependent interleukin-8 expression. *Virol. J.* **2008**, *5*, 17. [CrossRef]
143. Rajaiya, J.; Sadeghi, N.; Chodosh, J. Specific NFkappaB subunit activation and kinetics of cytokine induction in adenoviral keratitis. *Mol. Vis.* **2009**, *15*, 2879–2889.
144. Yousuf, M.A.; Zhou, X.; Mukherjee, S.; Chintakuntlawar, A.V.; Lee, J.Y.; Ramke, M.; Chodosh, J.; Rajaiya, J. Caveolin-1 associated adenovirus entry into human corneal cells. *PLoS ONE* **2013**, *8*, e77462. [CrossRef]
145. Rajaiya, J.; Yousuf, M.A.; Singh, G.; Stanish, H.; Chodosh, J. Heat shock protein 27 mediated signaling in viral infection. *Biochemistry* **2012**, *51*, 5695–5702. [CrossRef] [PubMed]
146. Lee, J.S.; Mukherjee, S.; Lee, J.Y.; Saha, A.; Chodosh, J.; Painter, D.F.; Rajaiya, J. Entry of Epidemic Keratoconjunctivitis-Associated Human Adenovirus Type 37 in Human Corneal Epithelial Cells. *Investig. Ophthalmol. Vis. Sci.* **2020**, *61*, 50. [CrossRef] [PubMed]

viruses

MDPI

Review

Genetic and Chemical Capsid Modifications of Adenovirus Vectors to Modulate Vector–Host Interactions

Denice Weklak †, Daniel Pembaur †, Georgia Koukou †, Franziska Jönsson, Claudia Hagedorn and Florian Kreppel *

Chair of Biochemistry and Molecular Medicine, Center for Biomedical Education and Research (ZBAF), Witten/Herdecke University, Stockumer Street 10, 58453 Witten, Germany; denice.weklak@uni-wh.de (D.W.); daniel.pembaur@uni-wh.de (D.P.); georgia.koukou@uni-wh.de (G.K.); franziska.joensson@uni-wh.de (F.J.); claudia.hagedorn@uni-wh.de (C.H.)

* Correspondence: florian.kreppel@uni-wh.de

† These authors contributed equally.

Abstract: Adenovirus-based vectors are playing an important role as efficacious genetic vaccines to fight the current COVID-19 pandemic. Furthermore, they have an enormous potential as oncolytic vectors for virotherapy and as vectors for classic gene therapy. However, numerous vector–host interactions on a cellular and noncellular level, including specific components of the immune system, must be modulated in order to generate safe and efficacious vectors for virotherapy or classic gene therapy. Importantly, the current widespread use of Ad vectors as vaccines against COVID-19 will induce antivector immunity in many humans. This requires the development of strategies and techniques to enable Ad-based vectors to evade pre-existing immunity. In this review article, we discuss the current status of genetic and chemical capsid modifications as means to modulate the vector–host interactions of Ad-based vectors.

Keywords: adenovirus; capsid modification; chemical modification; PEGylation; HPMA; COVID-19 vaccine

1. Introduction

Adenovirus (Ad) vectors are the most frequently used vectors in gene therapy trials [https://a873679.fmphost.com/fmi/webd/GTCT, accessed on 30 June 2021] and are currently mainly applied in cancer therapies (reviewed in [1]) and as novel vaccines (reviewed in [2]). During the ongoing global SARS-CoV2 pandemic, Ad-based vaccines were successfully tested in clinical trials, were approved as vaccines against COVID-19, and are being used as efficacious preventive vaccines [3–6].

The human Ad has a ~38 kb double-stranded DNA genome. The viral capsid is composed of three major capsid proteins (hexon, penton base, and fiber) and four minor proteins (IIIa, VI, VIII, and IX) that are organized in icosahedral symmetry. Of these major capsid proteins, hexon is both the largest and most abundant protein of the shell. Twelve hexon homotrimers form one capsid facet, of which 20 are present in the capsid. Penton consists of homopentamers of penton-base proteins located on each icosahedral edge (vertex) forming the base for the vertex's spike that consists of trimers of glycosylated fiber protein [7–9].

The recent success of Ad vectors in clinical trials is based on numerous advantageous features: Ad-based vectors exhibit broad tropism profiles, have large packaging capacities (up to ~36 kb), and persist episomally in infected cells. They can efficiently infect both dividing and quiescent cells, and the availability of scalable production systems enables large-scale vector production to high titers. However, despite these advantages the clinical use of Ad-based vectors in virotherapy and gene therapy still faces several hurdles. While local delivery, e.g., for genetic vaccination, is comparably unproblematic, systemic delivery of Ad bears some challenges mainly associated with widely pre-existing immunity, strong innate immune responses to the viral capsid, and robust adaptive immune responses to de novo synthesized viral and transgene gene products [10].

Upon systemic delivery, innate responses can occur within minutes to hours, leading to low systolic blood pressure, release of proinflammatory cytokines, thrombocytopenia, and transient liver toxicity. These adverse effects are triggered by interactions of vector particles with endothelial cells of the liver and vascular system, hepatocytes, Kupffer cells, platelets, macrophages, and dendritic cells (DC) [11]. Once in the blood stream, coagulation factor X (FX) binds to the hexon protein of the viral capsid [12], thus activating Toll-like receptor 4 on splenic macrophages resulting in a rapid clearance of vector particles from the spleen [13]. These innate immune responses are also observed for biologically inactive viral particles [14] or after administration of helper-dependent (HD) Ad vectors lacking all viral genes [15]. Therefore, innate immune responses seem to be largely independent from the expression of viral genes and instead relate to the viral capsid proteins themselves. Besides binding to coagulation factors, Ad capsid proteins are also bound by proteins of the complement system, such as C3 and C4BP, as well as natural immunoglobulin (IgM) antibodies. These interactions redirect vector particles from the blood stream to the liver [12,16] and mediate clearance through particle uptake by Kupffer cells [17,18].

The vector–host interactions of Ad known thus far are well described and build a basis for different capsid modification strategies. Mutating or blocking Ad vector interactions with cellular or noncellular host components is an efficient tool to detarget Ad from undesired off-target tissues. Moreover, safe and efficient therapeutic applications of Ad vectors require targeted cell entry opposing the broad tropism of adenoviruses. Genetic and chemical strategies that modify the interaction spectra of Ad capsid proteins may be used to retarget Ad vectors to new target cells or tissues.

Furthermore, pre-existing antibodies [19,20], originating either from previous Ad infections or vaccinations, can prevent the vectors from reaching target cells. For example, homologous prime-boost regimes using Ad-based vaccines, as applied in the recent pandemic [3], may elicit potent neutralizing antibody responses directed not only against the transgene itself but also against the vector. During boost vaccination, capsid-bound antibodies mediate vector sequestration by Fc receptor (FcR)-positive cells including DCs [21], neutrophils [22], and tissue-specific macrophages [23]. Antibody–vector complexes may further trigger the release of proinflammatory cytokines in macrophages by activating the intracellular antibody receptor TRIM21 [24–26]. Here, heterologous prime-boost regimes based on either different vector species [5,6] or platforms [27] could contribute to overcome this hurdle. However, with an emerging role of Ad-based vectors in vaccination strategies or virotherapy, these strategies will only postpone the initial problem. In the long-term, capsid modification strategies that shield vector particles from both innate and adaptive immune recognition may help to overcome this problem.

The above-described vector–host interactions and the challenges arising thereof highlight the necessity of viral capsid modifications in vector development to eliminate these obstacles. Since the majority of these interactions are interactions of the viral capsid with cellular and noncellular blood components, most strategies focus on modifications of the viral capsid, including genetic and chemical capsid modification approaches.

In the following, we will describe several mechanisms of vector detargeting to prevent unwanted interactions, vector targeting to direct vector particles to specific cell types, and combinations thereof.

2. Genetic Capsid Modifications

Genetic modification of capsid proteins is the most direct approach to modulate immune recognition and Ad tropism. This can be achieved by either the introduction of mutations into the Ad capsid proteins themselves or by the incorporation of foreign peptides into the capsid. To date, all Ad capsid proteins of the outer capsid surface have been genetically modified. This part of the review focuses on previously described genetic modification strategies for abrogation of Ad liver tropism and anti-Ad immune response (Figure 1).

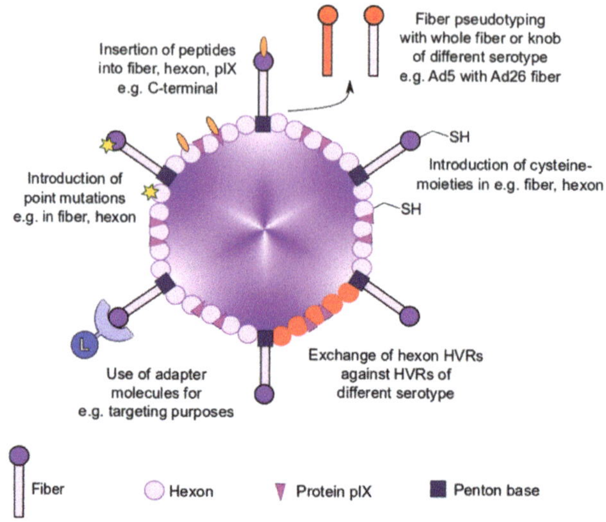

Figure 1. Overview of genetic modification strategies to modulate vector–host interaction. Genetic modification strategies for ablating viral tropism and anti-Ad immune responses by introduction of point mutations, insertion of peptides into capsid proteins, fiber pseudotyping, use of adapter molecules, or exchange of hexon hypervariable regions (HVRs).

2.1. Genetic Modifications to Ablate Vector Tropism

The targeted cell entry of Ad is mediated by two fundamental steps. The first step includes the high-affinity attachment of the fiber protein to its primary receptor, the coxsackie and Ad receptor (CAR) [28]. This binding increases the proximity between virus and cell surface and allows for the binding of cellular integrins such as $\alpha v\beta 3$ and $\alpha v\beta 5$ to an Arg-Gly-Asp (RGD) motif in the Ad penton base [29]. The penton RGD motif is exposed by bending of the fiber protein due to a flexible joint in the shaft domain in a process previously described as 'virus yoga' (reviewed in [30]). The Ad-integrin interaction promotes the internalization of viral particles by clathrin-mediated endocytosis [31]. After partial disassembly of the viral capsid in the endosome and release of virus into the cytoplasm [32], replication-defective vectors transport their genome into the nucleus [33].

Importantly, while CAR is expressed in some therapeutically relevant cell types, it is also expressed in human erythrocytes, which also express complement receptor 1 (CR1). Therefore, human erythrocytes influence blood circulation and infectivity of systemically delivered viral particles [34]. However, as erythrocytes do not express integrins, Ad vector particles can bind to erythroycytes but cannot be internalized, indicating that this interaction is reversible [35]. A feasible approach to detarget Ad vecors from CAR-expressing cells is the introduction of specific point mutations into the fiber protein. Kirby et al. proved abolished binding to CAR for Ad5 vectors bearing a S408E, P409K, T477A, or L485K point mutation in their fiber protein [36]. However, the binding of CAR-ablated vectors to erythrocytes was not prevented, as antibody-mediated attachment to CR1 still occurred [37]. Moreover, blocking fiber-CAR interactions did not prevent liver sequestration, as the transduction of hepatocytes was still mediated by blood components bound to capsid proteins other than fiber [38]. The high-affinity binding of blood coagulation factors such as FX and IX to hexon leads to the sequestration of virus in the liver, possibly hindering efficient transduction of target cells [39]. FX binds Ad5 in the presence of Ca^{2+}

and Mg^{2+} with picomolar affinity and increases heparan-sulfate-proteoglycan-mediated uptake into hepatocytes [12,39–41]. These observations demonstrate that although most detargeting approaches are straightforward, the interactions of Ad in vivo are far more complex and need to be better understood and carefully addressed to increase the safety and efficacy of Ad vectors in vaccinations and virotherapy.

2.1.1. Generation of Chimeric Ad Vectors by Fiber Pseudotyping

An approach for modifying Ad tropism is the generation of chimeric Ad vectors using fiber pseudotyping (Figure 1). Fiber pseudotyping describes the substitution of the knob domain or the entire fiber protein with its counterpart of other serotypes. While CAR is used as the primary receptor for cell entry by most Ad types, interactions with other attachment receptors, such as CD46 (e.g., Ad21, Ad35), desmoglein-2 (e.g., Ad3), GD1a, and sialic acid (e.g., Ad37), have been described (reviewed in [42]). As such, fiber pseudotyping may be used to trim CAR-dependent serotypes to different cell types. This particular method of altering viral tropism was established by Krasnykh et al., and since its inception in 1996, many different chimeric Ad vectors with a broad spectrum of infection profiles have been developed [43]. For example, Shayakhmetov et al. demonstrated that a chimeric Ad5/35 vector showed increased infection of $CD34^+$ hematopoietic cells [44]. In another example, the generation of an Ad5/3 chimera improved the infectivity of ovarian cancer cell lines allowing for the establishment of a retargeted oncolytic Ad [45,46]. Last but not least, in vivo experiments with pseudotyped Ad vectors uncovered a role of the fiber protein in Ad-induced thrombocytopenia. Mice infected with either wildtype Ad5 or Ad3 displayed a drop in platelet counts 2 days post infection. In contrast, the pseudotyping of Ad5 with either the complete fiber or fiber shaft of Ad3 prevented thrombocytopenia in infected mice [47]. These results indicate a role of fiber in Ad-induced thrombocytopenia that might be overcome by fiber-pseudotyping strategies. Importantly, while adverse blood clotting side effects have been observed for example with the chimpanzee Ad-vector-based COVID-19 vaccine, it is too early to speculate on the role of Ad fiber in this specific context since other factors such as the spike protein itself or fragments thereof could likely contribute to these rare observed events.

2.1.2. Insertion of Targeting Peptides into Ad Capsid Proteins

One of the most frequently applied method of altering Ad tropism is the incorporation of foreign peptides into Ad capsid proteins to either increase or shift target cell infection. Most suitable for the incorporation of these foreign ligands is the fiber protein (Figure 1). While the C-terminus presents itself as an obvious location for peptide insertion, another optimal location has been identified as a solvent exposed, flexible loop between β-sheets H and I, the so called HI loop [48]. Peptides with a length of nearly 90 amino acids were C-terminally incorporated, while a peptide length of 83 amino acids was reported for the HI loop insertion site [49,50]. Commonly, an RGD-peptide is incorporated at the HI loop to increase integrin-based interactions between virus and host cells allowing for higher transduction efficacy of cancer cell lines, a strategy pioneered by the research group around *D.T. Curiel* [51]. Combining the insertion of an RGD-peptide in the HI loop with the insertion of a heparan sulfate binding polylysine in the C-terminus amplified the transduction efficacy of CAR-positive and -negative cells compared to un- or singly modified vectors [52]. Moreover, Curiel et al. reported that these vectors demonstrated a loss of natural CAR binding and infected ovarian carcinoma cell lines with increased efficacy [53]. As such, the safety and efficacy of RGD-incorporated Ad vectors was tested in clinical phase I trials for the treatment of ovarian cancer patients [54,55]. MacLeod et al. generated an Ad5 vector in which the epidermal growth factor (EGF)-like domain of heregulin-α was inserted into the HI loop to increase the in vitro infection of HER3/ErbB3- and HER4/ErbB4-expressing breast cancer cells [56]. It should be noted that incorporation of foreign peptides into the fiber protein can interfere with its trimerization and ultimately prevent viral assembly. Thus, the insertion site as well as the length of the desired targeting ligand should be considered carefully and will differ between different Ad types [57].

Another possible peptide insertion site is the cement protein IX which can be modified at its surface-exposed C-terminus. The incorporation of a polylysine sequence with an upstream FLAG peptide allowed for increased heparan sulfate binding and knob-independent infection [58]. Furthermore, IX is suitable for modification with large polypeptides and proteins, as the incorporation of enhanced green fluorescent protein (EGFP) with a length of nearly 240 amino acids was demonstrated, and only minimal effects on thermostability and bioactivity were observed [59]. Therefore, they demonstrated that protein IX has a higher peptide insertion capacity than the fiber protein.

In addition to protein IX and fiber, ligand insertion was also reported for the Ad hexon protein. It could be proven that hexon hypervariable regions (HVR) 2, HVR3, HVR5, HVR6, and HVR7 were appropriate for peptide insertion after testing feasibility with His_6 epitopes [60]. While being rather limited in the length of the incorporated peptide, an example of hexon HVR5 accommodating a foreign RGD-peptide showed increased transduction of human vascular smooth muscle cells with only minor effects on virus viability and growth [61]. Moreover, the introduction of point mutations in the Ad hexon protein presents an elegant solution for escaping blood coagulation factor-mediated hepatocyte tropism. Doronin et al. demonstrated that a point mutation (T425A) within HVR7 of hexon abolished binding of FX to Ad5 and reduced hepatocyte infection [13]. Of note is that interfering with FX association can lead to natural IgM- and complement-mediated neutralization of Ad [62].

2.1.3. Adapter-Based Strategies to Ablate Ad Tropism

A different strategy for employing altered viral tropism is the use of adapters serving as fusion proteins between capsid structure proteins and cell-binding ligands (Figure 1). One of these adapter strategies utilized the fusion of single chain variable fragments (scFv) to peptides such as EGF to target EGF-receptor expressing cells, thus enhancing the efficacy of Ad gene therapy [63]. In a similar way, Barry et al. fused the hexon-binding domain of FX to scFv to retarget Ad vectors to cells expressing Her2, ATP-binding cassette protein G2, and EGF-receptor [64]. Parrott et al. also reported the use of another adapter strategy in which a biotin-acceptor peptide (BAP) was fused to the C-terminus of fiber. This peptide is metabolically biotinylated in producer cell lines and can be used for retargeting purposes by conjugation to biotinylated antibodies [50]. However, fusing BAP to protein IX failed retargeting when biotinylated antibodies were used, and retargeting was only achieved when biotinylated ligands instead of antibodies were used [50,65]. An alternative method was reported by Dimitriev et al. in which a soluble, truncated version of CAR was fused to EGF, resulting in an increased infection of EGF-receptor overexpressing cancer cell lines compared to untargeted Ad or EGF-receptor negative cells in vitro [66]. These adapter strategies are based on noncovalent protein–protein interactions. Due to this, natural CAR, FX, or antibody binding can compete with the adapter molecule binding, ultimately preventing vector retargeting.

2.2. Genetic Modifications to Circumvent Anti-Ad Immune Responses

Another difficulty of systemic delivery of Ad vectors is the high prevalence of anti-Ad immunity in the population. Neutralizing antibodies against common Ad types, such as Ad5, is relevant, and means of immune escape must be taken into consideration prior systemic delivery of Ad vectors, especially in regard to the current coronavirus pandemic. As mentioned above, worldwide vaccination with Ad-vector-based vaccines such as Vaxzevria® (AstraZeneca plc) and *Ad26.COV2-S* (Janssen Pharmaceutica N. V.) will increase the prevalence of neutralizing anti-Ad antibodies in the population and probably limit the efficacy of subsequent treatment with Ad vectors, including prime-boost regimens. The combinatorial utilization of different de- and retargeting strategies described above requires a detailed understanding of the interactions not only between Ad capsid proteins and target cells but also between Ad and (noncellular) blood components. For example, while the binding of FX to Ad hexon increases the transduction efficiency of hepatocytes, it also acts as a shield and thus prevents the binding of natural or pre-existing antibodies. Therefore, the ablation of FX binding (e.g., by introduction of point mutations in hexon HVR7) may result in increased IgM- and complement-mediated vector neutralization after systemic delivery [62].

As most of the neutralizing antibodies are directed against HVRs of Ad hexon [67], the generation of chimeric Ad variants by replacing these regions against HVRs of a less prevalent serotype is one method to hinder antibody-mediated neutralization (Figure 1). For example, Roberts et al. demonstrated that the exchange of all seven HVRs of Ad5 hexon for their Ad48 counterparts increased Ad immune escape in vivo [67]. However, replacing HVRs proved to be insufficient for anti-Ad immune escape [68]. Still, it should be noted that Roberts et al. reported a three- to fivefold lower production yield in comparison to the parental Ad5 vectors. In addition, most of their chimeric vectors were nonviable, indicating structural and biochemical constraints in regard to the manipulation of hexon HVRs [67]. In a different approach, Rojas et al. incorporated an albumin-binding peptide into hexon for shielding via binding to blood serum albumin after systemic delivery [69]. While it was shown that the virus was protected from neutralization by natural antibodies, Rojas et al. observed a decrease in the transduction efficacy of the albumin-binding vector [69]. Recently, another method for Ad escape from natural IgM- and complement-mediated neutralization has been reported. Atasheva et al. demonstrated that the deletion of a small stretch of negatively charged amino acids in the HVR1 in combination with an FX-ablating mutation could prevent the binding of natural IgM antibodies and complement components [70]. In summary, the research group generated an Ad vector that resisted inactivation by blood factors and liver cell sequestration and did not trigger hepatotoxicity. This is, to our knowledge, the most complete approach to genetic capsid modifications with substantial modulation of Ad vector–host interactions so far. However, the interaction of anti-Ad antibodies with the vector after systemic delivery requires further research.

2.2.1. Removal of Viral Genes from the Vector Genome

Ad-associated immunogenicity and toxicity can be decreased by the use of high-capacity (HC) or helper-dependent (HD) vectors, also termed 'gutless vectors'. HD-Ad vectors were generated by the deletion of all viral coding sequences but not the inverted terminal repeats (ITR) necessary for genome replication and the packaging signal (Ψ) (reviewed in [71,72]). The deletion of nearly all viral genes allows for a high cloning capacity of about 36 kilobases (kb). However, efficient DNA packaging requires a minimum vector size of 27 kb, and as such, stuffer DNA is required to reach minimum viral packaging size [73]. While the first DNA stuffers relied on the use of lambda phage, bacterial, or yeast DNA, the use of intronic sequences of human genes, such as hypoxanthine-guanine phosphoribosyltransferase (HPRT), has become more frequent. HD-Ad vectors containing lambda-phage stuffer DNA showed limited gene expression as well as an increased cytotoxic T cell (CTL) response in vivo [74], while vectors equipped with HPRT-derived stuffer DNA displayed prolonged transgene expression without inducing CTL responses [74]. Since the dsDNA itself may trigger intracellular innate immune responses via TLR signaling [75], stuffer DNA sequences should be designed carefully.

With all viral genes being deleted, HD-Ad vectors rely on a helper Ad supplying viral proteins in trans for propagation (reviewed in [71]). This risks contamination of, e.g., gene therapy preparations with helper Ad and replication-competent particles (reviewed in [76]). Due to the lack of viral gene expression, HD-Ad exhibit prolonged transgene expression and reduced toxicity. While the adaptive immune response to these vectors is dampened, acute inflammatory responses were still observed [15]. Despite this, HD-Ad vectors were successfully used for long-term expression of transgenes in nonhuman primates [77,78].

2.2.2. Geneti-Chemical Modification Strategies

Finally, another approach to prevent liver cell sequestration and antibody-mediated neutralization of Ad is the genetic introduction of cysteine moieties into Ad capsid proteins, allowing for site-specific chemical coupling of polymers (Figures 1 and 2) [79]. This so-called geneti-chemical approach enables researchers to shield specific capsid proteins from unwanted interactions, to attach ligand molecules to the capsid, and to study vector–host interactions after capsid modification in a position-dependent manner. As an example of this geneti-chemical modification approach, Prill et al. introduced a cysteine moiety in hexon HVR5 which allowed for chemical coupling of small maleimide-activated polyethylene glycol moieties, leading to a reduction in hepatocyte transduction after systemic delivery [80]. The incorporation of foreign cysteine moieties is not limited to the hexon protein, as modulation of the fiber HI loop and protein IX has also been shown to be successful [79,81]. Generally, the chemical and geneti-chemical coupling of polymers to various capsid proteins are promising approaches to evade anti-Ad immune responses and will be discussed in more detail in the following parts of this review.

All things considered, engineering vectors for anti-Ad immune escape by genetic modifications alone may prove difficult. HD-Ad vectors still elicit innate immune responses, while the exchange of hexon HVRs affects vector viability, and shielding with albumin reduces viral infectivity. The generation of chimeras between Ad types or with different viruses (e.g., by inserting the genome of small viruses into the Ad genome) may present itself as a future endeavor for escaping anti-Ad immunity as well as for improving target cell transduction efficacy and biodistribution (reviewed in [76]).

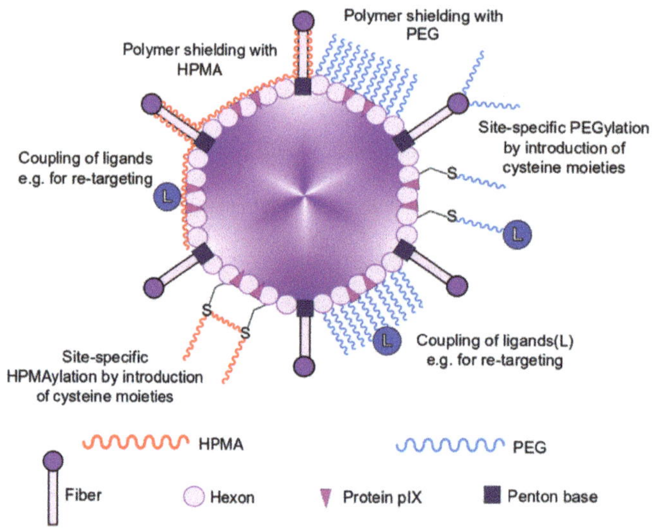

Figure 2. Overview of chemical modification strategies to modulate vector–host interaction. Chemical modification strategies for polymer capsid shielding with polyethylene glycol (PEG) or poly-*N*-(2-hydroxypropyl) methacrylamide (HMPA) and retargeting purposes by use of ligands.

Conclusively, genetic modifications of the Ad genome offer a robust toolkit for altering Ad tropism and may provide means of escaping anti-Ad immunity. While the possibilities seem nearly endless, the combination of different re- and detargeting approaches requires an excessive amount of knowledge about the interaction of Ad and blood components. It has been shown that the incorporation of peptides as well as the generation of chimeric Ad vectors can be rather challenging. Interference with Ad structural integrity can result in nonviable, less infectious vectors, and thus the choice of motif and site of insertion should be considered carefully. Furthermore, Ad-based vaccines are currently employed in the ongoing SARS-CoV2 pandemic, and therefore genetic modifications alone may be insufficient to prevent anti-Ad-mediated neutralization. Chemical modification of the Ad capsid by polymer shielding could provide additional help for the evasion of anti-Ad immunity.

3. Chemical Shielding of Ad Vectors

As mentioned above, genetic modifications of Ad vectors were used to evade from the immune system, to detarget particles from their natural tropism, and to retarget them to new receptors [82]. However, genetic modifications are limited by several obstacles, including structural capsid protein integrity and stability often in conjunction with impaired vector production [60]. In contrast, the chemical modification of the vector capsid ensures (i) a highly efficient vector production without impairment by protein structural changes, (ii) a comparably fast and easy capsid modification strategy, (iii) no safety concernsdue to inherited vector immune evasion properties, and (iv) heterogeneous capsid modifications given by a wide variety of appropriate molecules that allow for the chemical modification of the capsid under mild conditions [83]. These molecules are, in general, strongly hydrophilic polymers that are activated with amine- or thiol-reactive groups to bind to α- or ε-amino groups of the N-terminus and of lysine residues, respectively. In the case of thiol-reactive groups, these molecules react with reduced cysteine residues (Figure 2). Besides the possibility to define the binding site of the polymer, the binding mode can be also selected. Thereby, the type of reactive group determines whether the binding of the polymer is irreversible or bioresponsive, i.e., reversible under certain cellular conditions (e.g., reducing or acidic compartments) [84,85]. Bioresponsive modifications offer the option of evading immune components in the blood while ensuring a regular intracellular trafficking of the vectors by uncovering polymer modified proteins after cell entry. Further variety of the polymers is offered by the number of reactive groups per polymer. A polymer that possesses only one reactive group would protrude from the capsid whereas a polymer with multiple reactive groups would closely attach to the capsid surface (Figure 2). Thus, even the size of the vector capsid could be modified by the choice of the polymer, which could have the potential of altering vector tropism [82,86]. Chemical modification of vector capsids can be performed very fast (about 1 h) and under mild conditions (room temperature, physiological pH) and, furthermore, without impairment of vector production. Moreover, it offers endless options that allow for detargeting and even retargeting of the vectors by the introduction of new ligand binding sites. Here, we discuss key results from the two major polymers polyethylene glycol (PEG) and poly-*N*-(2-hydroxypropyl) methacrylamide (HPMA), which have successfully been used for chemical modification of Ad vector capsids for protection against immune and blood factors as well as for de- and retargeting (Table 1).

Table 1. Overview of the discussed shielding polymers and their main characteristics. Reactivity of the polymers was directed to amine- (–NH) or thiol- (–SH) groups, whereas the functionality describes the number of reactive groups per polymer (monofunctional = one reactive group; heterobifunctional = two different reactive groups; polyfunctional = many reactive groups). Bioresponsive polymers were coupled reversible (under physiological conditions) to the capsid.

Abbreviation	Polymer	Reactivity	Functionality	Bioresponsive
ONp-HPMA	4-nitrophenoxy-poly-*N*-(2-hydroxypropyl) methacrylamide	–NH	polyfunctional	no
mal-HPMA	maleimide-poly-*N*-(2-hydroxypropyl) methacrylamide	–SH	polyfunctional	no
OPSS-HPMA	orthopyridyl-disulfide-poly-*N*-(2-hydroxypropyl) methacrylamide	–SH	polyfunctional	yes
Next Generation HPMA	EC208	–NH	polyfunctional	yes
T-MPEG	tresyl-monomethoxypolyethylene glycol	–NH	monofunctional	no
CC-MPEG	cyanuric chloride monomethoxypolyethylene glycol	–NH	monofunctional	no

Table 1. *Cont.*

Abbreviation	Polymer	Reactivity	Functionality	Bioresponsive
SS-MPEG	succinimidyl succinate monomethoxypolyethylene glycol	–NH	monofunctional	no
T-MPEG-mal	tresyl-polyethylene-glycol-maleimide	–NH, –SH	heterobifunctional	no
mal-PEG	maleimide-polyethylene glycol	–SH	monofunctional	no
SPA-PEG	succinimidyl propionate polyethylene glycol	–NH	monofunctional	no

4. Shielding with HPMA

4.1. HPMA-Coating of Ad

HPMA is one of the important polymers to chemically modify the Ad capsid. The main advantage of HPMA polymers is their high biocompatibility and the low immunogenicity which were tested for a wide variety of different HPMA polymers (reviewed in [87]). Over two months, in vivo assays in mice showed no significant influence of the basic immune defense activity after treatment with a 25 kDa HPMA and a high dose of 2 g HPMA per kg body weight [88]. The tested basic immune activities included humoral immune responses, activation of complement pathways, phagocytotic activity, and bone marrow status. Furthermore, the examined doses of 2 g HPMA per kg body weight were much higher than the doses that would be used in clinical practice. Moreover, the size of the administered HPMA polymer was shown to be decisive regarding their immunogenicity [89]. Říhová et al. detected a two- to fivefold increase of the number of antibody-forming cells in the spleen in mice after injection of 150–200 kDa HPMA polymers in comparison to 5 kDa HPMA polymers [89].

Evasion from undesired vector–host interactions of Ad vectors can be achieved by modifying the vector capsid with diverse HPMA polymers and thus limit vector sequestration and reduce vector immunogenicity. One of the first attempts was presented in 2001 by Fisher et al. with an all-in-one approach [90]. In brief, an Ad type 5 vector was chemically modified with an amine-directed HPMA that was activated with ONp groups. An efficient coupling of the polymer to the vector was shown, and in vitro transduction assays with A549 cells revealed a significant detargeting of the shielded vectors from these CAR-positive cells. Furthermore, the coupling of fibroblast growth factor (FGF) or vascular endothelial growth factor (VEGF) to the HPMA-coated vectors resulted in a retargeting towards receptor-positive cells in vitro. Additionally, an enzyme-linked immunosorbent assay (ELISA) showed protection against neutralizing antibodies [90]. This all-in-one approach was a door opener to the field of HPMA-coated Ad vectors.

4.2. Chemistry of HPMA Coupling

The coupling of HPMA polymers can be generally performed at amine groups or genetically introduced thiol groups of solvent accessible residues on the Ad capsid surface (Figure 2). The conditions of HPMA coupling were reported to be very mild with a pH of 7.2–7.8 and at room temperature [84,90,91]. As reviewed by Říhová et al., another welcome feature of HPMA polymers is the low immunogenicity of the molecule [87]. HPMA polymers used for Ad capsid modification possess multiple reactive side chains that bind cooperatively to multiple sites at the capsid surface [92]. Kreppel and Kochanek reviewed that this multivalent reaction mode closely attached HPMA to the capsid surface, which is in contrast to the binding of monovalent PEGs which protrude with one free end from the capsid surface (see Figure 2) [83]. Thus, the size of HPMA-modified vectors increased only slightly when compared to the molecular weight of the used polymer, which commonly ranges from 12 to 150 kDa [34,84,90,91,93,94]. However, it was shown that by altering the initial concentration and the reaction temperature during polymer synthesis, the molecular weight of HPMA copolymers could be influenced [93]. In the following, we give an overview of the different coupling strategies using HPMA, and we further describe the chemical properties of HPMA polymers.

4.2.1. Amine-Directed HPMA Coupling

The amine-directed HPMA copolymers that are mostly used in chemical Ad modification allow for the binding of the polymer to multiple, solvent-exposed terminal α-amino groups and ε-amino groups of lysine residues, respectively [93,94]. Thus, the multivalency of HPMA copolymers facilitates a cooperative binding to the Ad capsid that provides about 18,000 possible binding sites for amine-directed HPMA coupling [90,95]. It was reported by Green et al. that up to 90% of these primary amines could be modified by HPMA coating [96]. The HPMAs used, were activated by 4-nitrophenoxy (ONp) side chains that covalently attach the polymer to the vector capsid (Figure 3A). of the different amine reactive HPMA copolymers are presented in Figure 3.

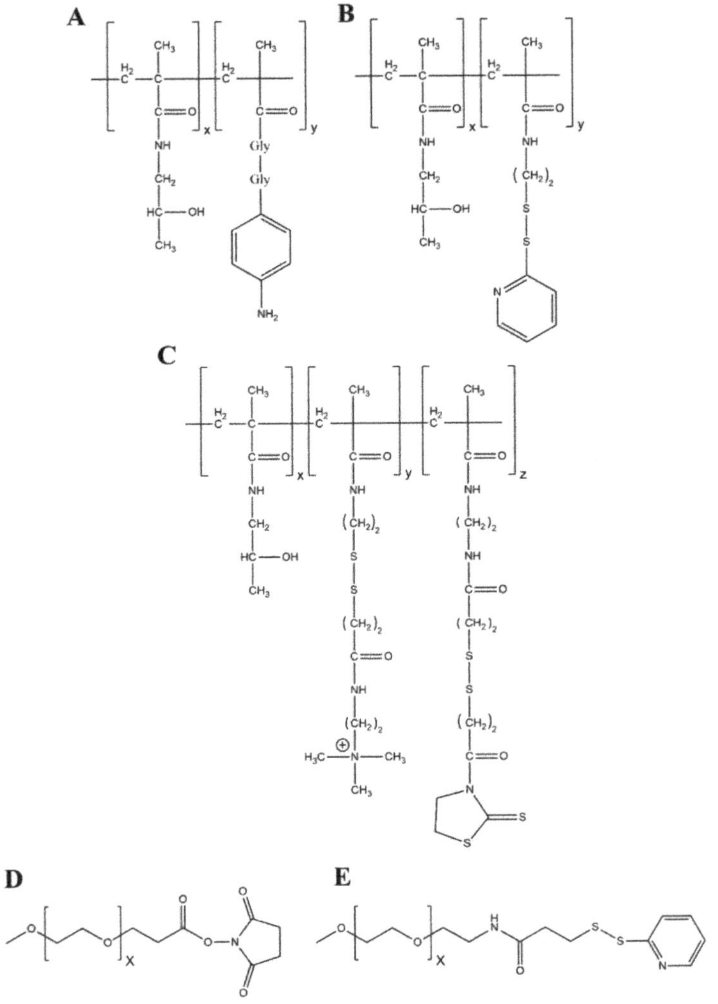

Figure 3. HPMA copolymer and PEG structures. (**A**) HPMA copolymer activated with a diglycine-linked ONp as presented by Fisher et al. [90]. ONp as an amine-reactive group will covalently bind to ε-amino groups of lysine residues on the capsid surface. Commonly used ratio for the monomers

is 9:1 (X:Y). (**B**) Orthopyridyl disulfide (OPSS)-activated HPMA as presented by Prill et al. [84]. This copolymer will bind to genetically introduced cysteines on the capsid surface and results in a disulfide bound (bioresponsive, reducible) HPMA-coated vector. (**C**) "Next generation HPMA copolymer" with disulfide linkers (bioresponsive, reducible) in the quaternary amine (QA) monomer and in the amine reactive carbonyl thiazolidine-2-thione (TT) monomer as presented by Subr et al. [91]. The amine reactive TT group will covalently attach the HPMA to ε-amino groups of lysine residues on the capsid surface, while the QA group will bind to the negatively charged surface of the capsid to enhance the efficacy of the covalent attachment. (**D**) Succinimidyl propionate (SPA)-activated polyethylene glycol (PEG). SPA is an amine-reactive group and will covalently and irreversibly bind to ε-amino groups of lysine residues on the capsid surface [97]. (**E**) OPSS-activated PEG. OPSS is a thiol-reactive group and will covalently, bioresponsively bind to genetically introduced cysteines [98].

4.2.2. Thiol-Directed HPMA Coupling

By the genetic introduction of cysteines into vector capsid proteins, solvent-exposed thiol groups can be introduced into the Ad capsid surface. This approach allowed for thiol-directed HPMA coupling at introduced cysteines in the hexon protein [84]. In comparison to amine-directed HPMA coupling, this method enables a position-specific coupling of the polymer to selected capsid proteins [98]. For thiol-directed HPMA coupling, the HPMA polymers were activated with maleimide or orthopyridyl disulfide (OPSS) groups (Figure 3B) to generate irreversible or bioresponsive bonds between HPMA and vectors, respectively [84]. Further properties of the thiol-directed HPMA coupling are presented below.

4.2.3. Next Generation HPMA Polymers—Bioresponsive HPMA Coupling

A wide diversity of HPMA copolymers can be synthesized by altering the reactive group or introducing charged side chains (e.g., quaternary amino groups). The choice of the type of the reactive group is crucial for the intracellular fate of the vector. As mentioned above, an activation of HPMAs by ONp side chains resulted in an irreversible covalent attachment of the polymer to the vector capsid (Figure 3A). As shown for HPMA-coated DNA/polyethyleneimine (PEI) complexes, the irreversible ONp-HPMA coating resulted in an ablation of transgene expression, while it could be restored with a disulfide-linked HPMA coating [99]. Based on these findings, Subr et al. presented the "next generation HPMA copolymers" that included a degradable disulfide bond in the spacer between the amine reactive carbonyl thiazolidine-2-thione group and the HPMA (which possessed additional quaternary amines for a higher coating efficiency, Figure 3C) [91]. Prill et al. reported with live cell imaging that an irreversible attachment of HPMA copolymers to specific sites on the hexon protein caused an intracellular trafficking impairment of the vectors [84]. Moreover, this trafficking impairment could be converted to a trafficking delay by using bioresponsive disulfide-bound HPMAs (Figure 3B). The authors could show that these biorepsonsive polymers can be shed from the particles in the reducing intracellular environment in a traceless manner, i.e., leaving no chemical residues on the capsid surface. Wang et al. reported that an electrostatic attachment of HPMAs to the vector capsid with biodegradable linkers enhanced transgene expression significantly compared to nonbiodegradable linkers [100]. Taken together, these results show that bioresponsive HPMAs are more suitable for chemical Ad vector modification, than the irreversible attached HPMAs. These results highlight the importance to maintain dynamic abilities of the vector particles after chemical modifications.

As mentioned above, the HPMAs presented by Subr et al. possessed quaternary amines (QA) as an additional modification of the HPMA copolymer that is not solely based on a chemically reactive group. The QAs imparted a positive charge to the HPMAs and enabled them to bind to the negative vector capsid surface, especially to the hexon protein [91]. Thus, they bind the vector capsid with higher affinity and at the same time convert the slightly negative surface (ζ) potential into a neutral or even minimal positive charge which—as discussed later (see Section 4.3)—may alter the vector tropism by the binding of the positively charged vector capsid to negatively charged surface molecules [34,84,100].

Another feature of vector-attached HPMAs is the excess of reactive groups which did not react with the capsid. By the addition of ligands with suitable binding sites (e.g., antibodies, growth factors), these ligands could be adhered to the HPMA-vector complex, allowing an easy retargeting of the whole complex [90,101]. Alternatively, the ligand could react with the HPMA prior to the coating of the vector capsid [102].

4.3. Detargeting and Immune Evasion of HPMAylated Ad Vectors

Intravenous delivery of Ad vectors remains a favorable route for virotherapy. In that case, several barriers must be overcome, such as rapid inactivation of the vectors by humoral and cellular immune and blood components. The wide range of cell types targeted by the wildtype Ad further leads to off-targeting in tumor therapy [103]. The intraperitoneal treatment of mice with Ad vectors was shown to result in peritoneal adhesion formation and bowel obstruction, likely caused by inflammatory effects that could be circumvented by altering the vector tropism with HPMA coating [102].

HPMA-modified vector capsids were shown to evade humoral immune and blood factors as well as the CAR- and FX-mediated binding to host cells. Subr et al. assessed three different amine reactive HPMA copolymers and demonstrated in vitro that exhaustive coating of Ad vectors greatly reduced CAR-mediated transduction and binding to human erythrocytes, respectively [91]. In addition, they demonstrated an ablation of FX-mediated transduction in in vitro experiments for the vectors modified with QA containing HPMAs and, moreover, a great reduction of CR1-mediated binding to human erythrocytes. These results indicated an ablation of the natural Ad tropism for the QA-containing HPMA-modified vectors and, likewise, partial protection against the complement and binding of antibodies. Green et al. showed that in in vivo experiments with BALB/c mice, Ad5 vectors modified with a 31.6 kDa thiazolidine-2-thione (9.5 mol%)-activated HPMA had an extended plasma circulation time compared to unmodified Ad5 and Ad11 [96]. Since Ad11 is unable to infect the cells of BALB/c mice (including hepatocytes) the extended circulation time of HPMA-modified vectors was therefore not only a result of an ablation of the natural tropism, but also due to a prevention of phagocytosis by macrophages. Furthermore, for the HPMA-coated vectors, Green et al. demonstrated a stable circulation for 30 min in BALB/c and melanoma tumor-bearing C57BL/6 mice after removal of Kupffer cells by preadministration of clodronate liposomes. In fact, Green et al. were able to show a complete prevention of tumor growth in HEPG2 xenograft-bearing mice by a 29-day treatment of HPMA-coated wildtype Ad5 and clodronate liposomes [96]. This extended plasma circulation time was crucial for a passive targeting of the vectors to the tumors, which was described by the enhanced permeability and retention (EPR) effect in the tumor area [104]. Fisher et al. first described that the EPR effect can also be exploited by using an ONp-HPMA-modified Ad5 vector and showed a targeting of subcutaneous AB22 mesothelioma tumors in BALB/c mice. Additionally, the initial vector dose found in the liver was greatly reduced compared to uncoated Ad5 vectors. Nevertheless, specific transgene expression (expression per vector particle) of coated vectors in tumor was diminished compared to uncoated vectors, and it was postulated that this might be caused by an altered infection pathway [92].

The altered tropism of HPMA-coated Ad5 vectors was also assessed by Prill et al. with geneti-chemically modified vectors (as described by Kreppel et al. in [79]). In their study, the selective binding of maleimide- or OPSS-activated HPMAs to introduced cysteines in the hexon protein was used to maintain CAR-mediated transduction capacity but, at the same time, to shield against FX-mediated transduction [84]. This approach resulted in decreased liver tropism and thus reduced transgene expression in the liver of vectors modified with uncharged HPMA. In contrast, positively charged HPMA mimicked the effect of FX-mediated liver transduction, underlining the necessity of a careful

HPMA design. Regardless of whether the vectors were modified with charged or uncharged HPMAs, the number of vector particles in the spleen, lung, and kidneys was markedly reduced compared to the number of unmodified vector particles [84]. Moreover, the plasma circulation time of coated vectors in comparison to unmodified vector particles increased significantly to 2.4-fold for an uncharged HPMA and 8.8-fold for a charged HPMA, respectively. The effect of positively charged HPMAs binding to negatively charged surface molecules (e.g., heparan sulfate proteoglycans, HSPGs) of cells was also described by Wang et al. [100].

Stevenson et al., who modified chicken embryonal lethal orphan virus (CELO, fowl Ad type 1) with HPMA, showed a reduced neutralization of coated human Ad5 vector particles in human serum, reflecting a partial evasion of the humoral immune defense by HPMA coating [105]. Furthermore, in a capture ELISA, Fisher et al. showed no detectable antibody binding with a fully HPMA-coated Ad5 vector [90]. Moreover, Carlisle et al. demonstrated an improved resistance against neutralizing antibodies and plasma components when using a retargeted Ad5 vector modified with bioresponsive QA-bearing HPMA. Additionally, an ablation of complement or fiber-mediated binding to erythrocytes was achieved [34]. However, the immune evasion of HPMA-coated Ad vectors in human serum is a major challenge that needs to be assessed with every new HPMA copolymer.

4.4. Retargeting of HPMAylated Ad Vectors

In addition to the passive tumor targeting of fully shielded Ad vectors, an active retargeting by the introduction of the appropriate ligands to the shielded vector capsid is a highly promising and versatile option for chemical Ad vector modification.

As described above, the excessive reactive groups of HPMA copolymers attached to Ad vectors could be used to introduce ligands. Fisher et al. detargeted HPMA-coated Ad5 vectors from CAR expressing cells, and after the introduction of fibroblast growth factor (FGF) or VEGF, vectors were retargeted to cells with corresponding receptors [90]. The FGF-retargeted vector was able to induce a 10-fold increase in transgene expression in SUIT-2 cells (pancreatic ductal carcinoma) in mice in vivo. The same retargeting strategy was used for a CELO virus with a successful retargeting in vitro [105]. However, in vivo, in BALB/c mice, a higher number of these vectors were found in all examined organs compared to unmodified Ad5 vectors. However, the transgene expression of the retargeted CELO vectors was observed to be greatly reduced. The authors stated that the high particle number in organs likely originates, on one hand, from a faster degradation of Ad5 compared to CELO vectors and, on the other hand, probably from the low-affinity binding of FGF to widely presented HSPGs. Nevertheless, the reduced transgene expression might be a result of an altered intracellular trafficking that was perhaps caused by differences between human and avian cells [105]. Importantly, targeting by FGF could lead to excessive binding to human erythrocytes and may therefore be unfavorable for systemic delivery [105].

EGF was also employed as a retargeting ligand [34,102]. Morrison et al. retargeted wildtype Ad5 to EGF receptor by coating the vector with thiazolidine-activated and EGF-bearing HPMAs. They showed detargeting from the CAR receptor and retargeting to the EGF receptor. The complex of EGF bound to its receptor was endocytosed, and the EGF still colocalized to its receptor in perinuclear areas. Furthermore, a murine xenograft model with intraperitoneal implanted SKOV-3 cells was used to demonstrate the efficacy of the retargeted viruses. The intraperitoneal injection of detargeted (HPMA-coated) virus showed no effect on survival, whereas the retargeted (EGF-HPMA-coated) virus and, in a similar way, the uncoated virus resulted in longer survival of the mice compared to the PBS control. However, fewer side effects (peritoneal adhesions and bowel formation) were observed for coated virus in comparison to uncoated virus [102]. Similar effects were reported for the same HPMA-coated vector when retargeted with the monoclonal antibody Cetuximab. Cetuximab is directed to the EGF receptor and is used for colorectal cancer therapy [101,106].

Another interesting approach for the retargeting of HPMA-coated Ad5 vectors was presented by Li et al. who used an activable cell penetrating peptide (ACPP) to retarget the coated vectors to matrix metalloproteinase (MMP)-overexpressing cells [107]. ACPP consists of a polycationic (polyarginine) domain, which is able to penetrate cell membranes and is linked via an MMP-cleavable linker to an inhibitory, polyanionic domain. The polyanionic domain can be cleaved in MMP-overexpressing tissues from the ACPP, allowing the polycationic domain to bind to and enter cells. Li et al. showed that the loss of transduction efficiency of a fully detargeted HPMA-coated vector could be regained by linkage of the ACPP to the HPMA-coated vector. Furthermore, reduced neutralization of the coated vector by neutralizing antibodies was shown. The selectivity of the ACPP retargeting was demonstrated by monitoring transgene expression in vitro. While human bronchial epithelial cells exhibited only diminished transgene expression, three assessed tumor cell lines showed significantly higher expression [107].

Conclusively, HPMA copolymers were shown to successfully and efficiently detarget Ad vectors from their natural target cells. More importantly, HPMA polymer shielding was able to protect the vectors against neutralizing blood components while introducing possibilities for vector retargeting by the introduction of appropriate ligands.

Nevertheless, HPMA coating must be designed carefully as it is able to coat the vector particles, resulting in a fully passive targeting that might not be sufficient for a successful tumor therapy [108]. In contrast, incomplete HPMA modification of the vectors that still ensures maintenance of the natural tropism by the binding of vector proteins to host cell receptors (e.g., fiber knob binding to CAR) likewise offers an unfavorable possibility of neutralization of the vectors by immune components [84]. Therefore, the most promising solution for vector design should combine both genetic and chemical modifications. While the polymer shielding is best for the repeated delivery of the vectors without triggering immune responses against the vector, genetical modification of the capsid could close the remaining gaps in the shielding or retargeting by polymers. This may also offer an option for a two-step targeting for oncolytic vectors. These vectors may lose their polymer coating upon transduction mediated by the first chemically introduced targeting molecule to present a second genetically introduced targeting molecule after replication in tumor cells [82].

5. Shielding with Polyethylene Glycol (PEG)

5.1. Chemistry of PEGylation

An alternative to HPMA is polyethylene glycol (PEG). PEG molecules are synthetic yet biocompatible compounds of repeating ethylene glycol units, widely used in cosmetic and pharmaceutical industries. They are hydrophilic and can be produced in various lengths up to 40 kDa in branched or linear forms. PEG molecules are able to attach to proteins with covalent or noncovalent bonds. Due to their low toxicity and immunogenicity, the conjugated protein acquires new potentials such as an increased half-life and solubility when administered in vivo without loss of its function [109,110]. Thus, protein PEGylation has been proven, for many years now, to be a good and robust approach in the field of targeted drug delivery systems. To date, numerous drugs with PEGylated compounds have been approved. For example, PEGylated interferons have been used for therapies of multiple sclerosis [111] and hepatitis C [112]. Pegademase bovine has been applied in enzyme replacement therapy for severe combined immunodeficiency [113], while PEGylated peptides have been used for acromegaly [114,115].

The chemistry behind the coupling of PEG moieties to proteins is comparable to the coupling of HPMA moieties described above. Here, reactive groups such as maleimide (mal-) or succinimidyl propionate (SPA-) covalently bound to PEG molecules can couple such reactive PEG molecules to ε-amino groups of naturally occurring lysine residues or -SH groups of genetically introduced cysteine residues within the viral capsid. Additionally, linkers can contribute in the vector's performance in vivo [83]. PEG molecules can be engineered as monofunctional moieties with one reactive group or as bifunctional moieties offering two different reactive groups at both ends. Bifunctional PEGs can be further considered as homobifunctional when the same reactive groups are used or heterobifunctional if these are different [83]. An overview of chemical modification strategies with PEG is shown in Figure 2. Since the structure and composition of the Ad capsid are described in great detail, the density of the covalently coupled PEG shield can easily be regulated by the number of PEG molecules used.

In general, PEGylation, such as HPMAylation, can be performed under mild conditions, such as room temperature and at a pH of 7.4–8.2 [83]. Since the PEGylation of therapeutic protein compounds has been shown to significantly reduce immunogenicity and antigenicity [116,117], coupling PEG molecules to viral capsid proteins might also contribute to viral vector gene therapy. By PEGylating the major capsid proteins fiber, penton and hexon, vectors can be shielded from various interactions with blood and immune components, thus overcoming the most prominent obstacles in their application in gene therapy.

One of the most pivotal attempts of chemical capsid modification via PEGylation was performed by O' Riordan et al. in 1999 [118]. In fact, that was one of the very first attempts to chemically alter the Ad capsid structure. The authors engineered a whole capsid shield using covalently attached tresyl-monomethoxypolyethylene glycol (T-MPEG) molecules to ε-amino groups of capsid lysine residues in first generation Ad5 vectors. The T-MPEG-shielded vectors remained bioactive in comparison to cyanuric chloride-activated MPEG (CC-MPEG) or succinimidyl succinate MPEG (SS-MPEG) vectors, and, in in vitro studies, T-MPEG vectors reached higher gene expression levels in the presence of neutralizing antibodies. After intranasal immunization of mice with T-MPEG Ad2-βgal, MPEG Ad2-βgal, or unshielded Ad2-βgal, expression of the transgene product in the lung tissue cells of mice treated with T-MPEG vectors was highest, also indicating the efficient evasion of Ad neutralization in vivo due to this PEGylation strategy [118].

Bioresponsive PEGylation Approaches

While PEGylation protects Ad vectors from host immune and blood components (see Section 5.3), chemical modification of the capsid may interfere with the biological functionality of the vectors. Hence, the reactive group of a PEG molecule defines not only its reactivity towards functional groups on the vector capsid, but also the bioresponsiveness of the PEG–vector capsid bond. A well-designed reversible bond can thereby be bioresponsive, i.e., reversible under certain physiological conditions (e.g., in low pH in endosomes or in highly reducing cell compartments). The environmentally triggered release of the PEGs can restore the initial biological function of the capsid after delivery of the protected PEGylated vectors to the tissue of interest.

As described above, this concept was presented by Carlisle et al. for HPMA coated DNA–PEI complexes and was engineered with a reducible disulfide bond between HPMA and the DNA–PEI complex [99]. Moreover, other research groups developed a bioresponsive HPMA coating of Ad vector capsids which facilitated an extracellular shielding of vector particles (see Section 4.2), whereas the natural intracellular transport of the vector particles was ensured by shedding of the HPMA shield upon cell entry [84,91]

Thus far, only little is known about bioresponsive PEGylated Ad vectors and their ability to shield against immune and blood components. Espenlaub et al. presented an in vitro laser scanning microscopy assay with fluorescent dyes. In this assay, the PEGs or dyes used were either irreversibly or pH- or redox-sensitively bound to Ad capsids, and their intracellular fate was followed. While irreversibly bound PEGs or dyes coupled to amine groups impaired intracellular trafficking, pH-sensitive hydrazone-bound PEGs had no influence on the intracellular trafficking of the Ad-vectors [98].

The field of bioresponsive PEGylation of Ad vectors is poorly studied especially with regard to shielding against immune and blood components. Bioresponsive PEGylation should efficiently protect Ad vectors extracellularly from host blood components while ensuring vector transport intracellularly and preserving the natural transduction efficiency.

5.2. Detargeting and Retargeting with PEGylation

Many research groups used PEGylation strategies not only for detargeting but also for retargeting PEGylated Ad vectors to tissues or cells of choice, such as tumor cells. Tissue- or organ-specific targeting is a major challenge towards Ad vector-mediated gene therapy or virotherapy.

The group of Lanciotti et al. retargeted vectors PEGylated with heterobifunctional T-MPEG-mal coupled with FGF-2 protein to tumor cells via the FGF receptor [119]. In vitro experiments showed high transduction levels in cells of the ovarian cancer cell line SKOV-3, which express the FGF-receptor. Further in vivo experiments showed similar results: vectors were successfully detargeted from off-target organs such as the liver and spleen and retargeted to intraperitoneal tumor cells. The immunological response to PEGylated vectors was decreased, as in vitro; PEGylated vectors were neutralized less; and after treatment of BALB/c mice with PEGylated vectors, lower Th-1 and -2 responses of splenocytes were detected compared to the administration of un-PEGylated vectors [119].

However, PEGylation goes beyond masking vectors from antibodies and immune cells. Hofherr et al. showed that PEGylated vectors reduced the adverse effect of thrombocytopenia and disseminated intravascular coagulation as well as the elevation of D-dimers in mice [120].

As mentioned in the first part of this review (see Section 2.2.2), the idea of the geneti-chemical approach was first established by Kreppel et al. [79,83]. A chemically reactive group is introduced at specific solvent-exposed sites of one or more major capsid proteins via minor genetic modifications. The added reactivity can be further used to specifically mask the mutated protein with PEG moieties. This approach can be used for detargeting as well as for retargeting purposes.

A great example of site-specific PEGylation was achieved with the introduction of a peptide containing a central cysteine residue into the HI loop of fiber. The cysteine -SH groups could be further specifically and covalently attached to mal-PEG. Moreover, this study showed that retargeting of the vector to the transferrin receptor, which is highly expressed in tumor cells, was possible by coupling full-length transferrin to the other end of the heterobifunctional PEG [79].

The introduction of a cysteine residue in HVR5 of hexon, followed by thiol-based coupling with small PEG moieties, prevented the binding of coagulation FX to the viral capsid and showed decreased hepatotropism. The vector was then successfully retargeted to liver cells by chemically coupling transferrin to hexon [80].

In a study of Krutzke et al., first-generation Ad vectors were generated with a mutation in HVR7 and in fiber in order to ablate the vector's ability to bind to the coagulation FX and CAR receptor, respectively [37]. Another point mutation resulting in a cysteine residue in HVR1 enabled site-specific PEGylation of this HVR. Although FX binding was ablated, hepatocyte transduction by 2 kDa- and 5 kDa-PEGylated Ad-HVR1-ΔCAR-ΔFX was shown in BALB/c mice as well as in antibody-deficient JHD mice. These data indicated an FX and natural antibody-independent mechanism of liver transduction. This study also provided evidence that the specific PEGylation of Ad-HVR1-ΔFX replaced the FX shield and protected the viral particles from natural antibodies and complement factors. Furthermore, PEGylated Ad-HVR1-ΔFX prevented the binding of C3b, thus protecting against opsonization through complement as well as the uptake by macrophages through scavenger receptors [37].

Site-specific mutations and further PEGylation of capsid proteins, particularly hexon, elucidated new capabilities of this specific protein and its significance in the vector's pathogenicity and biology. Future studies applying or even improving this technique can enlighten the still unknown interactions between humans and Ads and develop new approaches for overcoming major obstacles in Ad gene therapy.

5.3. Immune Evasion with PEGylation and PEGylated Ad Vectors as Vaccines

The main goal of capsid shielding is to prevent interactions of vectors with components of the immune system, such as pre-existing neutralizing antibodies, natural IgM, and opsonization, through complement and macrophages. Adenoviral immunity in the human population, due to natural infections, is one of the main obstacles in adeno-based systemic applications. Furthermore, since globally a large fraction of the human population is currently vaccinated with Ad-based vectors, strategies for reapplication of these vectors will be mandatory. Therefore, the need for optimizing PEGylation is highly important, and promising results have been shown from various research groups.

Of great importance in this field are the studies of Croyle et al. [121,122] and, as mentioned above, the studies of O'Riordan et al. [118]. Croyle et al. demonstrated that fully MPEGylated vectors with different reactive groups, such as CC-MPEG, SS-MPEG, and T-MPEG, resulted in reduced cytotoxic lymphocyte responses in vivo. Although shielded vectors evaded humoral and adaptive immune responses, the readministration of the same PEGylated vector showed less efficient gene expression, indicating that natural antibodies against Ad vectors could potentially still become an additional obstacle for Ad gene therapy. However, the researchers furthermore showed that using different activation groups for PEG coupling resulted in the same gene expression levels after readminstration [118,121,122].

Wortmann et al. modified capsids of Ad vectors expressing the hepatitis B surface antigen (HBsAg) with SPA-PEG 20 kDa [123]. Surprisingly, these excessively PEGylated vaccine vectors induced strong anti-transgene product-directed (anti HBsAg) humoral and cellular responses, despite detargeting vectors from CAR. Even in the presence of antibodies against Ad5 after plasma transfer, the PEGylation allowed for evasion from these anti-Ad immunity in vivo. PEGylated vectors when compared to their un-PEGylated counterparts, achieved a significantly higher number of HBsAg $CD8^+/IFN\text{-}\gamma^+$ positive T-cells. These findings were consistent with the results of Mok et al. [97]. In this study, SPA-PEGylated vectors appeared to be less phagocytosed in vitro by Raw264 macrophages, evaluated via qPCR. Furthermore, in vivo analyses regarding liver specific macrophages showed that the abovementioned vectors retained the uptake through hepatocytes but with a decreased uptake by liver specific macrophages like Kupffer cells [97].

An important study by Khare et al. using geneti-chemical shielding revealed more information about the contribution of the different hexon HVRs to the virus' biology [124]. They introduced cysteine residues in all hexon HVRs and blocked them by PEGylation to investigate the function of this important capsid protein. Transduction in vitro was not affected, but the authors provided evidence that the PEGylation of HVR 1,2,5,7 enhanced liver transduction in mice in conjunction with reduced recognition by Kupffer cells [124].

The immunoevasion potentials of PEGylated vectors for priming and boosting vaccination were also investigated by Weaver and Barry et al. [125]. Ad-immune mice were injected with PEGylated Ads (PEG sizes upt to 35 kDa) intramuscular (IM) or intranasally. Although the vectors initially appeared inactive in vitro, transduction and transgene delivery in vivo were possible, thus confirming the findings by Wortmann et al. [123]. Additional experiments examined the effects of PEGylated vectors after boosting immunization. PEGylated vectors elicited high titers of antibodies against the transgene product similar to those achieved by the unmodified vector after priming, indicating once more that this polymer shield has the ability to medate evasion of the vector particles from neutralizing antibodies. After priming immunization of Ad naïve mice, T-cell responses were also assessed. Intranasally administered 35 kDa PEG-Ad vaccines generated higher levels of neutralizing antibodies when compared to the 5 kDa PEG-Ad. Interestingly, further analyses using 5 kDa PEG-Ad for priming demonstrated that after boosting with either unmodified or modified vectors, strong T-cell responses could be elicited. IM administration of all vectors reached high neutralizing antibody levels, but after boosting vaccination, strong T-cell responses were also achieved. In general, the authors suggested that T-cell and antibody responses could be enhanced after boosting immunization with PEGylated vectors and that priming vaccination with modified vectors is able to induce higher immune responses when compared with unmodified vectors [125].

5.4. Preventing Ad-Mediated Toxicity with PEGylation

Ad vectors can be utilized in all fields of virotherapy. They can successfully deliver genes for gene therapy and are able to induce significant immunity against pathogens when used as genetic vaccines. However, the immediate Ad toxicity remains a limiting factor for their use. Cell damage causes not only tissue damage, but in vivo experiments have shown that this cell damage elicited the activation of the complement cascade and therefore further immune responses [126].

The value of PEGylation in reducing tissue-related toxicity was investigated by Mok et al. [97]. In this study, whole capsid SPA-PEGylated first-generation vectors (FG) resulted in low transduction levels in vitro. This was in contrast with in vivo data, which demonstrated similar distribution and transduction levels of the SPA-PEGylated and unmodified vectors. Furthermore, FG-PEGylated vectors resulted in lower interleukine-6 (IL) levels, measured at different time points, both in vivo and in vitro in a macrophage cell line. In particular, IL-6 release was reduced by 10-fold in the first 48 h of infection. Moreover, in mice, the administration of PEGylated and non-PEGylated vectors did not cause an elevation of the liver enzymes alanine aminotransferase (ALT) and aspartate aminotransferase (AST), which are highly released in the case of liver damage, indicating that PEG is not liver toxic.

Interestingly, the researchers also looked into the PEGylation of helper-dependent HD-Ad vectors IL-6 responses and especially liver damage were reduced when using PEGylated HD-vectors suggesting that these vectors could be safer for in vivo administration [97]. Similarly, another study showed that the secretion of other cytokines, such as IL-2, tumor necrosis factor-α, and IL-6, was also significantly lower after vector administration, and Ad-induced thrombocytopenia in mice was not observed [127]. These data underline the safety profile of PEGylated HD-vectors.

6. Conclusions

To date, a wide variety of technologies to modify Ad vector capsids has been developed based on scientific knowledge about desirable and undesirable vector–host interactions. Ad vectors are already being successfully used as genetic vaccines; however, their use as classic gene transfer vectors is still hampered because none of the technologies summarized here was sufficient to completely overcome the hurdles imposed by the complex network of vector–host interactions. In particular, the dose-dependent toxicity of systemically delivered Ad vectors and problems arising from pre-existing anti-vector immunity have not been solved so far. Presumably, only a combination of genetics and chemistry might allow for the development of safe and efficient Ad-based vectors in the future.

Genetic modifications typically yield highly defined and uniform particles and can be used to generate vectors detargeted from their primary and secondary receptors. However, for each adenovirus type, the precise sites of receptor interactions and the precise sites involved in unwanted interactions (for example, with blood components) must be identified to the amino acid level prior to successful genetic detargeting. The same holds true for potential insertion sites to retarget the vector particles by peptide ligands. Thus far, insertion sites have been characterized to great detail for Ad5-based vectors, but only very little is known about such sites for rare Ad types. We think that in the near future an in-depth characterization of rare Ad types, including the description of insertion sites/detargeting mutations, will be required to increase the chances for clinically successful Ad vector development. The example of Atasheva et al. [70] discussed above is pioneering for Ad5 and demonstrates that the substantial modulation of Ad vector tropism in vivo is feasible by multiple genetic modifications. However, the problem of adaptive antibody responses remains unsolved.

In addition, it seems likely that in the near future, novel producer cell lines will be required in order to produce such genetically modified de- and retargeted vector particles. It is well known that the generation of cell lines suitable for Ad vector production is anything but trivial, and research projects with a focus on the generation of cell lines for the production of next generation Ad vectors should be fostered not only on an industrial but also on an academic basis.

It is currently unclear if genetic modifications alone will suffice to circumvent the barriers for repeated and systemic Ad vector delivery to patients. Here, chemical modifications offer the possibility of shielding the vector particles. The advantage of shielding by chemical modifications is that conventional producer cell lines can be used, and the sites to be shielded do not need to be described in detail. However, there is a risk that the shielding of larger capsid areas will interfere with biological vector functions, such as intracellular capsid transport and/or nuclear import of the vector DNA. It has become clear in the recent years that dense and tight shielding bears the risk of generating inactive particles that cannot undergo the dynamic structural changes that are required for a successful transduction process. Therefore, bioresponsive shielding modes have been developed that allow a reversible shield to be generated that is removed from the capsid upon cell entry (for example). In addition, a more fine-tuned shielding, such as the thiol-directed geneti-chemical capsid shielding, has been shown to be advantageous, simply due to modifying fewer capsid sites compared to amine-directed large area shielding. However, similar to successful genetic modifications, successful geneti-chemical modifications depend on the precise knowledge of interaction sites on the capsid and might suffer from incomplete shielding against antibodies. We think that special focus should be put on the development of novel bioresponsive coupling modes, which currently seem to be a potential solution for pre-existing humoral immunity. Obviously, both the coupling mode and the nature of the shielding molecule itself are very important. While PEG and HPMA are important examples, novel shielding moieties based, for example, on carbohydrates or even lipids might be developed in the near future.

The last decades of Ad vector development have taught us that there will not be one "magic bullet" that fits all needs. The capsids of second-generation Ad vector-based vaccines to be delivered intramuscularly will likely look different than those of future oncolytic Ads for virotherapy that are delivered intravenously. The further development of existing technologies and novel combinations are most likely to have success in specific applications, may it be in vaccine development, virotherapy or even classic gene therapy.

Author Contributions: D.W., D.P., and G.K. contributed equally to the manuscript. D.W., D.P., and G.K. wrote the manuscript. D.W. created Figures 1 and 2. D.P. created Figure 3. G.K. created Table 1. C.H. wrote the introduction. F.K. outlined the structure of the manuscript and wrote the conclusion and the abstract. F.J., C.H. and F.K. revised the manuscript. All authors have read and agreed to the published version of the manuscript.

Funding: This research received no external funding.

Institutional Review Board Statement: Not applicable.

Informed Consent Statement: Not applicable.

Data Availability Statement: Not applicable.

Acknowledgments: This review was supported by the Center of Biomedical Education and Research (ZBAF) of the University of Witten/Herdecke. D.W. and G.K. received a fellowship for their PhD studies of biomedicine at the University Witten/Herdecke.

Conflicts of Interest: The authors declare no conflict of interest.

References

1. Shaw, A.R.; Suzuki, M. Immunology of Adenoviral Vectors in Cancer Therapy. *Mol. Ther. Methods Clin. Dev.* **2019**, *15*, 418–429. [CrossRef] [PubMed]
2. Bulcha, J.T.; Wang, Y.; Ma, H.; Tai, P.W.L.; Gao, G. Viral vector platforms within the gene therapy landscape. *Signal Transduct. Target. Ther.* **2021**, *6*, 1–24. [CrossRef]
3. Folegatti, P.M.; Ewer, K.J.; Aley, P.K.; Angus, B.; Becker, S.; Belij-Rammerstorfer, S.; Bellamy, D.; Bibi, S.; Bittaye, M.; Clutterbuck, E.A.; et al. Safety and immunogenicity of the ChAdOx1 nCoV-19 vaccine against SARS-CoV-2: A preliminary report of a phase 1/2, single-blind, randomised controlled trial. *Lancet* **2020**, *396*, 467–478. [CrossRef]
4. Ewer, K.J.; The Oxford COVID Vaccine Trial Group; Barrett, J.R.; Belij-Rammerstorfer, S.; Sharpe, H.; Makinson, R.; Morter, R.; Flaxman, A.; Wright, D.; Bellamy, D.; et al. T cell and antibody responses induced by a single dose of ChAdOx1 nCoV-19 (AZD1222) vaccine in a phase 1/2 clinical trial. *Nat. Med.* **2021**, *27*, 270–278. [CrossRef]
5. Logunov, D.Y.; Dolzhikova, I.V.; Zubkova, O.V.; Tukhvatullin, A.I.; Shcheblyakov, D.V.; Dzharullaeva, A.S.; Grousova, D.M.; Erokhova, A.S.; Kovyrshina, A.V.; Botikov, A.G.; et al. Safety and immunogenicity of an rAd26 and rAd5 vector-based heterologous prime-boost COVID-19 vaccine in two formulations: Two open, non-randomised phase 1/2 studies from Russia. *Lancet* **2020**, *396*, 887–897. [CrossRef]
6. Logunov, D.Y.; Dolzhikova, I.V.; Shcheblyakov, D.V.; Tukhvatulin, A.I.; Zubkova, O.V.; Dzharullaeva, A.S.; Kovyrshina, A.V.; Lubenets, N.L.; Grousova, D.M.; Erokhova, A.S.; et al. Safety and efficacy of an rAd26 and rAd5 vector-based heterologous prime-boost COVID-19 vaccine: An interim analysis of a randomised controlled phase 3 trial in Russia. *Lancet* **2021**, *397*, 671–681. [CrossRef]
7. Van Oostrum, J.; Burnett, R.M. Molecular composition of the adenovirus type 2 virion. *J. Virol.* **1985**, *56*, 439–448. [CrossRef]
8. Stewart, P.; Fuller, S.; Burnett, R. Difference imaging of adenovirus: Bridging the resolution gap between X-ray crystallography and electron microscopy. *EMBO J.* **1993**, *12*, 2589–2599. [CrossRef]
9. Stewart, P.L.; Burnett, R.M.; Cyrklaff, M.; Fuller, S.D. Image reconstruction reveals the complex molecular organization of adenovirus. *Cell* **1991**, *67*, 145–154. [CrossRef]
10. Shirley, J.L.; de Jong, Y.P.; Terhorst, C.; Herzog, R.W. Immune Responses to Viral Gene Therapy Vectors. *Mol. Ther.* **2020**, *28*, 709–722. [CrossRef]
11. Atasheva, S.; Yao, J.; Shayakhmetov, D.M. Innate immunity to adenovirus: Lessons from mice. *FEBS Lett.* **2019**, *593*, 3461–3483. [CrossRef]
12. Shayakhmetov, D.M.; Gaggar, A.; Ni, S.; Li, Z.-Y.; Lieber, A. Adenovirus Binding to Blood Factors Results in Liver Cell Infection and Hepatotoxicity. *J. Virol.* **2005**, *79*, 7478–7491. [CrossRef]
13. Doronin, K.; Flatt, J.; Di Paolo, N.C.; Khare, R.; Kalyuzhniy, O.; Acchione, M.; Sumida, J.P.; Ohto, U.; Shimizu, T.; Akashi-Takamura, S.; et al. Coagulation Factor X Activates Innate Immunity to Human Species C Adenovirus. *Science* **2012**, *338*, 795–798. [CrossRef]
14. Schnell, M.A.; Zhang, Y.; Tazelaar, J.; Gao, G.-P.; Yu, Q.; Qian, R.; Chen, S.-J.; Varnavski, A.N.; LeClair, C.; Raper, S.E.; et al. Activation of Innate Immunity in Nonhuman Primates Following Intraportal Administration of Adenoviral Vectors. *Mol. Ther.* **2001**, *3*, 708–722. [CrossRef]
15. Muruve, D.A.; Cotter, M.J.; Zaiss, A.K.; White, L.R.; Liu, Q.; Chan, T.; Clark, S.A.; Ross, P.J.; Meulenbroek, R.A.; Maelandsmo, G.M.; et al. Helper-Dependent Adenovirus Vectors Elicit Intact Innate but Attenuated Adaptive Host Immune Responses In Vivo. *J. Virol.* **2004**, *78*, 5966–5972. [CrossRef]
16. Allen, R.J.; Byrnes, A.P. Interaction of adenovirus with antibodies, complement, and coagulation factors. *FEBS Lett.* **2019**, *593*, 3449–3460. [CrossRef]
17. Alemany, R.; Suzuki, K.; Curiel, D.T. Blood clearance rates of adenovirus type 5 in mice. *J. Gen. Virol.* **2000**, *81*, 2605–2609. [CrossRef]
18. Xu, Z.; Tian, J.; Smith, J.S.; Byrnes, A.P. Clearance of Adenovirus by Kupffer Cells Is Mediated by Scavenger Receptors, Natural Antibodies, and Complement. *J. Virol.* **2008**, *82*, 11705–11713. [CrossRef]
19. Moskalenko, M.; Chen, L.; van Roey, M.; Donahue, B.A.; Snyder, R.O.; McArthur, J.G.; Patel, S.D. Epitope Mapping of Human Anti-Adeno-Associated Virus Type 2 Neutralizing Antibodies: Implications for Gene Therapy and Virus Structure. *J. Virol.* **2000**, *74*, 1761–1766. [CrossRef]
20. Zhi, Y.; Figueredo, J.; Kobinger, G.P.; Hagan, H.; Calcedo, R.; Miller, J.R.; Gao, G.; Wilson, J.M. Efficacy of Severe Acute Respiratory Syndrome Vaccine Based on a Nonhuman Primate Adenovirus in the Presence of Immunity Against Human Adenovirus. *Hum. Gene Ther.* **2006**, *17*, 500–506. [CrossRef]
21. Perreau, M.; Pantaleo, G.; Kremer, E.J. Activation of a dendritic cell–T cell axis by Ad5 immune complexes creates an improved environment for replication of HIV in T cells. *J. Exp. Med.* **2008**, *205*, 2717–2725. [CrossRef]
22. Cotter, M.J.; Zaiss, A.K.; Muruve, D.A. Neutrophils Interact with Adenovirus Vectors via Fc Receptors and Complement Receptor. *J. Virol.* **2005**, *79*, 14622–14631. [CrossRef]
23. Zaiss, A.K.; Vilaysane, A.; Cotter, M.J.; Clark, S.A.; Meijndert, H.C.; Colarusso, P.; Yates, R.M.; Petrilli, V.; Tschopp, J.; Muruve, D.A. Antiviral Antibodies Target Adenovirus to Phagolysosomes and Amplify the Innate Immune Response. *J. Immunol.* **2009**, *182*, 7058–7068. [CrossRef]

24. McEwan, W.A.; Tam, J.; Watkinson, R.E.; Bidgood, S.; Mallery, D.L.; James, L.C. Intracellular antibody-bound pathogens stimulate immune signaling via the Fc receptor TRIM21. *Nat. Immunol.* **2013**, *14*, 327–336. [CrossRef]
25. Khare, R.; Hillestad, M.L.; Xu, Z.; Byrnes, A.P.; Barry, M.A. Circulating Antibodies and Macrophages as Modulators of Adenovirus Pharmacology. *J. Virol.* **2013**, *87*, 3678–3686. [CrossRef]
26. Bottermann, M.; Foss, S.; van Tienen, L.M.; Vaysburd, M.; Cruickshank, J.; O'Connell, K.; Clark, J.; Mayes, K.; Higginson, K.; Hirst, J.C.; et al. TRIM21 mediates antibody inhibition of adenovirus-based gene delivery and vaccination. *Proc. Natl. Acad. Sci. USA* **2018**, *115*, 10440–10445. [CrossRef]
27. Groß, R.; Zanoni, M.; Seidel, A.; Conzelmann, C.; Gilg, A.; Krnavek, D.; Erdemci-Evin, S.; Mayer, B.; Hoffmann, M.; Pöhlmann, S.; et al. Heterologous ChAdOx1 NCoV-19 and BNT162b2 Prime-Boost Vaccination Elicits Potent Neutralizing Antibody Responses and T Cell Reactivity. *medRxiv* **2021**. [CrossRef]
28. Coyne, C.B.; Bergelson, J.M. CAR: A virus receptor within the tight junction. *Adv. Drug Deliv. Rev.* **2005**, *57*, 869–882. [CrossRef]
29. Wickham, T.J.; Mathias, P.; Cheresh, D.A.; Nemerow, G.R. Integrins $\alpha v\beta 3$ and $\alpha v\beta 5$ promote adenovirus internalization but not virus attachment. *Cell* **1993**, *73*, 309–319. [CrossRef]
30. Wu, E.; Nemerow, G.R. Virus yoga: The role of flexibility in virus host cell recognition. *Trends Microbiol.* **2004**, *12*, 162–169. [CrossRef]
31. Meier, O.; Boucke, K.; Hammer, S.V.; Keller, S.; Stidwill, R.P.; Hemmi, S.; Greber, U.F. Adenovirus triggers macropinocytosis and endosomal leakage together with its clathrin-mediated uptake. *J. Cell Biol.* **2002**, *158*, 1119–1131. [CrossRef] [PubMed]
32. Wiethoff, C.M.; Wodrich, H.; Gerace, L.; Nemerow, G.R. Adenovirus Protein VI Mediates Membrane Disruption following Capsid Disassembly. *J. Virol.* **2005**, *79*, 1992–2000. [CrossRef] [PubMed]
33. Greber, U.F.; Fornerod, M. Nuclear Import in Viral Infections. *Curr. Top. Microbiol. Immunol.* **2004**, *285*, 109–138. [CrossRef]
34. Carlisle, R.C.; Di, Y.; Cerny, A.M.; Sonnen, A.F.-P.; Sim, R.; Green, N.K.; Subr, V.; Ulbrich, K.; Gilbert, R.; Fisher, K.D.; et al. Human erythrocytes bind and inactivate type 5 adenovirus by presenting Coxsackie virus-adenovirus receptor and complement receptor. *Blood* **2009**, *113*, 1909–1918. [CrossRef]
35. Rojas, L.A.; Moreno, R.; Calderón, H.; Alemany, R. Adenovirus coxsackie adenovirus receptor-mediated binding to human erythrocytes does not preclude systemic transduction. *Cancer Gene Ther.* **2016**, *23*, 411–414. [CrossRef]
36. Kirby, I.; Davison, E.; Beavil, A.J.; Soh, C.P.C.; Wickham, T.J.; Roelvink, P.W.; Kovesdi, I.; Sutton, B.J.; Santis, G. Identification of Contact Residues and Definition of the CAR-Binding Site of Adenovirus Type 5 Fiber Protein. *J. Virol.* **2000**, *74*, 2804–2813. [CrossRef]
37. Krutzke, L.; Prill, J.; Engler, T.; Schmidt, C.; Xu, Z.; Byrnes, A.; Simmet, T.; Kreppel, F. Substitution of blood coagulation factor X-binding to Ad5 by position-specific PEGylation: Preventing vector clearance and preserving infectivity. *J. Control. Release* **2016**, *235*, 379–392. [CrossRef]
38. Alemany, R.; Curiel, D.T. CAR-binding ablation does not change biodistribution and toxicity of adenoviral vectors. *Gene Ther.* **2001**, *8*, 1347–1353. [CrossRef]
39. Waddington, S.N.; McVey, J.; Bhella, D.; Parker, A.L.; Barker, K.; Atoda, H.; Pink, R.; Buckley, S.M.; Greig, J.A.; Denby, L.; et al. Adenovirus Serotype 5 Hexon Mediates Liver Gene Transfer. *Cell* **2008**, *132*, 397–409. [CrossRef]
40. Kalyuzhniy, O.; Di Paolo, N.C.; Silvestry, M.; Hofherr, S.E.; Barry, M.A.; Stewart, P.L.; Shayakhmetov, D.M. Adenovirus serotype 5 hexon is critical for virus infection of hepatocytes in vivo. *Proc. Natl. Acad. Sci. USA* **2008**, *105*, 5483–5488. [CrossRef]
41. Vigant, F.; Descamps, D.; Jullienne, B.; Esselin, S.; Connault, E.; Opolon, P.; Tordjmann, T.; Vigne, E.; Perricaudet, M.; Benihoud, K. Substitution of Hexon Hypervariable Region 5 of Adenovirus Serotype 5 Abrogates Blood Factor Binding and Limits Gene Transfer to Liver. *Mol. Ther.* **2008**, *16*, 1474–1480. [CrossRef]
42. Stasiak, A.C.; Stehle, T. Human adenovirus binding to host cell receptors: A structural view. *Med. Microbiol. Immunol.* **2019**, *209*, 325–333. [CrossRef]
43. Krasnykh, V.N.; Mikheeva, G.V.; Douglas, J.T.; Curiel, D.T. Generation of recombinant adenovirus vectors with modified fibers for altering viral tropism. *J. Virol.* **1996**, *70*, 6839–6846. [CrossRef]
44. Shayakhmetov, D.M.; Papayannopoulou, T.; Stamatoyannopoulos, G.; Lieber, A. Efficient Gene Transfer into Human CD34(+) Cells by a Retargeted Adenovirus Vector. *J. Virol.* **2000**, *74*, 2567–2583. [CrossRef]
45. Kanerva, A.; Mikheeva, G.V.; Krasnykh, V.; Coolidge, C.J.; Lam, J.T.; Mahasreshti, P.J.; Barker, S.D.; Straughn, M.; Barnes, M.N.; Alvarez, R.D.; et al. Targeting adenovirus to the serotype 3 receptor increases gene transfer efficiency to ovarian cancer cells. *Clin. Cancer Res.* **2002**, *8*, 275–280.
46. Kanerva, A.; Zinn, K.R.; Chaudhuri, T.R.; Lam, J.T.; Suzuki, K.; Uil, T.G.; Hakkarainen, T.; Bauerschmitz, G.J.; Wang, M.; Liu, B.; et al. Enhanced therapeutic efficacy for ovarian cancer with a serotype 3 receptor-targeted oncolytic adenovirus. *Mol. Ther.* **2003**, *8*, 449–458. [CrossRef]
47. Raddi, N.; Vigant, F.; Wagner-Ballon, O.; Giraudier, S.; Custers, J.; Hemmi, S.; Benihoud, K. Pseudotyping Serotype 5 Adenovirus with the Fiber from Other Serotypes Uncovers a Key Role of the Fiber Protein in Adenovirus 5-Induced Thrombocytopenia. *Hum. Gene Ther.* **2016**, *27*, 193–201. [CrossRef]
48. Krasnykh, V.; Dmitriev, I.; Mikheeva, G.; Miller, C.R.; Belousova, N.; Curiel, D.T. Characterization of an Adenovirus Vector Containing a Heterologous Peptide Epitope in the HI Loop of the Fiber Knob. *J. Virol.* **1998**, *72*, 1844–1852. [CrossRef]
49. Belousova, N.; Krendelchtchikova, V.; Curiel, D.T.; Krasnykh, V. Modulation of Adenovirus Vector Tropism via Incorporation of Polypeptide Ligands into the Fiber Protein. *J. Virol.* **2002**, *76*, 8621–8631. [CrossRef]

50. Parrott, M.; Adams, K.E.; Mercier, G.T.; Mok, H.; Campos, S.K.; Barry, M.A. Metabolically biotinylated adenovirus for cell targeting, ligand screening, and vector purification. *Mol. Ther.* **2003**, *8*, 688–700. [CrossRef]
51. Dmitriev, I.; Krasnykh, V.; Miller, C.R.; Wang, M.; Kashentseva, E.; Mikheeva, G.; Belousova, N.; Curiel, D.T. An Adenovirus Vector with Genetically Modified Fibers Demonstrates Expanded Tropism via Utilization of a Coxsackievirus and Adenovirus Receptor-Independent Cell Entry Mechanism. *J. Virol.* **1998**, *72*, 9706–9713. [CrossRef] [PubMed]
52. Wu, H.; Seki, T.; Dmitriev, I.; Uil, T.; Kashentseva, E.; Han, T.; Curiel, D.T. Double Modification of Adenovirus Fiber with RGD and Polylysine Motifs Improves Coxsackievirus–Adenovirus Receptor-Independent Gene Transfer Efficiency. *Hum. Gene Ther.* **2002**, *13*, 1647–1653. [CrossRef] [PubMed]
53. Wu, H.; Han, T.; Lam, J.T.; A Leath, C.; Dmitriev, I.; Kashentseva, E.; Barnes, M.N.; Alvarez, R.D.; Curiel, D.T. Preclinical evaluation of a class of infectivity-enhanced adenoviral vectors in ovarian cancer gene therapy. *Gene Ther.* **2004**, *11*, 874–878. [CrossRef] [PubMed]
54. Kimball, K.J.; Preuss, M.A.; Barnes, M.N.; Wang, M.; Siegal, G.P.; Wan, W.; Kuo, H.; Saddekni, S.; Stockard, C.R.; Grizzle, W.E.; et al. A Phase I Study of a Tropism-Modified Conditionally Replicative Adenovirus for Recurrent Malignant Gynecologic Diseases. *Clin. Cancer Res.* **2010**, *16*, 5277–5287. [CrossRef] [PubMed]
55. Kim, K.H.; Dmitriev, I.; O'Malley, J.P.; Wang, M.; Saddekni, S.; You, Z.; Preuss, M.A.; Harris, R.D.; Aurigemma, R.; Siegal, G.P.; et al. A Phase I Clinical Trial of Ad5.SSTR/TK.RGD, a Novel Infectivity-Enhanced Bicistronic Adenovirus, in Patients with Recurrent Gynecologic Cancer. *Clin. Cancer Res.* **2012**, *18*, 3440–3451. [CrossRef]
56. MacLeod, S.H.; Elgadi, M.M.; Bossi, G.; Sankar, U.; Pisio, A.; Agopsowicz, K.; Sharon, D.; Graham, F.L.; Hitt, M.M. HER3 targeting of adenovirus by fiber modification increases infection of breast cancer cells in vitro, but not following intratumoral injection in mice. *Cancer Gene Ther.* **2012**, *19*, 888–898. [CrossRef]
57. Behr, M.; Kaufmann, J.K.; Ketzer, P.; Engelhardt, S.; Muck-Hausl, M.; Okun, P.M.; Petersen, G.; Neipel, F.; Hassel, J.C.; Ehrhardt, A.; et al. Adenoviruses Using the Cancer Marker EphA2 as a Receptor In Vitro and In Vivo by Genetic Ligand Insertion into Different Capsid Scaffolds. *PLoS ONE* **2014**, *9*, e95723. [CrossRef]
58. Dmitriev, I.P.; Kashentseva, E.A.; Curiel, D.T. Engineering of Adenovirus Vectors Containing Heterologous Peptide Sequences in the C Terminus of Capsid Protein IX. *J. Virol.* **2002**, *76*, 6893–6899. [CrossRef]
59. Le, L.P.; Everts, M.; Dmitriev, I.P.; Davydova, J.G.; Yamamoto, M.; Curiel, D.T. Fluorescently Labeled Adenovirus with PIX-EGFP for Vector Detection. *Mol. Imaging* **2004**, *3*, 105–116. [CrossRef]
60. Wu, H.; Han, T.; Belousova, N.; Krasnykh, V.; Kashentseva, E.; Dmitriev, I.; Kataram, M.; Mahasreshti, P.J.; Curiel, D.T. Identification of Sites in Adenovirus Hexon for Foreign Peptide Incorporation. *J. Virol.* **2005**, *79*, 3382–3390. [CrossRef]
61. Vigne, E.; Mahfouz, I.; Dedieu, J.-F.; Brie, A.; Perricaudet, M.; Yeh, P. RGD Inclusion in the Hexon Monomer Provides Adenovirus Type 5-Based Vectors with a Fiber Knob-Independent Pathway for Infection. *J. Virol.* **1999**, *73*, 5156–5161. [CrossRef]
62. Xu, Z.; Qiu, Q.; Tian, J.; Smith, J.S.; Conenello, G.M.; Morita, T.; Byrnes, A.P. Coagulation factor X shields adenovirus type 5 from attack by natural antibodies and complement. *Nat. Med.* **2013**, *19*, 452–457. [CrossRef]
63. Watkins, S.; Mesyanzhinov, V.; Kurochkina, L.; Hawkins, R. The 'Adenobody' Approach to Viral Targeting: Specific and Enhanced Adenoviral Gene Delivery. *Gene Ther.* **1997**, *4*, 1004–1012. [CrossRef]
64. Chen, C.Y.; May, S.M.; Barry, M.A. Targeting Adenoviruses with Factor X–Single-Chain Antibody Fusion Proteins. *Hum. Gene Ther.* **2010**, *21*, 739–749. [CrossRef]
65. Campos, S.K.; Parrott, M.B.; Barry, M.A. Avidin-based targeting and purification of a protein IX-modified, metabolically biotinylated adenoviral vector. *Mol. Ther.* **2004**, *9*, 942–954. [CrossRef]
66. Dmitriev, I.; Kashentseva, E.; Rogers, B.E.; Krasnykh, V.; Curiel, D.T. Ectodomain of Coxsackievirus and Adenovirus Receptor Genetically Fused to Epidermal Growth Factor Mediates Adenovirus Targeting to Epidermal Growth Factor Receptor-Positive Cells. *J. Virol.* **2000**, *74*, 6875–6884. [CrossRef]
67. Roberts, D.M.; Nanda, A.; Havenga, M.J.E.; Abbink, P.; Lynch, D.M.; Ewald, B.A.; Liu, J.; Thorner, A.R.; Swanson, P.E.; Gorgone, D.A.; et al. Hexon-chimaeric adenovirus serotype 5 vectors circumvent pre-existing anti-vector immunity. *Nature* **2006**, *441*, 239–243. [CrossRef]
68. Bradley, R.R.; Maxfield, L.F.; Lynch, D.M.; Iampietro, M.J.; Borducchi, E.N.; Barouch, D.H. Adenovirus Serotype 5-Specific Neutralizing Antibodies Target Multiple Hexon Hypervariable Regions. *J. Virol.* **2011**, *86*, 1267–1272. [CrossRef]
69. Rojas, L.A.; Condezo, G.N.; Moreno, R.; Fajardo, C.A.; Arias-Badia, M.; Martín, C.S.; Alemany, R. Albumin-binding adenoviruses circumvent pre-existing neutralizing antibodies upon systemic delivery. *J. Control. Release* **2016**, *237*, 78–88. [CrossRef]
70. Atasheva, S.; Emerson, C.C.; Yao, J.; Young, C.; Stewart, P.L.; Shayakhmetov, D.M. Systemic cancer therapy with engineered adenovirus that evades innate immunity. *Sci. Transl. Med.* **2020**, *12*, eabc6659. [CrossRef]
71. Alba, R.; Bosch, A.; Chillon, M. Gutless adenovirus: Last-generation adenovirus for gene therapy. *Gene Ther.* **2005**, *12*, S18–S27. [CrossRef]
72. Rosewell, A.; Vetrini, F.; Ng, P. Helper-Dependent Adenoviral Vectors. *J. Genet. Syndr. Gene Ther.* **2011**, (Suppl. 5), 1. [CrossRef]
73. Parks, R.J.; Graham, F.L. A helper-dependent system for adenovirus vector production helps define a lower limit for efficient DNA packaging. *J. Virol.* **1997**, *71*, 3293–3298. [CrossRef]
74. Parks, R.J.; Bramson, J.L.; Wan, Y.; Addison, C.L.; Graham, F.L. Effects of Stuffer DNA on Transgene Expression from Helper-Dependent Adenovirus Vectors. *J. Virol.* **1999**, *73*, 8027–8034. [CrossRef]

75. Cerullo, V.; Seiler, M.P.; Mane, V.; Brunetti-Pierri, N.; Clarke, C.; Bertin, T.K.; Rodgers, J.R.; Lee, B. Toll-like Receptor 9 Triggers an Innate Immune Response to Helper-dependent Adenoviral Vectors. *Mol. Ther.* **2007**, *15*, 378–385. [CrossRef]
76. Liu, J. Helper virus-free gutless adenovirus (HF-GLAd): A new platform for gene therapy. *BMB Rep.* **2020**, *53*, 565–575. [CrossRef]
77. Brunetti-Pierri, N.; Ng, T.; Iannitti, D.; Cioffi, W.; Stapleton, G.; Law, M.; Breinholt, J.; Palmer, D.; Grove, N.; Rice, K.; et al. Transgene Expression up to 7 Years in Nonhuman Primates Following Hepatic Transduction with Helper-Dependent Adenoviral Vectors. *Hum. Gene Ther.* **2013**, *24*, 761–765. [CrossRef]
78. Rastall, D.P.W.; Seregin, S.S.; Aldhamen, Y.A.; Kaiser, L.; Mullins, C.; Liou, A.; Ing, F.; Pereria-Hicks, C.; Godbehere-Roosa, S.; Palmer, D.; et al. Long-term, high-level hepatic secretion of acid α-glucosidase for Pompe disease achieved in non-human primates using helper-dependent adenovirus. *Gene Ther.* **2016**, *23*, 743–752. [CrossRef]
79. Kreppel, F.; Gackowski, J.; Schmidt, E.; Kochanek, S. Combined Genetic and Chemical Capsid Modifications Enable Flexible and Efficient De- and Retargeting of Adenovirus Vectors. *Mol. Ther.* **2005**, *12*, 107–117. [CrossRef]
80. Prill, J.-M.; Espenlaub, S.; Samen, U.; Engler, T.; Schmidt, E.; Vetrini, F.; Rosewell, A.; Grove, N.; Palmer, D.; Ng, P.; et al. Modifications of Adenovirus Hexon Allow for Either Hepatocyte Detargeting or Targeting With Potential Evasion From Kupffer Cells. *Mol. Ther.* **2011**, *19*, 83–92. [CrossRef]
81. Corjon, S.; Wortmann, A.; Engler, T.; Van Rooijen, N.; Kochanek, S.; Kreppel, F. Targeting of Adenovirus Vectors to the LRP Receptor Family with the High-affinity Ligand RAP via Combined Genetic and Chemical Modification of the pIX Capsomere. *Mol. Ther.* **2008**, *16*, 1813–1824. [CrossRef] [PubMed]
82. Barry, M.A.; Rubin, J.D.; Lu, S. Retargeting adenoviruses for therapeutic applications and vaccines. *FEBS Lett.* **2020**, *594*, 1918–1946. [CrossRef] [PubMed]
83. Kreppel, F.; Kochanek, S. Modification of Adenovirus Gene Transfer Vectors With Synthetic Polymers: A Scientific Review and Technical Guide. *Mol. Ther.* **2008**, *16*, 16–29. [CrossRef] [PubMed]
84. Prill, J.-M.; Šubr, V.; Pasquarelli, N.; Engler, T.; Hoffmeister, A.; Kochanek, S.; Ulbrich, K.; Kreppel, F. Traceless Bioresponsive Shielding of Adenovirus Hexon with HPMA Copolymers Maintains Transduction Capacity In Vitro and In Vivo. *PLoS ONE* **2014**, *9*, e82716. [CrossRef]
85. Fella, C.; Walker, G.F.; Ogris, M.; Wagner, E. Amine-reactive pyridylhydrazone-based PEG reagents for pH-reversible PEI polyplex shielding. *Eur. J. Pharm. Sci.* **2008**, *34*, 309–320. [CrossRef]
86. Hofherr, S.E.; Shashkova, E.V.; Weaver, E.A.; Khare, R.; Barry, M.A. Modification of Adenoviral Vectors with Polyethylene Glycol Modulates In Vivo Tissue Tropism and Gene Expression. *Mol. Ther.* **2008**, *16*, 1276–1282. [CrossRef]
87. Říhová, B.; Kovar, M. Immunogenicity and immunomodulatory properties of HPMA-based polymers. *Adv. Drug Deliv. Rev.* **2010**, *62*, 184–191. [CrossRef]
88. Volfova, I.; Říhová, B.; VetviČKa, V.; Rossmann, P.; Ulbrich, K. Biocompatibility of Biopolymers. *J. Bioact. Compat. Polym.* **1992**, *7*, 175–190. [CrossRef]
89. Říhová, B.; Ulbrich, K.; Kopeček, J.; Mančal, P. Immunogenicity of N-(2-hydroxypropyl)-methacrylamide copolymers—Potential hapten or drug carriers. *Folia Microbiol.* **1983**, *28*, 217–227. [CrossRef]
90. Fisher, K.; Stallwood, Y.; Green, N.K.; Ulbrich, K.; Mautner, V.; Seymour, L.W. Polymer-coated adenovirus permits efficient retargeting and evades neutralising antibodies. *Gene Ther.* **2001**, *8*, 341–348. [CrossRef]
91. Subr, V.; Kostka, L.; Selby-Milic, T.; Fisher, K.; Ulbrich, K.; Seymour, L.W.; Carlisle, R.C. Coating of Adenovirus Type 5 with Polymers Containing Quaternary Amines Prevents Binding to Blood Components. *J. Control. Release* **2009**, *135*, 152–158. [CrossRef]
92. Fisher, K.D.; Green, N.K.; Hale, A.; Subr, V.; Ulbrich, K.; Seymour, L.W. Passive tumour targeting of polymer-coated adenovirus for cancer gene therapy. *J. Drug Target.* **2007**, *15*, 546–551. [CrossRef]
93. Šubr, V.; Ulbrich, K. Synthesis and properties of new N-(2-hydroxypropyl)methacrylamide copolymers containing thiazolidine-2-thione reactive groups. *React. Funct. Polym.* **2006**, *66*, 1525–1538. [CrossRef]
94. Ulbrich, K. Polymeric Drugs Based on Conjugates of Synthetic and Natural Macromolecules I. Synthesis and Physico-Chemical Characterisation. *J. Control. Release* **2000**, *64*, 63–79. [CrossRef]
95. Espenlaub, S.; Wortmann, A.; Engler, T.; Corjon, S.; Kochanek, S.; Kreppel, F. Reductive amination as a strategy to reduce adenovirus vector promiscuity by chemical capsid modification with large polysaccharides. *J. Gene Med.* **2008**, *10*, 1303–1314. [CrossRef]
96. Green, N.K.; Hale, A.; Cawood, R.; Illingworth, S.; Herbert, C.; Hermiston, T.; Subr, V.; Ulbrich, K.; Van Rooijen, N.; Seymour, L.W.; et al. Tropism ablation and stealthing of oncolytic adenovirus enhances systemic delivery to tumors and improves virotherapy of cancer. *Nanomedicine* **2012**, *7*, 1683–1695. [CrossRef]
97. Mok, H.; Palmer, D.J.; Ng, P.; Barry, M.A. Evaluation of polyethylene glycol modification of first-generation and helper-dependent adenoviral vectors to reduce innate immune responses. *Mol. Ther.* **2005**, *11*, 66–79. [CrossRef]
98. Espenlaub, S.; Corjon, S.; Engler, T.; Fella, C.; Ogris, M.; Wagner, E.; Kochanek, S.; Kreppel, F. Capsomer-Specific Fluorescent Labeling of Adenoviral Vector Particles Allows for Detailed Analysis of Intracellular Particle Trafficking and the Performance of Bioresponsive Bonds for Vector Capsid Modifications. *Hum. Gene Ther.* **2010**, *21*, 1155–1167. [CrossRef]
99. Carlisle, R.C.; Etrych, T.; Briggs, S.S.; Preece, J.A.; Ulbrich, K.; Seymour, L.W. Polymer-coated polyethylenimine/DNA complexes designed for triggered activation by intracellular reduction. *J. Gene Med.* **2004**, *6*, 337–344. [CrossRef]

100. Wang, C.-H.K.; Chan, L.W.; Johnson, R.N.; Chu, D.S.; Shi, J.; Schellinger, J.G.; Lieber, A.; Pun, S.H. The transduction of Coxsackie and Adenovirus Receptor-negative cells and protection against neutralizing antibodies by HPMA-co-oligolysine copolymer-coated adenovirus. *Biomaterials* **2011**, *32*, 9536–9545. [CrossRef]
101. Morrison, J.; Briggs, S.S.; Green, N.K.; Thoma, C.; Fisher, K.; Kehoe, S.T.; Seymour, L.W. Cetuximab retargeting of adenovirus via the epidermal growth factor receptor for treatment of intraperitoneal ovarian cancer. *Hum. Gene Ther.* **2008**, *15*. [CrossRef]
102. Morrison, J.; Briggs, S.S.; Green, N.; Fisher, K.; Subr, V.; Ulbrich, K.; Kehoe, S.; Seymour, L.W. Virotherapy of Ovarian Cancer with Polymer-cloaked Adenovirus Retargeted to the Epidermal Growth Factor Receptor. *Mol. Ther.* **2008**, *16*, 244–251. [CrossRef]
103. Jönsson, F.; Kreppel, F. Barriers to systemic application of virus-based vectors in gene therapy: Lessons from adenovirus type 5. *Virus Genes* **2017**, *53*, 692–699. [CrossRef]
104. Matsumura, Y.; Maeda, H. A new concept for macromolecular therapeutics in cancer chemotherapy: Mechanism of tumoritropic accumulation of proteins and the antitumor agent smancs. *Cancer Res.* **1986**, *46*, 6387–6392.
105. Stevenson, M.; Boos, E.; Herbert, C.; Hale, A.; Green, N.; Lyons, M.; Chandler, L.; Ulbrich, K.; Van Rooijen, N.; Mautner, V.; et al. Chick embryo lethal orphan virus can be polymer-coated and retargeted to infect mammalian cells. *Gene Ther.* **2005**, *13*, 356–368. [CrossRef]
106. Jonker, D.J.; O'Callaghan, C.J.; Karapetis, C.S.; Zalcberg, J.R.; Tu, D.; Au, H.-J.; Berry, S.R.; Krahn, M.; Price, T.; Simes, R.J.; et al. Cetuximab for the Treatment of Colorectal Cancer. *N. Engl. J. Med.* **2007**, *357*, 2040–2048. [CrossRef]
107. Li, S.; Chen, J.; Xu, H.; Long, J.; Xie, X.; Zhang, Y. The Targeted Transduction of MMP-Overexpressing Tumor Cells by ACPP-HPMA Copolymer-Coated Adenovirus Conjugates. *PLoS ONE* **2014**, *9*, e100670. [CrossRef]
108. Sun, Y.; Lv, X.; Ding, P.; Wang, L.; Sun, Y.; Li, S.; Zhang, H.; Gao, Z. Exploring the functions of polymers in adenovirus-mediated gene delivery: Evading immune response and redirecting tropism. *Acta Biomater.* **2019**, *97*, 93–104. [CrossRef]
109. Veronese, F.M. Peptide and protein PEGylation: A review of problems and solutions. *Biomaterials* **2001**, *22*, 405–417. [CrossRef]
110. Veronese, F.M.; Saccà, B.; De Laureto, P.P.; Sergi, M.; Caliceti, P.; Schiavon, O.; Orsolini, P. New PEGs for peptide and protein modification, suitable for identification of the PEGylation site. *Bioconjugate Chem.* **2001**, *12*, 62–70. [CrossRef] [PubMed]
111. Pepinsky, R.B.; Lepage, D.J.; Gill, A.; Chakraborty, A.; Vaidyanathan, S.; Green, M.; Baker, D.P.; Whalley, E.; Hochman, P.S.; Martin, P. Improved pharmacokinetic properties of a polyethylene glycol-modified form of interferon-beta-1a with preserved in vitro bioactivity. *J. Pharmacol. Exp. Ther.* **2001**, *297*, 1059–1066. [PubMed]
112. Wang, Y.-S.; Youngster, S.; Bausch, J.; Zhang, R.; McNemar, C.; Wyss, D.F. Identification of the Major Positional Isomer of Pegylated Interferon Alpha-2b. *Biochemistry* **2000**, *39*, 10634–10640. [CrossRef] [PubMed]
113. Booth, C.; Gaspar, H.B. Pegademase bovine (PEG-ADA) for the treatment of infants and children with severe combined immunodeficiency (SCID). *Biol. Targets Ther.* **2009**, *3*, 349–358.
114. Thorner, M.O. The Discovery of Growth Hormone-Releasing Hormone. *J. Clin. Endocrinol. Metab.* **1999**, *84*, 4671–4676. [CrossRef]
115. Thorner, M.O.; Strasburger, C.J.; Wu, Z.; Straume, M.; Bidlingmaier, M.; Pezzoli, S.S.; Zib, K.; Scarlett, J.C.; Bennett, W.F. Growth Hormone (GH) Receptor Blockade with a PEG-Modified GH (B2036-PEG) Lowers Serum Insulin-Like Growth Factor-I but Does Not Acutely Stimulate Serum GH. *J. Clin. Endocrinol. Metab.* **1999**, *84*, 2098–2103. [CrossRef]
116. Delgado, C.; Francis, G.E.; Fisher, D. The uses and properties of PEG-linked proteins. *Crit. Rev. Ther. Drug Carr. Syst.* **1992**, *9*, 249–304.
117. Parveen, S.; Sahoo, S.K. Nanomedicine. *Clin. Pharmacokinet.* **2006**, *45*, 965–988. [CrossRef]
118. O'Riordan, C.R.; Lachapelle, A.; Delgado, C.; Parkes, V.; Wadsworth, S.C.; Smith, A.E.; Francis, G.E. PEGylation of Adenovirus with Retention of Infectivity and Protection from Neutralizing Antibody in Vitro and in Vivo. *Hum. Gene Ther.* **1999**, *10*, 1349–1358. [CrossRef]
119. Lanciotti, J.; Song, A.; Doukas, J.; Sosnowski, B.; Pierce, G.; Gregory, R.; Wadsworth, S.; O'Riordan, C. Targeting adenoviral vectors using heterofunctional polyethylene glycol FGF2 conjugates. *Mol. Ther.* **2003**, *8*, 99–107. [CrossRef]
120. Hofherr, S.E.; Mok, H.; Gushiken, F.C.; Lopez, J.A.; Barry, M.A. Polyethylene Glycol Modification of Adenovirus Reduces Platelet Activation, Endothelial Cell Activation, and Thrombocytopenia. *Hum. Gene Ther.* **2007**, *18*, 837–848. [CrossRef]
121. Croyle, M.A.; Chirmule, N.; Zhang, Y.; Wilson, J.M. "Stealth" Adenoviruses Blunt Cell-Mediated and Humoral Immune Responses against the Virus and Allow for Significant Gene Expression upon Readministration in the Lung. *J. Virol.* **2001**, *75*, 4792–4801. [CrossRef]
122. Croyle, M.A.; Chirmule, N.; Zhang, Y.; Wilson, J.M. PEGylation of E1-Deleted Adenovirus Vectors Allows Significant Gene Expression on Readministration to Liver. *Hum. Gene Ther.* **2002**, *13*, 1887–1900. [CrossRef]
123. Wortmann, A.; Vöhringer, S.; Engler, T.; Corjon, S.; Schirmbeck, R.; Reimann, J.; Kochanek, J.; Kreppel, F. Fully Detargeted Polyethylene Glycol-coated Adenovirus Vectors Are Potent Genetic Vaccines and Escape from Pre-existing Anti-adenovirus Antibodies. *Mol. Ther.* **2008**, *16*, 154–162. [CrossRef]
124. Khare, R.; Reddy, V.S.; Nemerow, G.R.; Barry, M.A. Identification of Adenovirus Serotype 5 Hexon Regions That Interact with Scavenger Receptors. *J. Virol.* **2011**, *86*, 2293–2301. [CrossRef]
125. Weaver, E.A.; Barry, M.A. Effects of Shielding Adenoviral Vectors with Polyethylene Glycol on Vector-Specific and Vaccine-Mediated Immune Responses. *Hum. Gene Ther.* **2008**, *19*, 1369–1382. [CrossRef]

126. Tian, J.; Xu, Z.; Smith, J.S.; Hofherr, S.E.; Barry, M.A.; Byrnes, A.P. Adenovirus Activates Complement by Distinctly Different Mechanisms In Vitro and In Vivo: Indirect Complement Activation by Virions In Vivo. *J. Virol.* **2009**, *83*, 5648–5658. [CrossRef]
127. Croyle, M.A.; Le, H.T.; Linse, K.D.; Cerullo, V.; Toietta, G.; Beaudet, A.; Pastore, L. PEGylated helper-dependent adenoviral vectors: Highly efficient vectors with an enhanced safety profile. *Gene Ther.* **2005**, *12*, 579–587. [CrossRef]

viruses

MDPI

Review

Understanding Post Entry Sorting of Adenovirus Capsids; A Chance to Change Vaccine Vector Properties

Coralie F. Daussy †, Noémie Pied † and Harald Wodrich *

Microbiologie Fondamentale et Pathogénicité, MFP CNRS UMR 5234, University of Bordeaux, 146 rue Leo Saignat, CEDEX, 33076 Bordeaux, France; coralie.daussy@u-bordeaux.fr (C.F.D.); noemie.pied@u-bordeaux.fr (N.P.)

* Correspondence: harald.wodrich@u-bordeaux.fr

† These authors contributed equally to this work.

Abstract: Adenovirus vector-based genetic vaccines have emerged as a powerful strategy against the SARS-CoV-2 health crisis. This success is not unexpected because adenoviruses combine many desirable features of a genetic vaccine. They are highly immunogenic and have a low and well characterized pathogenic profile paired with technological approachability. Ongoing efforts to improve adenovirus-vaccine vectors include the use of rare serotypes and non-human adenoviruses. In this review, we focus on the viral capsid and how the choice of genotypes influences the uptake and subsequent subcellular sorting. We describe how understanding capsid properties, such as stability during the entry process, can change the fate of the entering particles and how this translates into differences in immunity outcomes. We discuss in detail how mutating the membrane lytic capsid protein VI affects species C viruses' post-entry sorting and briefly discuss if such approaches could have a wider implication in vaccine and/or vector development.

Keywords: adenovirus; virus entry; vaccine vector; autophagy; innate immunity; adaptive immunity; intracellular trafficking; SARS-CoV-2

1. Introduction

Viral infections, unlike bacterial infections that can be treated with antibiotics, remain a threat to health as very few antiviral remedies exist. Therefore, vaccination is the only widely applicable strategy to protect human (and animal) populations from viral infections. Until recently, vaccination against viral infections was mainly considered in the context of preventing childhood diseases or as the annual "flu" shot for influenza infection prevention. With the outbreak of the SARS-CoV-2 health crisis in 2020, vaccination has returned to the center of public attention as it is now seen as the only sustainable way out of this pandemic. In this context, adenovirus (Ad) vector-based vaccines have emerged as one of two major successful vaccination strategies. Following emergency approval, Ad vaccines are now being injected to a large proportion of the earth's population in an unprecedented effort to install planetary immunity against SARS-CoV-2.

Adenovirus vector vaccines were not widely used before the COVID-19 outbreak, but their success is not unexpected. Viruses are in general very suitable vaccine platforms. They are biologically optimized and adapted for nucleic acid transport, and they elicit immune responses. Adenoviruses have many features that make them particularly ideal vaccine candidates. They have a low and well-characterized pathogenic profile paired with a high infectivity. They are technologically approachable and have been extensively used for vectorization. Their high production yield and stability allow industrial production under good manufacturing practice (GMP) conditions and provide practical features (i.e., storage) for regional distribution at a reduced cost [1]. Furthermore, Ad cell delivery into antigen-presenting cells is efficient, and they naturally provoke an innate immune response that acts as an adjuvant to boost vaccination success [2,3].

Understanding Ad biology has been crucial in developing Ad-based (vaccine) vectors. For example, the use of non-human or rare Ad serotypes in some of the currently marketed SARS-CoV-2 vaccines was deliberately chosen to prevent pre-existing vector immunity from reducing vaccine efficiency. The use of viral gene-deleted vectors provides improved vector safety. However, there are still many aspects of the viral life cycle that have not been considered when developing Ad-based vaccine vectors, and knowledge on rare or non-human genotypes is limited, including the ones currently used as SARS-CoV-2 vaccines. With this review, we would like to raise awareness for potential modifications that could be exploited when developing Ad vectors for medical purposes including vaccination. We focus on the viral capsid and how the choice of genotypes influences the uptake und subsequent subcellular sorting. We describe how understanding capsid properties such as stability during the entry process can change the fate of the entering particles and what different immunity outcomes are observed. As an example of how capsid stability and fate can be changed, we discuss in detail how mutating the membrane lytic capsid protein VI affects species C viruses. Finally, we briefly discuss if such approaches have the potential to be exploited for vaccine and/or vector development.

1.1. *Adenoviruses Constitute a Diverse Family of Infectious Pathogens*

Historically, Ad was first isolated in 1953 from adenoids of patients with respiratory infections [4], and soon after, the term "adenovirus" was universally adopted [5]. The family of *Adenoviridae* has more than 120 members, divided into five genera—depending on whether they infect mammals (e.g., *Mastadenoviridae*), birds (e.g., *Aviadenoviridae*), reptiles (e.g., *Atadenoviridae*), amphibians (*Siadenoviridae*), or fish (*Ichtadenoviridae*) [6,7]. The genus *Mastadenoviridae* also accounts for more than 70 types of human Ads [8,9], which are classified into seven species (A–G), according to their morphological, biological, and physicochemical properties, as listed in Table 1 [10,11]. Furthermore, Ads can also spontaneously recombine, thereby generating new species [12], as it was the case for AdE4, which is the only human species E virus and is likely to have resulted from the recombination between human species B and a simian Ad [13]. Although infections in humans with non-human Ads are naturally rare, they are not impossible [14,15]. The capacity of a variety of non-human Ads to infect (or transduce) human cells makes them attractive (vaccine) vector platforms. Among them are Ads derived from chimpanzees [16–18], gorilla [19], sheep [20], cow [21], dog [21,22], or the new world monkey [23].

Adenovirus infections can cause different pathologies depending on their cellular or tissue tropism including respiratory, ocular, urinary, or gastrointestinal diseases (see Table 1). Tissue tropism is often the result of distinct receptor usage by different Ads and may also be related to intrinsic particle stability [24]. Most infections with Ads are mild or asymptomatic in an immune-competent host. Primary Ad infections are mostly occurring during childhood and result in strong protective immunity. In contrast, in immune-suppressed individuals, Ads can cause uncontrolled, severe, and life-threatening systemic infections leading to serious cell toxicity, which can result in excessive inflammation and multiple organ failure, causing high mortality [12]. The best-studied human Ads are the species C viruses type 2 and 5, which are predominantly found in patients with upper respiratory tract and gastrointestinal infections [25] and which will be in the major focus of this review. Adenoviruses can be responsible for occasional outbreaks, such as species B viruses causing pneumonia (e.g., Ad3, 7, and 14) among immune-competent individuals [26] or Ad40/41 of species F and Ad12, 18, and 31 of species A, which infect the gastrointestinal tract (as reviewed in [26,27]). Species D viruses are a very diverse group and prone to recombination and are a leading cause of eye infections [28].

Table 1. A simplified overview of human Ads and non-human Ads and their uses for vaccination purposes. Species of human Ads, their receptor, and main tropism.

Species	Type	Tropism	Receptor	Remarks
A	12, 18 and 31	Intestinal	CAR	
B	B1: 3, 7, 16, 21 and 50 B2: 11, 14, 34 and 35	Respiratory and ocular	CD46/DSG-2	
C	1, 2, 5 and 6	Respiratory	CAR	Ad5: used as vaccine against SARS-CoV-2 and in clinical trial against Ebola and HIV-1 (but prematurely stopped)
D	8-10, 13, 15, 17, 19, 20, 22-30, 32, 33, 36-39, 42-49	Ocular and intestinal	CAR	Ad26: used vaccine against Ebola and SARS-CoV-2
E	4	Respiratory and ocular	CAR	
F	40,41	Intestinal	CAR	Oral vaccine
G	52	Intestinal	CAR/sialic acid	
Non-human (Chimpanzee) (Canine)	ChAd3 ChAdOx1 CAV-2	—	CAR	ChAd3 : used as a vaccine in a clinical trial against Ebola ChAdOx1 : used as a vaccine against SARS-CoV-2 CAV-2: can target neuronal cells

1.2. Adenovirus Capsids Are Metastable Structures

Despite their diversity, all Ads follow a similar structural and organizational framework. Adenoviruses are non-enveloped DNA viruses containing a relatively large, 26–45 kb linear double-stranded DNA genome [29–31]. The capsid is ~70–90 nm in diameter and has the form of an icosahedron with 20 facets and 12 vertices [32]. The capsid consists of three major structural proteins (hexon, penton, and fiber) and a variety of minor cement proteins (i.e., IIIa, VI, VIII, and IX in species C viruses) that stabilize the structure. The high-resolution structure of several Ad capsids has been obtained by cryo-electron microscopy and X-ray crystallography [24,33–36]. The icosahedral capsid has a pseudo T25 structure with 720 hexons assembled in different trimers depending on their location in the capsid. The 20 facets of the capsid are composed of nine assembled hexons (so called GON for Group of Nine), whereas at the 12 vertices, hexons are assembled in GOS (for Group of Six) (see [37,38] for more details). The pentameric penton base is inserted into the 12 vertices and serves as the basis for the protruding trimeric fiber [39,40]. The interaction between major and minor capsid proteins stabilizes the capsid, mainly through the interaction with hexons. For example, GON are stabilized by IX [41]; IX and IIIa are themselves stabilized by their interaction with VIII, which links them with GOS and GON, respectively [37]. VIII molecules outline the GON and help to stabilize hexons on the inside of the capsid. Protein VI also stabilizes hexon by binding the hexon trimer central cavity exposed to the inner viral lumen [42–44].

The capsid surrounds and protects the viral core composed of the viral genome, the viral core proteins (i.e., V, VII, Mu in species C viruses), and the terminal protein (TP), which is covalently bound to each 5′ end of the Ad genome. Minor capsid proteins interact with core proteins establishing a connection between capsid and genome; however, their spatial organization in relation to the core proteins remains poorly understood. It has been shown that V maintains DNA inside the capsid by interacting with VI (capsid side) and VII (genome side) [45–47] and probably contributes to genome release at the nuclear pore [48]. VII is the main capsid protein associated with viral DNA and promotes its condensation inside the capsid [49–51]. VII also protects the viral genome from immune detection through cellular factors [50,52–54]. Interestingly, capsid protein IX and protein

V are only present in *Mastadenoviridae* [55,56], and viral capsids can be produced without IX [57] or protein VII [58], showing some flexibility in the capsid structure.

The genome of human Ads encodes ~45 proteins (including the structural proteins), organized into temporally regulated transcription units [59]. The transcription factor E1A is the first expressed protein and is necessary to initiate viral gene expression [60]. The genome also harbors a 5′ encapsidation sequence and inverted terminal repeats (ITR) at each extremity which are required for replication [59,61–64]. Adenoviruses encode an adenoviral protease (AVP) which is packed into the particle and processes several proteins (i.e., IIIa, VI, VII, VIII, μ, TP, and 52.55K in species C viruses) [65–67]. This step is required for the formation of mature infectious particles and induces the formation of metastable (i.e., less stable) capsids primed for capsid disassembly upon cell attachment and entry [68,69].

There are also species-specific differences in the structure of human Ads, e.g., in the fiber length and flexibility, which directly affects receptor binding [70–72]. Differences can also be observed in penton assembly. Penton of species B viruses can form inter-penton contacts, resulting in alternative and less stable virus-like particles that are devoid of genomes and occasionally without fibers. Such particles are called dodecahedron and are studied for the cell transfer of plasmids or peptides as an alternative to Ad vectors. Dodecahedrons better expose a RGD short peptide motif present in penton of most Ad species, which is required to bind target cells [73,74]. Differences in capsid stability, as discussed below, may be an important property that determines the entry fate of Ad particles.

1.3. Adenoviruses Follow a Lytic Life Cycle

The Ad infection cycle has been extensively reviewed elsewhere [75–77]. Here, we briefly focus on the entry part of the life cycle, and, unless stated otherwise, we will refer to species C viruses, which are the best studied (Figure 1). Adenoviruses use the capsid fiber as the primary cell attachment molecule. Most Ads, including species C viruses, use the coxsackievirus and Ad receptor (CAR) to bind target cells [78–81]. In contrast, some species B viruses preferentially bind the CD46 receptor [82] or desmoglein 2 [83], and some species D viruses were shown to interact with sialic acids [84, 85] or directly with $\alpha v\beta$ integrins (see Table 1) [86]. Thus, fiber switching between species is an attractive strategy to change tropism [87–89]. The primary role of the fiber molecule is to mediate the physical association with target cells, which is followed by a second interaction between penton and integrins, such as $\alpha v\beta 5$ [90]. This interaction is mediated by the RGD peptide motif in the penton sequence [91]. Absence of this motif, as in species F viruses, is associated with less efficient cell entry [92]. Integrin binding results in integrin clustering at the cell surface [93] that triggers a signaling cascade, leading to the reorganization of the actin cytoskeleton [94,95] and the endocytic uptake via clathrin-mediated endocytosis or micropinocytosis [96,97]. RGD binding may also destabilize the capsid [98] by decreasing contacts between penton and hexon, helping the capsid to uncoat [99]. As a consequence, Ad particles lose their fiber, and one can find dissociated hexon and penton inside the endosome [100–102]. The endosome is a dynamic organelle primed to mature into degradative lysosomes, and Ads have to escape rapidly into the cytosol. Endosome acidification may accelerate the escape process by further weakening the capsid [100,103]. The endosome penetration process relies on the release of the internal capsid protein VI [103,104]. The released VI is characterized by a N-terminal amphipathic helix that binds to the inner leaflet of the endosome and induces positive membrane curvature, resulting in locally confined membrane rupture [75,103,105,106]. Cells respond to the virus-induced membrane rupture and activate a localized autophagy response to clear the damaged organelle [107–109]. To avoid degradation, Ads stall autophagy via a short PPxY peptide motif in protein VI until the particle has reached the safety of the cytosol [104,108]. Species C viruses were shown to escape from early endosomes [110]. Other species including A, B, and D first traffic to the lysosome before escaping into the cytosol [111–113], which may affect the immune activation triggered by those viruses [114].

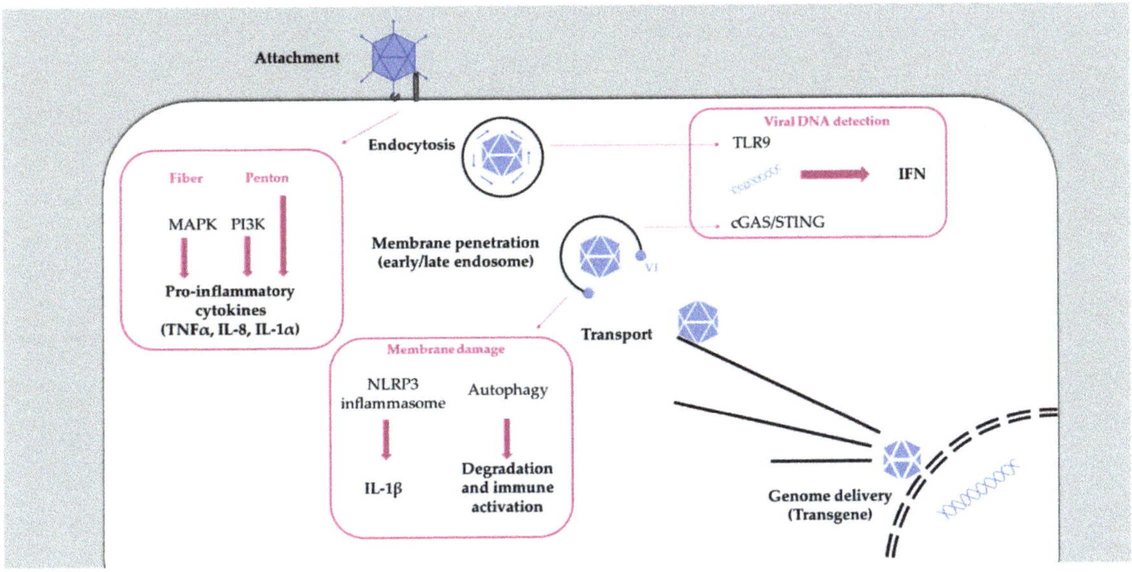

Figure 1. Adenovirus cycle and associated immune response. Along its journey to the nucleus, Ad faces several cellular immune restrictions. Interaction of fiber and penton are required for initial attachment of Ad to cellular receptors; this interaction also induces a signaling cascade leading to the production of pro-inflammatory cytokines. Following entry, Ad is trapped inside an endosome where viral genomes can be detected via TLR9. The membrane lytic function of the viral protein VI allows viral penetration of the endosomal membrane, permitting viral escape into the cytosol. Endosomal escape occurs from within early or late endosome, depending on the Ad species. Membrane penetration of the viral capsid allows exposure of viral DNA in the cytosol that can be sensed by cGAS/STING. Membrane damage is also detected as a danger signal and induces both inflammasome and autophagy activation.

Once Ads have reached the cytosol, they engage with dynein motors and use retrograde transport along microtubules to reach the microtubule organizing center [115–117]. Motor binding is probably mediated by hexon [115] and may require acidic priming [118]. In the vicinity of the nucleus, capsids switch transport directionality, probably by binding to kinesin motor proteins, and accumulate at the nuclear envelope [115,117,119,120]. They dock at the nuclear pore complex, then capsids are completely disassembled, and genomes are released and imported into the nucleus [51,121–124]. Once inside the nucleus, the cycle continues by transcription activation of the genome and expression of the immediate early E1A gene, which serves as a transcription factor for all other early transcription units [60,125–130]. Following expression of the replication enzymes, viral genomes are replicated and accumulate at the periphery of replication centers [131]. Then, structural proteins are expressed from the late expression unit under the control of the major late promoter [132,133]. Expression of late genes culminates in production and nuclear accumulation of all structural proteins where they assemble into progeny and package newly synthesized genomes. Nuclear assembly of the next generation of Ads then results in particle egress and cell lysis [42,134–136]. As mentioned above, assembly of the capsids is followed by maturation, i.e., the proteolytic cleavage of several capsid proteins by the AVP to produce infectious particles [66]. The importance of this process is exemplified by the temperature-sensitive mutant, *Ts*1, initially identified in a mutagenesis screen in Ad2. At the non-permissive temperature, this mutant fails to package the AVP, and assembled particles do not undergo maturation, resulting in hyperstable particles [137]. The responsible AVP mutation was genetically introduced into Ad5 retaining this phenotype. *Ts1* capsids are still able to attach to cell receptors and to be endocytosed, but then remain trapped in the endosome due to their failure to liberate protein VI [103,138,139].

2. Immune Detection of Adenovirus

Adenoviruses are widespread, and a large majority of the population has been confronted with this virus, mainly during childhood. Adenoviruses are highly immunogenic, and every encounter provokes the development of specific anti-Ad immunity, both innate and adaptive, that can last a lifetime. This is important in the fight against the virus and renders Ads less harmful to the immune-competent host, but it can be an obstacle when Ads are to be used as a vector in vaccinology or gene therapy [140,141]. Understanding how Ad (and its vector derivatives) activates the immune system will allow us to improve the effectiveness and safety of vaccines. Observations in vivo, from patients participating in clinical trials, clearly showed that Ad-based vectors have the ability to induce potent immune activation [142,143], sometimes with fatal consequences [144]. In addition, various studies carried out in vitro or in animal models [145] suggest a key role of the viral particle itself (i.e., capsid and incorporated DNA) in activating the initial immune response. Here, we briefly focus on immune responses associated with the entering capsid; for a more in-depth overview, see previously published reviews [146–149].

2.1. *Adenoviruses Trigger Cell Intrinsic Immunity*

With the development of Ads as vectors, it became apparent that cell intrinsic or innate immunity induced by Ads is a response to the invading virion. Interestingly, both replicative and non-replicative viruses (i.e., UV-inactivated virus, non-replicative vectors, or empty capsids) activate immune responses in the infected cells, showing that the capsid has a key role in this response [145,150–156].

In general, Ads are sensed by cells at several steps throughout the entire viral life cycle and activate innate immunity pathways (Figure 1). This activation starts with the Ad fiber binding to the cell surface receptor CAR, which activates the mitogen-activated protein kinase (MAPK) signaling that leads to the activation of the transcription factor NF-κB. This trigger results in the production of pro-inflammatory cytokines (IL-6, IL-8) [157,158]. However, depending on Ad species, other receptors usage may modulate the strength of the signaling cascade involved in cytokine expression [159]. The interaction between viral penton and cellular integrin is a further trigger of immune activation. A study in murine macrophages showed that this binding activates the expression of IL-1α via the RGD motif [160]. This might be restricted to immune cells because in other cell models (e.g., HeLa, epithelial cells line), the role of the integrin/RGD interaction in chemokines and pro-inflammatory cytokines production was not confirmed [151,161,162]. This suggests a more important role of the internalization process itself in immune activation than the interaction with integrins. Indeed, penton binding to integrin also activates the phosphoinositide 3-kinase (PI3K) signaling to increase viral internalization [163]. Moreover, PI3K activation has been shown upon Ad infection to trigger the production of pro-inflammatory cytokines, such as TNFα [164], and the internalization process has been described as required for immune activation [165]. These studies suggest that immune activation mainly occurs at a post-internalization step, highlighting the importance of post-entry sorting (reviewed in [166]).

Once in the endosome, Ads are exposed to intraluminal pathogen-recognition receptors (PRR) such as TLR9, which is a double-stranded DNA sensor restricted to immune cells. Murine macrophages can sense Ad vectors via TLR9 to induce the production of pro-inflammatory cytokines [167]. TLR9 was also strictly required for IFNα/β production when murine plasmacytoid dentritic cells (pDC) were challenged with Ad [168]. However, in TLR9 knock-out mouse models, IFNα/β was still produced, suggesting that, in cells other than pDC, IFN production occurs independently from TLR9 signaling [168,169].

Viral genomes (or vector genomes) become accessible to sensing during or after the endosomal escape of the virus. Once they reach the cytosol, they are sensed by the cytosolic double-stranded DNA sensor cGAS [168,170]. Upon viral genome recognition, cGAS promotes phosphorylation and nuclear translocation of the transcription factor IRF3 to drive IFNα/β expression, inducing an antiviral state. Endosomal passage requires virus-induced endosomal membrane damage to facilitate endosomal escape of viral particles. Membrane rupture thus considerably contributes to immune activation, highlighted by the fact that the escape defective *Ts1* mutant fails to activate a complete and efficient immune response [161,162,170,171]. A pivotal role in the inflammatory response upon Ad membrane penetration is played by the tank-binding kinase, TBK1, suggested to be part of a down-stream cytosolic sensing of the viral genome via the cGAS/STING pathway [168,170].

Adenovirus membrane damage also results in the release of cathepsin B from the endo-lysosomal compartment, causing oxidative stress, which activates the NLRP3 inflammasome, resulting in IL-1β maturation [114]. Species C viruses escape from early endosomes, while species B viruses traffic until late endosomes, presumably due to differences in receptor usage. Acidification in late endosome/lysosome is more likely to activate lysosomal acid hydrolases. Both species induce inflammasome activation [172], but, probably owed to the residing time in the endosome, the extent of their responses differ. Globally, species B viruses induce a stronger innate and adaptive immune activation than species C viruses [113,173,174]. This may explain why species B viruses are more pathogenic, causing outbreaks in immune competent hosts [26]. It further suggests that cells are able to discriminate the penetration compartment (i.e., endosome vs. lysosome, [109]) and adapt the efficiency of immunity. Endosomal membrane damage during the Ad escape process can also be a danger signal in itself. As discussed in detail below, membrane damage results in the cytosolic exposure of intraluminal glycans, which are detected by the cell as danger signals (as reviewed in [109]). The detection of exposed glycans by galectins subsequently activates autophagy, triggering a second branch of antiviral immunity [108].

2.2. Adenoviruses Provoke an Aadaptive Immune Response and Subvert Antiviral Autophagy upon Cell Entry

Innate immune responses provoked by Ads or their vectors subsequently translates into adaptive immunity through chemokine secretion that attracts immune cells (neutrophils, natural killer cells, and macrophages) and reinforces potent antigen presentation [153,154]. Adenoviruses are highly immunogenic, and their main immunogens are the major capsid proteins (hexon, penton, and fiber) that trigger the species-specific production of neutralizing antibodies by B cells [175–177]. Neutralizing antibodies are mostly directed against the hyper-variable loops of the hexon protein [178–180] and represent a major limitation for the use of Ads as vectors, if derived from the same serotype. This problem has been circumvented by the use of less seroprevalent or non-human Ads [181–185]. Despite this anti-vector immunity, when Ads are used as vaccine platforms (e.g., to elicit immune protection against Ebola in clinical trials), they still induce a strong B cell response, allowing the production of antibodies against the desired vaccine immunogen that can last for up to 6 months [186]. In addition, pre-existing vector immunity could be overcome with higher vaccine vector doses, as in the case of the SARS-CoV-2 vaccine vector based on the highly seroprevalant AdC5 [187].

As with neutralizing antibodies, Ad infections trigger a persistent T-cell immunity, mainly through CD4+ activation. This protection can still be found in adults, suggesting a long-lasting immunity from childhood [188–190]. In addition, activation of CD8+ is also described to maintain a cytotoxic response [191–193]. This T-cell activation is not necessarily serotype-specific and allows cross-protection against different Ad subgroups and serotypes [192–194]. Moreover, in stem-cell-transplanted patients infected by Ad, a strong activation of CD4+ and CD8+ has been described [195]. This activation is directed against hexon peptides and leads to a clearance of Ad viremia, emphasizing the major role of both CD4+ and CD8+ T-cell activation in immune protection. While this T-cell-based immunity is directed against the Ad particle, Ads also trigger strong cellular immunity based on the T-cell response against the vaccine antigen, which is an advantage for long-lasting immunity when using recombinant Ad vectors as vaccine platforms [196,197]. Importantly, data from clinical trials have confirmed that when Ad is used as a vaccine platform (with the desired antigen as transgene), it induces both cytotoxic CD8+ T-cell and CD4+ T-cell activation, leading to memory immunity against the expressed transgene [184,186,198–202].

T-cell activation relies on the presentation of viral peptides through the major histocompatibility complex (MHC). Briefly, antigen presentation by MHC class I results from proteasomal degradation, and the peptides obtained will induce CD8+ T-cell activation, allowing priming and proliferation of a cytotoxic T-cell response. MHC class II antigen presentation results from lysosomal proteolysis, and the resulting peptides will induce CD4+ T-cell activation upon presentation, triggering the subsequent activation of B cells [203]. It is now well characterized that autophagy promotes peptide presentation to the MHC class II molecules [204]. Not surprisingly, autophagy is one of the oldest defense mechanisms against infection and an important part of the cellular repertoire against invading pathogens, participating in both innate and adaptive immune responses. It can directly sequester a pathogen for degradation (xenophagy) and participates in their presentation to the immune system [205–207]. Macroautophagy (more commonly called autophagy) is a conserved lysosomal degradation pathway that participates in many fundamental physiological processes such as homeostasis and innate and adaptive immunity [208–210]. Autophagy is characterized by the formation of a double membrane vesicle called "autophagosome" that engulfs cytoplasmic cargoes (i.e., organelles, aggregates, and pathogens) destined for degradation (Figure 2). During this process, cargo-containing autophagosomes fuse with lysosomes in order to be degraded and recycled.

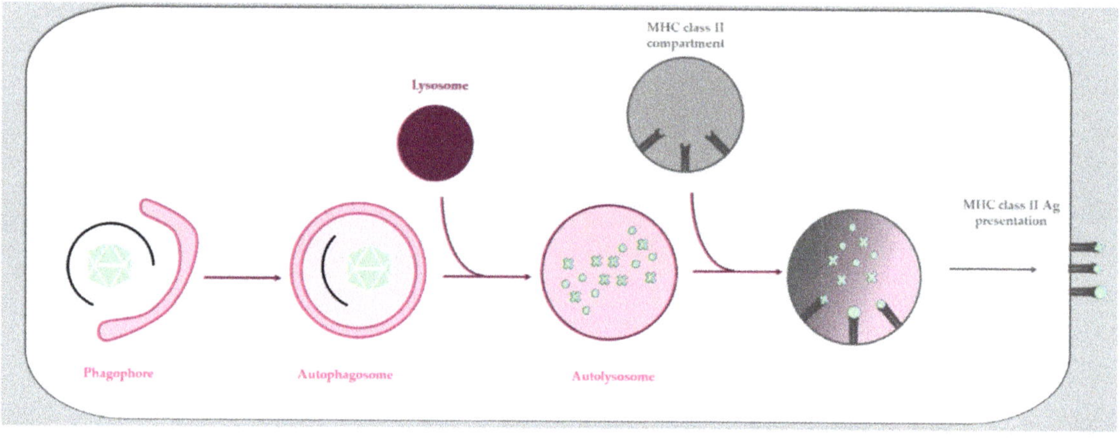

Figure 2. MHC class II presentation by autophagy. Autophagy starts with the formation of a double membrane structure called phagophore that will elongate around the cargo to form a double-membrane vesicle called autophagosome. After fusion with lysosome, the cargo will be degraded in a so-called autolysosome. Further fusion of this vesicle with MHC class II compartment allows the loading of peptides for the ensuing MHC class II antigen presentation.

Antigen presentation via autophagy has been described for both MHC class I and class II [204]. After induction of autophagy, intracellular antigens are engulfed in autophagosomes, fusing with MHC class II containing compartments to provide them with external pathogen material (Figure 2) [211]. The role of autophagy in this particular MHC class II antigen presentation was first described for Epstein-Barr virus infection, where inhibition of autophagy led to decreased CD4+ T-cell activation [212,213]. To date, examples include several viruses including herpes simplex virus [214], influenza virus [215], and human immunodeficiency virus type 1 (HIV-1) [216], and probably extends to Ads (see below). MHC class II presentation is consistent with the process of autophagy as it is a cytosolic defense mechanism.Autophagy is also involved in MHC class II cross-presentation of extracellular antigens [214]. Indeed, the MHC presentation of extracellular ovalbumin is more efficient when the autophagy machinery (and especially ATG5) is functional, and the uptake of extracellular material was shown to involve a non-canonical form of autophagy called LC3-associated phagocytosis (LAP) [217]. Once present in the LAP-osome, extracellular antigens can be exposed through MHC class II presentation [218]. Autophagy is also involved to some degree in MHC class I presentation, although the vast majority of the epitopes are presented via a mechanism strictly dependent on the transporter associated with the antigen processing (TAP) complex [219]. TAP-independent presentation mediated by autophagy was so far shown only under TAP depletion conditions [219,220]. If this mode of presentation truly exists remains controversial [221] and is discussed elsewhere [204,222,223].

To prevent MHC presentations, Ads dedicate a great deal of their genome, namely the E3 region, to undermine the MHC presentation system in cells [224–228]. In most (vaccine) vectors, these genes are absent, and recent work suggests that increasing antigen presentation by MHC could improve Ad vaccine efficiency [229,230]. Autophagy plays a crucial role in processing Ad antigens for MHC class II presentation in Ad-induced immunity [108,229].

Montespan et al. showed that Ads induce autophagy upon endosome penetration and that entering particles cleared by autophagy in this process are efficiently presented to the immune system [108]. This observation is consistent with the study published by Klein et al. showing that in the context of oncolytic Ads, the viral antigens are matured by a JNK-dependent form of autophagy [229]. However, autophagy in the context of cancer plays a dual role. As a homeostasis keeper, autophagy limits tumor initiation [231] while in advanced cancer autophagy acts as a survival process to favor tumor growth [232,233]. Thus, in the context of oncolytic vectors, tumor-specific autophagy features should be carefully evaluated.

3. Adenoviruses as (Vaccine) Vectors

Vectorization of a virus involves the exploitation of its natural properties, while simultaneously minimizing the associated biological risk. In these terms, Ads are excellent vector candidates for vaccine development [234]. Adenovirus vectors are stable, are easy to produce to a high titer, and have large cloning capacities. This allows insertion of a transgene, i.e., to express an antigen of choice. They have a broad infectivity spectrum and transduce non-dividing cells, and their genome is non-integrative [1]. As described above, they also provoke an inflammatory response in antigen-presenting cells (APCs) that serves an adjuvant function to amplify the immune response necessary to vaccinate efficiently [2,3,142,235,236]. Moreover, Ad-derived vectors have been extensively used in preclinical and clinical trials to prove their safety [237,238]. To eliminate the unwanted part of the replication cycle, viral replication needs to be suppressed to prevent cell lysis and viral propagation. Adenovirus vector development started during the early 1990s as a platform for a large range of therapeutic approaches, such as gene therapy, oncolytic vectors, and vaccine development. Most of the early vector developments were modified versions of species C viruses.

3.1. Adenovirus Vector Development, a Generational Approach

The first generation of an Ad vector was made by deleting the E1 and E3 region from the viral genome, notably encoding the immediate early transcription factor, preventing these vectors from replicating. These vectors had a cloning capacity of around 8 kb to insert a transgene with regulatory sequences [239–241]. The first-generation vectors were produced in human embryonic kidney (HEK) cells stably transformed with a part of the Ad5 viral genome that complements the E1 region in *trans* [242–247]. One of the major drawbacks of this system is that it allows the reconstitution of replication-competent genomes after spontaneous recombination [248]. The additional deletion of E3, which is not necessary for vector production, increases the size of the transgene cassette [243], and promotes a better immune response [240]. Unfortunately, first-generation Ad vectors were not completely transcriptionally inert and still elicited a strong immune response against viral capsid proteins that were expressed at baseline levels [141,249,250]. This anti-vector immunity contributed to vector toxicity and short-term elimination of transduced cells [145]. The second generation of Ad vectors additionally had the E2 and E4 regions removed from the vector genome. E2 encodes the replication enzymes and E4 additional regulatory proteins [251–253]. This allowed space for 10.5 kb of transgene sequences, including up to four expression cassettes [254]. Likewise, the supplemental deletions strongly reduced unwanted recombination and prolonged transgene expression [255]. This was probably the result of a reduced aberrant viral gene expression that further reduces anti-vector immunity [141,250,251]. Adenoviral vectors of the third generation are conceptually different. They consist of a packaged vector genome and a non-packaged viral genome (helper virus) that produces vector particles. Such vectors are also known as "helper-dependent", "gutless", or "high-capacity" (HC-Ad) vectors. They have the highest possible cloning capacity and can carry sequences of up to 36 kb [256–258]. High-capacity Ads lack any viral coding sequence except ITRs and a short encapsidation signal that permits genome packaging and which is absent from the helper virus [259–262]. This strategy has allowed us to finally overcome most anti-vector immunity, permitting the long-term expression of transgenes [263–267]. However, HC-Ads are not part of the current vaccine vector repertoire.

3.2. Adenovirus as a Vaccine Vector

Before exploring how Ad vectors have been exploited, it is noteworthy to mention that vaccination against Ad species B viruses type 4 and type 7 themselves was developed in the early 1970s [268]. This oral live vaccine is mandatory since 2011 for military recruits in the USA and dramatically decreased infection rates, reaching virtually zero [269–271]. This highlighted the intrinsic efficiency and safety of Ad as a vaccine and opened the field to use Ad as a vaccination platform. To date, over 200 Ad-based vaccines entered clinical trials [272]. Many of them are directed against infectious pathogens such as HIV-1 and Ebola [230,273,274]. Adenovirus-based vaccines are not without drawbacks. A recently developed anti-HIV vaccine, based on the species C virus Ad5, used a mixture of vectors expressing the HIV-1 gag, pol, and nef genes [275,276]. The vectors were safe, immunogenic, and well-tolerated. Unfortunately, the resulting STEP clinical trial had to be prematurely stopped [277–279]. Not only has this vaccine been shown not to protect against HIV infection, but epidemiological data suggested that vaccination with Ad in this case increased the HIV infection risk [280–282]. Having a pre-existing immunity against Ad5 was identified as a potential post-vaccination risk factor [283,284]. However, this observation was not confirmed in similar studies [285,286] and remains mechanistically unclear [287,288]. Next to HIV, Ad-based vector vaccines have been developed in the fight against Ebola (Ad-EBOV), a recurrent threat in Western Africa [289]. All Ad-EBOV vaccines use the Ebola glycoprotein (GP) as the vaccine antigen. Initial preclinical trials in rodents and primates used Ad5 and showed protection over 3 months with production of neutralizing antibodies and CD8+ activation [273,290–292]. This vaccine went through phase II clinical trial; however, pre-existing immunity to Ad5 in the vaccinated population has led to variation in the level of elicited neutralizing antibodies [293–297]. To overcome this issue, vaccines based on the less prevalent species D Ad26 have been developed [186,298,299]. However, Ad26 is less immunogenic in humans, resulting in decreased vaccination efficiency, and required a prime boost using a measles virus-based vaccine (MVA). This vaccination regime has reached phase III clinical trials [289,300,301]. Use of non-human Ad vectors has also been under development [184]. Use of chimpanzee ChAd3 as a vaccine platform against EBOV showed protection in rodents and in chimpanzees [302,303], and clinical trials employing those vectors are at different steps of progress [17,300,301,304–306]. Until recently, HIV-1 and EBOV vaccines were the two main examples for the application of Ad vector-based vaccines, but they are not the only ones. Different attempts have been made to employ Ad vaccine vectors against various pathogens such as *Mycobacterium tuberculosis* [307–310], dengue virus [311–313], hepatitis B and C [314–320], rabies [321–324], or influenza [325–328]. However, their efficacy was always below expectations (for more details, see [18,142,300,329]). The EBOV vaccine candidate has remained the only practically applied Ad vaccine and is used with some restrictions in the Democratic Republic of Congo. This vaccine is based on the Ad26 type (Ad26-ZEBOV/MVA-BN-FILO), and, as of May 2020, 20,339 people received the first dose of vaccination, and 9560 were fully vaccinated [230,301,330].

Adenovirus-vectorized vaccines have received a major public boost in response to the SARS-CoV-2 pandemic. The urgency of the situation has made it possible to capitalize on their development, including experience gained with vaccines raised against coronaviruses that caused previous regionally confined SARS-CoV-1 outbreaks [331]. The coronavirus spike (S) glycoprotein was identified as the best vaccine antigen for immunity to coronavirus infections [332,333]. To date, four Ad vector vaccines, all encoding the spike antigen, are in an advanced state and have been granted for emergency use in large parts of the world. The applied vector strategies have been slightly different, but essentially involve the use of first-generation vectors. They range from using species C Ad5-based vectors (Ad5-nCoV), overcoming pre-existing immunity with a high vector dose (CanSino) [187,334], to the use of less prevalent species D virus Ad26, using a stabilized spike glycoprotein (*JNJ-78436735/Ad26.COV2.S*) [335]. An alternative approach uses ChAdOx1-S-(AZD1222), an Ad vector from chimpanzees [336,337]. The only vaccine in use based on a heterologous prime-boost strategy to overcome vector immunity (*Gam-COVID-Vac/Sputnik V*) applies first an Ad26-based vector and then in the booster injection uses an Ad5 vector [338,339]. In all cases, clinical trial data have been very promising, showing high levels of induced protective immunity, and emergency use has been granted in several countries for the use of Ad-based vaccines. Furthermore, epidemiological surveys in the ongoing vaccination campaigns in different countries suggest an overwhelming success in the use of Ad-based SARS-CoV-2 vaccines.

4. Modulating Adenovirus (Vaccine) Vector Efficacy, the Capsid Leads the Way

Adenovirus (vaccine) vector design has to take into account a number of factors. These include maintaining a long and sustained expression of the encoded antigen and the activation of an immunological context (adjuvant effect) that permits efficient translation of the antigen expression into a persistent adaptive immunity directed against the immunogen but not against the vector itself. The immunological context is provided by the innate immune activation that the vaccine vector elicits during administration. In this context, the efficiency with which innate immune sensors detect the vector is important, and will in turn determine the amplitude of the adaptive response. As described in the sections above, innate immune sensing is mainly taking place at the post-entry level, concomitant with Ad endosome penetration. The efficient expression of the (vaccine) antigen depends then on the balance between the level of vector clearance vs. successful delivery and expression of vector genomes to the nucleus. With the development of capsid display vectors, expression of the transgene is not required to elicit immune activation. This strategy is based on the insertion of an epitope of interest into Ad capsid proteins, allowing its direct exposure to the immune system. Even if this approach has some advantages, most vaccine vectors rely on genetically encoded genes. For further reading on capsid display vectors, please see these reviews, [340,341].

Pre-existing immunity against a viral vector (e.g., 30–90% against the species C virus Ad5, depending on the geographic location) can reduce vector uptake and decrease the antigen expression and the resulting immune response [342–344]. Several strategies have been developed to decrease anti-vector immunity and have been recently reviewed [249]. Briefly, the neutralizing antibodies mainly recognize the hypervariable loop of hexons and as a consequence are serotype-specific [178–180]. These hypervariable loops can therefore be replaced by those of another, less prevalent serotype [178,345]. This prevents the vector from being neutralized and inactivated. As Ad5 is the most seroprevalent type, using rare serotypes such as Ad26, Ad35, or non-human Ads would also avoid this issue [177,346–350]. However, a major drawback with the use of rare types is their lower immunogenicity that makes them less efficient regarding vaccine development [298]. Therefore, strategies to decrease anti-vector immunity should be developed in parallel to strategies to enhance vector immunogenicity in order to obtain the most efficient vector. Even if the initial level of neutralizing antibodies against the vector is low, vaccination with a given vector type may trigger a response that prevents its future use. Priming of the anti-vector response also takes place at the post-entry level, when vector particles are sorted and processed as antigens themselves. Eliciting an immunologically favorable context for vaccination success may come at the expense of increased priming for vector immunity. Understanding the parameters that determine the fate of Ad vectors after uptake may provide an opportunity for vector design.

4.1. Stability Lies in the Species

Receptor-mediated uptake and escape from the endo-lysosomal compartment are hallmarks of Ad infection and vector transduction. The initial uptake is determined by the interaction of the virus fiber and penton molecules with cell receptors. Fiber and penton are common to all Ads, but, according to their species, they engage with different cellular receptors (Table 1). Species C viruses use CAR, and entering particles escape from an early endosomal compartment [79]. If species C viruses (Ad5) are engineered to contain a fiber molecule from a species B virus (Ad16), the resulting particle (Ad5F16) then escapes from a lysosomal compartment (as shown by costaining with the lysosomal marker LAMP1) instead of the early endosome [351]. This delayed escape from lysosomes is a feature shared with all tested species B viruses [111,352,353]. Furthermore, the hybrid Ad5F16 as well as species B viruses elicit a stronger pro-inflammatory response than species C viruses [159,351]. This is attributed to increased detection via intraluminal TLR9 in immune cells [354]. In addition, lysosomal membrane damage, unlike early endosome damage, causes strong oxidative stress. Lysosome rupture co-releases lysosomal hydrolases such as cathepsin B that can damage mitochondria, thus amplifying the damage response through inflammasome activation [355–357]. The Ad uptake process destabilizes the capsid, allowing the release of the viral protein VI [103,106]. Released protein VI then penetrates the membrane from within the endo-lysosomal compartment to allow the virus to gain access to the cytosol. Comparing species C and B shows that both viruses are equally efficient in cytosol entry and genome delivery, but differ in their escape kinetics [111]. The fact that a simple fiber swap can transfer the entry and immunogenic properties from species B to a species C virus is remarkable. It provides a proof of concept that hybrid vectors with specific properties can be designed. The underlying difference seems to be that protein VI release is delayed in species B viruses (or the hybrid virus) and may require either acidification or another, unknown disassembly trigger. Since protein VI release is delayed in this case, this could reflect an increased capsid stability. Indeed, species C Ad5 viruses were shown to release the fiber molecule upon cell binding, a step assumed to weaken the capsid and prime the disassembly process for protein VI release. In this context, fiberless Ad5 were shown to be less stable than native particles [358]. In contrast, non-matured, hyperstable *Ts1* Ad2/5 do not release fiber or protein VI and are poor inducers of inflammation, since they are neither triggering TLR9 nor causing membrane damage [114,138,359,360]. Whether fiber swapping with species B prevents initial fiber release and renders the species C capsid more stable has not been investigated but is probable. Capsid stability maybe an essential determinant of post-entry capsid fate and, by extension, a feature that could be exploited for vector design. Species B viruses may be more stable due to the use of a different receptor or through a unique fiber-penton interaction. In contrast, species F viruses Ad40/41 infect the gastrointestinal tract and have increased capsid stability due to their tropism and their adaptation to the harsh gut environment [24,361]. The gut has an intrinsic temperature that is a few degrees above the upper respiratory tract targeted by species C viruses (37 °C vs. 33 °C) [362]. Thus, Ad40/41 are intrinsically more thermostable, supporting the infection process [24], making them also difficult to propagate [24,361,363]. In addition, they lack a RGD motif in the penton molecule, resulting in a further reduced cell entry [92]. Whether these unique capsid properties translate into escape from a lysosomal compartment or partial disassembly/protein VI release defects upon entry is not known. In contrast, these properties may provide a context for the development of oral vaccine applications. Unlike species B fibers, species F fiber-harboring vectors have a decreased immune activation [364] when applied at standard environmental conditions. If a given human Ad species is now taken out of context and employed as a vaccine vector, it could be worthwhile to look at the natural tropism of this vector to look for clues as to predict its behavior. The currently used vaccine vector based on Ad26 is a species D virus, and its natural tropism is the eyes and the gastrointestinal tract [365]. Species D viruses including Ad26 are less immunogenic than Ad5 [366] and were shown to traffic to the late endosomal compartment, probably involving delayed protein VI release. Consequently, blocking acidification of the late endosomal compartment further reduced its immunogenicity [113]. In contrast to human Ads, the knowledge on non-human Ads also used in vaccination and vectorization is very limited. Chimpanzee-derived Ads including the ChAd3 are closely related to species D viruses [367,368]. Interestingly, Ad5 vectors are more immunogenic than chimpanzee-derived Ad vectors and have a higher transduction potential. This observation was not linked to differences in the penton RGD motif or the fiber, and its mechanistic reason remains unknown [369]. Potential differences in capsid stability have not been investigated. Another non-human Ad exploited as a vector is the canine Ad type 2 (CAV-2). This vector has the remarkable property to transduce neurons selectively, but not surrounding glia cells, unlike an Ad5 control vector, despite both vectors using the same receptor [370–372]. This ability may be linked to an increased capsid stability, allowing uncoating only in neuronal cells. Dogs have a body temperature that is ~2 °C higher than humans, which potentially results in elevated CAV-2 capsid stability [373], analogous to the species F viruses. CAV-2 poorly elicits an innate immune response, consistent with a lack of efficient disassembly in target cells [372]. Thus, CAV-2 vectors might be another example where a biophysical property linked to its natural tropism or host has the potential to provide novel vector properties.

4.2. Fine-Tuning the Capsid Structure, the Example of Protein VI in Species C Ad2/5

If capsid stability and protein VI release determine the downstream fate of the viral capsid, the question arises whether this can be modified in Ad vectors, i.e., to attribute novel properties to viral vectors (Figure 3). Most of our understanding of how Ads penetrate the endo-lysosomal compartment comes from the species C viruses Ad2/5. The discovery of protein VI as a membrane lytic factor was essential to understand this process [103,106]. Protein VI encodes a N-terminal amphipathic helix that is the actual membrane lytic part of the protein. During virus production, protein VI is expressed as a precursor protein that associates with hexon trimers via the amphipathic helix [42,374]. The complex is imported into the nucleus via transport signals encoded in the N- and C-terminus of protein VI, which are removed during virus assembly and maturation by the viral protease. The C-terminal 11 amino acid peptide also activates the protease, providing a smart way to link virus assembly with shielding of the amphipathic helix. The *Ts1* Ad2/5 mutant virus has a packaging defect for the protease, and none of the capsid proteins, including protein VI, is processed. Resulting particles are hyperstable, retain fibers, and do not deploy protein VI. As consequence, *Ts1* particles are sorted into lysosomes upon receptor-mediated endocytosis. The exact location of protein VI in the mature virus is not clear, but the cleaved N-terminus localizes to the inner hexon surface of peripentonal hexons [37,104,375]. Two point mutations in protein VI, G33A and S28C, were shown to increase the Ad5 capsid stability (Figure 3A) [376,377]. The first mutant precedes the N-terminal protease cleavage site in protein VI and impairs the processing. Resulting particles are partially immature and are characterized by an endosome lysis defect, suggesting that the N-terminus of protein VI stabilizes intra-virion protein-protein interactions [376]. The second mutant was chosen for its ability to form intra-molecular disulfide bridges [377]. The virus contained partially unprocessed proteins, potentially including protein VII, but surprisingly did not show any infectivity defects, despite its increased stability. Whether either of the viruses has an altered immune activation profile is not known. Another point mutation in protein VI that affects the virus fate is the protein VI L40Q mutation, directly situated in the amphipathic helix (Figure 3A) [378,379]. Viruses with this mutation mature normally but are partially defective for the endosome lysis step. In contrast to the above mentioned mutants, L40Q virions are less stable than their wild type counterparts. Instead of a delay, protein VI as well as penton are released prematurely from the L40Q capsid. L40Q virions are less infectious and subject to partial lysosome sorting. However, the L40Q mutation does not completely abolish the membrane lysis, and some viruses escape to the cytosol, resulting in an overall ~four-fold reduced infectivity of a L40Q vector [379,380]. The observations suggest that the amphipathic helix contributes with its interaction to the particle stability. It is not known if and how the L40Q mutation alters the immunogenic property of Ad5 vectors. Recent observations suggest that release of protein VI from the capsid is triggered by internal competition with protein VII for the same or overlapping binding sites in hexons [375]. Consistently, a mutant virus devoid of protein VII fails to escape from the endosome and has a defect in endosome lysis [58]. Further understanding the structural and dynamic organization of the internal capsid (i.e., protein VI) and core proteins (i.e., protein V and VII) in the particle and during cell entry will greatly help to understand what determines (reversible) capsid stability.

Controlling the efficiency of immune activation of Ad vectors is a major interest for the development of vaccines, and expanding MHC presentation of antigens would improve vaccine development [381]. As mentioned previously, autophagy regulates immune activation and feeds peptides to the MHC. Autophagy is one of the oldest cellular defense mechanisms against infection, yet Ads, as adapted human pathogens, have not evolved mechanisms to avoid autophagy upon entry. Instead, they seem to embrace autophagy activation and to divert the response to promote infection. Protein VI plays a central role in controlling the autophagy response of the cell. Adenovirus membrane lysis by protein VI leads to the recruitment of galectin 3 and galectin 8, which are recognized by the cell as danger signals (Figure 3B) [108,109,382]. Detection of Ad penetration sites by galectins activates selective autophagy in the infected cell, as illustrated by the recruitment of autophagy receptor (e.g., NDP52 andp62) and LC3 punctae formation [108]. Interestingly, Ad5 vector particles manage to escape from the ruptured endosome and are able to traffic to the nucleus. Inhibiting autophagy does not affect viral infectivity, but delays their genome delivery to the nucleus, showing that Ads hijack the cellular machinery for their own profit [108]. Escape from the endosome therefore allows the virus to avoid degradation by autophagy via an active process. This process was discovered by yet another protein VI mutant, the M1 mutant. All human Ads (and several non-human Ads) encode a highly conserved PPxY peptide motif in protein VI. In the M1 mutant, this motif was changed to PGAA (Figure 3) [104]. Both wild type and the M1 mutant efficiently release protein VI upon entry and trigger autophagy after rupture of the endosomal membrane. However, the M1 mutant virus is unable to escape efficiently into the cytosol, and EM images have shown that M1 mutant viruses are trapped in ruptured endosomes, targeted by autophagy. Consequently, the M1 mutant virus was shown to have a strong infectivity defect, which is restored upon the pharmacological and genetic inhibition of autophagy. The study showed that wild type viruses use their PPxY motif to recruit the ubiquitin ligase NEDD4.2 and limit the maturation of autophagosomes long enough to escape to the cytosol. Furthermore, this property allows the virus to limit its antigenic presentation [108]. Importantly, the M1 mutant vector injected into mice resulted in much stronger anti-vector immunity and a reduced CD8+ response to the encoded transgene, showing that a simple capsid alteration can have a profound impact on the immunogenicity of the Ad5 capsid. Whether any of the other protein VI mutations affects immunogenicity is not known. However, the presence of the PPxY motif in the L40Q virus may explain why the vector retains a relatively high infectivity despite the membrane lysis defect [379,380]. Because the endosome lysis is less efficient, and L40Q vectors end up in part in the lysosomal compartment, it is conceivable that the elicited immune response is yet different from either wild type, *Ts1*, or M1 Ad5 capsids.

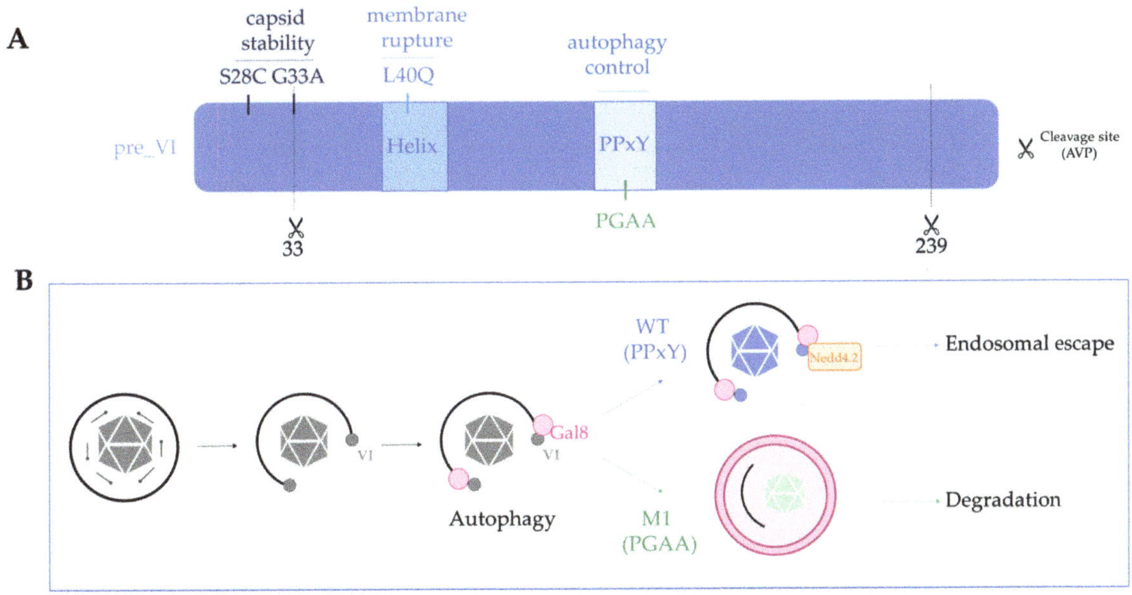

Figure 3. Protein VI, a multifunctional viral protein determining the post-entry fate of the particle. (**A**) Domain organization of the membrane lytic protein VI. Protein VI is produced as a precursor form (pre_VI) and processed by the viral protease (AVP) at two sites. A role for the N-terminal part in capsid stability has been shown by two-point mutation (S28C and G33A). An amphipathic helix is responsible for the membrane lytic activity of protein VI, and a point mutation in this domain (L40Q) reduces the lytic activity. Protein VI encodes a conserved PPxY motif required for viral control of the cellular autophagy response against membrane penetration. Mutating the motif (PGAA) reduces viral infectivity. (**B**) Adenovirus membrane penetration via protein VI is recognized by the cell through glycan binding galectin 8 followed by autophagy induction. WT virus uses the PPxY motif in protein VI to recruit the ubiquitin ligase NEDD4.2 to suppress or halt autophagy until endosomal escape is completed. A mutation in this motif (PGAA) abolishes viral autophagy control and renders the mutant virus (M1) particle susceptible to autophagic degradation and enhanced antigenic presentation.

5. Conclusions

As demonstrated in this review, development of Ad-based (vaccine) vectors with specific and desired properties is in part driven by a comparison between available Ad species without necessarily understanding the mechanistics behind existing differences. The lack of data concerning the underlying biology for most vectors, other than species C, means that relying on rare or non-human serotypes bears a risk of missing out on properties or overlooking problems that can arise. We strongly advocate to ameliorate this lack of knowledge by pairing vector development and application with investigating the underlying vector and virus biology. In the meantime, we believe that the existing knowledge base on Ad biology, especially for species C viruses, could help with the design of new Ad vectors, including rare or non-human species, using rational design. In this review, we highlighted the link between Ad post-entry sorting and the ensuing innate immune response and provided examples of how Ad post-entry sorting differs between species and has been modulated. The non-exhaustive list of possible capsid modifications exemplified on capsid protein VI and restricted to Ad2/5 shows the great potential that still lies buried in the Ad capsid, waiting to be exploited. Lysosomal membrane damage, such as early endosomal membrane damage, also triggers autophagy, but in addition results in oxidative stress that alters the innate immune response of the cell [383]. Altering the endo-lysosomal escape process of a species B, species D, or non-human Ad by introducing desirable mutations into the respective protein VI gene is only one of the possible options that come to mind. Other capsid proteins may hold additional functions related to transport or immune modulation. Extending rational capsid modifications to last-generation HC-Ad vectors and optimized antigen expression cassettes hold great promise for the next generation of Ad-based vaccine vectors.

Author Contributions: C.F.D., N.P. and H.W. wrote, reviewed, and edited the paper. C.F.D. and N.P. contributed equally to this manuscript. All authors have read and agreed to the published version of the manuscript.

Funding: C.F.D. received support from the Agence Nationale de la recherche (project VirMEDACO grant number ANR-19-CE15-0013-01) and N.P. received support from the Fondation pour la Recherche Médicale en France FRM DEQ20180339229. H.W. is an INSERM fellow.

Acknowledgments: We acknowledge the help of Sarah Cookson for manuscript preparation.

Conflicts of Interest: The authors declare no conflict of interest.

References

1. Ghosh, S.S.; Gopinath, P.; Ramesh, A. Adenoviral vectors: A promising tool for gene therapy. *Appl. Biochem. Biotechnol.* **2006**, *133*, 9–29. [CrossRef]
2. Adams, W.; Loré, K. Recombinant adenovirus vector infection of human dendritic cells. *Viral Gene Ther.* **2011**. [CrossRef]
3. Mercier, S.; Gahéry-Segard, H.; Monteil, M.; Lengagne, R.; Guillet, J.-G.; Eloit, M.; Denesvre, C. Distinct roles of adenovirus vector-transduced dendritic cells, myoblasts, and endothelial cells in mediating an immune response against a transgene product. *J. Virol.* **2002**, *76*, 2899–2911. [CrossRef]
4. Rowe, W.P.; Huebner, R.J.; Gilmore, L.K.; Parrott, R.H.; Ward, T.G. Isolation of a cytopathogenic agent from human adenoids undergoing spontaneous degeneration in tissue culture. *Proc. Soc. Exp. Biol. Med.* **1953**, *84*, 570–573. [CrossRef]
5. Enders, J.F.; Bell, J.A.; Dingle, J.H.; Francis, T.; Hilleman, M.R.; Huebner, R.J.; Payne, A.M.M. "Adenoviruses": Group name proposed for new respiratory-tract viruses. *Science* **1956**, *124*, 119–120. [CrossRef]
6. Lefkowitz, E.J.; Dempsey, D.M.; Hendrickson, R.C.; Orton, R.J.; Siddell, S.G.; Smith, D.B. Virus taxonomy: The database of the International Committee on Taxonomy of Viruses (ICTV). *Nucleic Acids Res.* **2018**, *46*, D708–D717. [CrossRef]
7. Harrach, B.; Tarján, Z.L.; Benkő, M. Adenoviruses across the animal kingdom: A walk in the zoo. *FEBS Lett.* **2019**, *593*, 3660–3673. [CrossRef] [PubMed]
8. Hage, E.; Liebert, U.G.; Bergs, S.; Ganzenmueller, T.; Heim, A. Human mastadenovirus type 70: A novel, multiple recombinant species D mastadenovirus isolated from diarrhoeal faeces of a haematopoietic stem cell transplantation recipient. *J. Gen. Virol.* **2015**, *96*, 2734–2742. [CrossRef]
9. Sarantis, H.; Johnson, G.; Brown, M.; Petric, M.; Tellier, R. Comprehensive detection and serotyping of human adenoviruses by PCR and sequencing. *J. Clin. Microbiol.* **2004**, *42*, 3963–3969. [CrossRef] [PubMed]
10. Crenshaw, B.J.; Jones, L.B.; Bell, C.R.; Kumar, S.; Matthews, Q.L. Perspective on adenoviruses: Epidemiology, pathogenicity, and gene therapy. *Biomedicines* **2019**, *7*, 61. [CrossRef] [PubMed]
11. Ghebremedhin, B. Human adenovirus: Viral pathogen with increasing importance. *Eur. J. Microbiol. Immunol.* **2014**, *4*, 26–33. [CrossRef]
12. Lion, T. Adenovirus infections in immunocompetent and immunocompromised patients. *Clin. Microbiol. Rev.* **2014**, *27*, 441–462. [CrossRef] [PubMed]
13. Dehghan, S.; Seto, J.; Liu, E.B.; Walsh, M.P.; Dyer, D.W.; Chodosh, J.; Seto, D. Computational analysis of four human adenovirus type 4 genomes reveals molecular evolution through two interspecies recombination events. *Virology* **2013**, *443*, 197–207. [CrossRef]

14. Dehghan, S.; Seto, J.; Liu, E.B.; Ismail, A.M.; Madupu, R.; Heim, A.; Jones, M.S.; Dyer, D.W.; Chodosh, J.; Seto, D. A zoonotic adenoviral human pathogen emerged through genomic recombination among human and nonhuman simian hosts. *J. Virol.* **2019**, *93*. [CrossRef] [PubMed]
15. Medkour, H.; Amona, I.; Akiana, J.; Davoust, B.; Bitam, I.; Levasseur, A.; Tall, M.L.; Diatta, G.; Sokhna, C.; Hernandez-Aguilar, R.A.; et al. Adenovirus infections in African humans and wild non-human primates: Great diversity and cross-species transmission. *Viruses* **2020**, *12*, 657. [CrossRef]
16. Hodgson, S.H.; Ewer, K.J.; Bliss, C.M.; Edwards, N.J.; Rampling, T.; Anagnostou, N.A.; de Barra, E.; Havelock, T.; Bowyer, G.; Poulton, I.D.; et al. Evaluation of the efficacy of ChAd63-MVA vectored vaccines expressing circumsporozoite protein and ME-TRAP against controlled human malaria infection in malaria-naive individuals. *J. Infect. Dis.* **2015**, *211*, 1076–1086. [CrossRef]
17. Tapia, M.D.; Sow, S.O.; Lyke, K.E.; Haidara, F.C.; Diallo, F.; Doumbia, M.; Traore, A.; Coulibaly, F.; Kodio, M.; Onwuchekwa, U.; et al. Use of ChAd3-EBO-Z Ebola virus vaccine in Malian and US adults, and boosting of Malian adults with MVA-BN-Filo: A phase 1, single-blind, randomised trial, a phase 1b, open-label and double-blind, dose-escalation trial, and a nested, randomised, double-bli. *Lancet Infect. Dis.* **2016**, *16*, 31–42. [CrossRef]
18. Guo, J.; Mondal, M.; Zhou, D. Development of novel vaccine vectors: Chimpanzee adenoviral vectors. *Hum. Vaccines Immunother.* **2018**, *14*, 1679–1685. [CrossRef]
19. Hollingdale, M.R.; Sedegah, M.; Limbach, K. Development of replication-deficient adenovirus malaria vaccines. *Expert Rev. Vaccines* **2017**, *16*, 261–271. [CrossRef]
20. Kümin, D.; Hofmann, C.; Rudolph, M.; Both, G.W.; Löser, P. Biology of ovine adenovirus infection of nonpermissive cells. *J. Virol.* **2002**, *76*, 10882–10893. [CrossRef]
21. Rasmussen, U.B.; Benchaibi, M.; Meyer, V.; Schlesinger, Y.; Schughart, K. Novel human gene transfer vectors: Evaluation of wild-type and recombinant animal adenoviruses in human-derived cells. *Hum. Gene Ther.* **1999**, *10*, 2587–2599. [CrossRef]
22. De Vleeschauwer, A.R.; Zhou, X.; Lefebvre, D.J.; Garnier, A.; Watier, F.; Pignon, C.; Lacour, S.A.; Zientara, S.; Bakkali-Kassimi, L.; de Clercq, K.; et al. A canine adenovirus type 2 vaccine vector confers protection against foot-and-mouth disease in guinea pigs. *Vaccine* **2018**, *36*, 2193–2198. [CrossRef]
23. Chen, E.C.; Yagi, S.; Kelly, K.R.; Mendoza, S.P.; Maninger, N.; Rosenthal, A.; Spinner, A.; Bales, K.L.; Schnurr, D.P.; Lerche, N.W.; et al. Cross-species transmission of a novel adenovirus associated with a fulminant pneumonia outbreak in a new world monkey colony. *PLoS Pathog.* **2011**, *7*, e1002155. [CrossRef]
24. Pérez-Illana, M.; Martínez, M.; Condezo, G.N.; Hernando-Pérez, M.; Mangroo, C.; Brown, M.; Marabini, R.; Martín, C.S. Cryo-EM structure of enteric adenovirus HAdV-F41 highlights structural variations among human adenoviruses. *Sci. Adv.* **2021**, *7*, 1–15. [CrossRef]
25. Berciaud, S.; Rayne, F.; Kassab, S.; Jubert, C.; Faure-Della Corte, M.; Salin, F.; Wodrich, H.; Lafon, M.E.; Barat, P.; Fayon, M.; et al. Adenovirus infections in Bordeaux University Hospital 2008–2010: Clinical and virological features. *J. Clin. Virol.* **2012**, *54*, 302–307. [CrossRef] [PubMed]
26. Khanal, S.; Ghimire, P.; Dhamoon, A.S. The repertoire of adenovirus in human disease: The innocuous to the deadly. *Biomedicines* **2018**, *6*, 30. [CrossRef] [PubMed]
27. Lynch, J.P.; Kajon, A.E. Adenovirus: Epidemiology, global spread of novel serotypes, and advances in treatment and prevention. *Semin. Respir. Crit. Care Med.* **2016**, *37*, 586–602. [CrossRef] [PubMed]
28. Robinson, C.M.; Seto, D.; Jones, M.S.; Dyer, D.W.; Chodosh, J. Molecular evolution of human species D adenoviruses. *Infect. Genet. Evol.* **2011**, *11*, 1208–1217. [CrossRef] [PubMed]
29. Greber, U.F. Disassembly: The adenovirus. *Rev. Med. Virol.* **1998**, *222*, 213–222. [CrossRef]
30. Chroboczek, J.; Bieber, F.; Jacrot, B. The sequence of the genome of adenovirus type 5 and its comparison with the genome of adenovirus type 2. *Virology* **1992**, *186*, 280–285. [CrossRef]
31. El Bakkouri, M.; Seiradake, E.; Cusack, S.; Ruigrok, R.W.H.; Schoehn, G. Structure of the C-terminal head domain of the fowl adenovirus type 1 short fibre. *Virology* **2008**, *378*, 169–176. [CrossRef]
32. Nemerow, G.R.; Stewart, P.L.; Reddy, V.S. Structure of human adenovirus. *Curr. Opin. Virol.* **2012**, *2*, 115–121. [CrossRef] [PubMed]
33. Stewart, P.L.; Burnett, R.M.; Cyrklaff, M.; Fuller, S.D. Image reconstruction reveals the complex molecular organization of adenovirus. *Cell* **1991**, *67*, 145–154. [CrossRef]
34. Fabry, C.M.S.; Rosa-Calatrava, M.; Conway, J.F.; Zubieta, C.; Cusack, S.; Ruigrok, R.W.H.; Schoehn, G. A quasi-atomic model of human adenovirus type 5 capsid. *EMBO J.* **2005**, *24*, 1645–1654. [CrossRef] [PubMed]
35. Liu, H.; Jin, L.; Koh, S.B.S.; Atanasov, I.; Schein, S.; Wu, L.; Zhou, Z.H. Atomic structure of human adenovirus by Cryo-EM reveals interactions among protein networks. *Science* **2010**, *329*, 1038–1043. [CrossRef] [PubMed]
36. Vassal-Stermann, E.; Effantin, G.; Zubieta, C.; Burmeister, W.; Iseni, F.; Wang, H.; Lieber, A.; Schoehn, G.; Fender, P. CryoEM structure of adenovirus type 3 fibre with desmoglein 2 shows an unusual mode of receptor engagement. *Nat. Commun.* **2019**, *10*, 1–7. [CrossRef]
37. Martín, C.S. Latest insights on adenovirus structure and assembly. *Viruses* **2012**, *4*, 847–877. [CrossRef]
38. Reddy, V.S.; Barry, M.A. Structural organization and protein-protein interactions in human adenovirus capsid. In *Subcellular Biochemistry*; Springer Science and Business Media: Berlin, Germany, 2021; Volume 96, pp. 503–518.

39. Cao, C.; Dong, X.; Wu, X.; Wen, B.; Ji, G.; Cheng, L.; Liu, H. Conserved fiber-penton base interaction revealed by nearly atomic resolution cryo-electron microscopy of the structure of adenovirus provides insight into receptor interaction. *J. Virol.* **2012**, *86*, 12322–12329. [CrossRef]
40. Zubieta, C.; Schoehn, G.; Chroboczek, J.; Cusack, S. The structure of the human adenovirus 2 penton. *Mol. Cell* **2005**, *17*, 121–135. [CrossRef] [PubMed]
41. Furcinitti, P.S.; van Oostrum, J.; Burnett, R.M. Adenovirus polypeptide IX revealed as capsid cement by difference images from electron microscopy and crystallography. *EMBO J.* **1989**, *8*, 3563–3570. [CrossRef] [PubMed]
42. Wodrich, H.; Guan, T.; Cingolani, G.; von Seggern, D.; Nemerow, G.; Gerace, L. Switch from capsid protein import to adenovirus assembly by cleavage of nuclear transport signals. *EMBO J.* **2003**, *22*, 6245–6255. [CrossRef]
43. San Martín, C. Transmission electron microscopy and the molecular structure of icosahedral viruses. *Arch. Biochem. Biophys.* **2015**, *581*, 59–67. [CrossRef] [PubMed]
44. Dai, X.; Wu, L.; Sun, R.; Zhou, Z.H. Atomic structures of minor proteins VI and VII in human adenovirus. *J. Virol.* **2017**, *91*. [CrossRef]
45. Russell, W.C. Adenoviruses: Update on structure and function. *J. Gen. Virol.* **2009**, *90*, 1–20. [CrossRef]
46. Chatterjee, P.K.; Vayda, M.E.; Flint, S.J. Interactions among the three adenovirus core proteins. *J. Virol.* **1985**, *55*, 379–386. [CrossRef] [PubMed]
47. Perez-Vargas, J.; Vaughan, R.C.; Houser, C.; Hastie, K.M.; Kao, C.C.; Nemerow, G.R. Isolation and characterization of the DNA and protein binding activities of adenovirus core protein V. *J. Virol.* **2014**, *88*, 9287–9296. [CrossRef]
48. Bauer, M.; Gomez-Gonzalez, A.; Suomalainen, M.; Hemmi, S.; Greber, U.F. The E3 ubiquitin ligase Mind bomb 1 enhances nuclear import of viral DNA by inactivating a virion linchpin protein that suppresses exposure of virion pathogen-associated molecular patterns. *bioRxiv* **2020**. [CrossRef]
49. Wang, I.H.; Suomalainen, M.; Andriasyan, V.; Kilcher, S.; Mercer, J.; Neef, A.; Luedtke, N.W.; Greber, U.F. Tracking viral genomes in host cells at single-molecule resolution. *Cell Host Microbe* **2013**, *14*, 468–480. [CrossRef]
50. Komatsu, T.; Dacheux, D.; Kreppel, F.; Nagata, K.; Wodrich, H. A method for visualization of incoming adenovirus chromatin complexes in fixed and living cells. *PLoS ONE* **2015**, *10*, e0137102. [CrossRef]
51. Greber, U.F.; Suomalainen, M.; Stidwill, R.P.; Boucke, K.; Ebersold, M.W.; Helenius, A. The role of the nuclear pore complex in adenovirus DNA entry. *EMBO J.* **1997**, *16*, 5998–6007. [CrossRef]
52. Karen, K.A.; Hearing, P. Adenovirus core protein VII protects the viral genome from a DNA damage response at early times after infection. *J. Virol.* **2011**, *85*, 4135–4142. [CrossRef] [PubMed]
53. Avgousti, D.C.; Herrmann, C.; Kulej, K.; Pancholi, N.J.; Sekulic, N.; Petrescu, J.; Molden, R.C.; Blumenthal, D.; Paris, A.J.; Reyes, E.D.; et al. A core viral protein binds host nucleosomes to sequester immune danger signals. *Nature* **2016**, *535*, 173–177. [CrossRef]
54. Avgousti, D.C.; della Fera, A.N.; Otter, C.J.; Herrmann, C.; Pancholi, N.J.; Weitzman, M.D. Adenovirus core protein VII downregulates the DNA damage response on the host genome. *J. Virol.* **2017**, *91*. [CrossRef] [PubMed]
55. Parks, R.J. Adenovirus protein IX: A new look at an old protein. *Mol. Ther.* **2005**, *11*, 19–25. [CrossRef] [PubMed]
56. Gorman, J.J.; Wallis, T.P.; Whelan, D.A.; Shaw, J.; Both, G.W. LH3, a "homologue" of the mastadenoviral E1B 55-kDa protein is a structural protein of atadenoviruses. *Virology* **2005**, *342*, 159–166. [CrossRef]
57. Colby, W.W.; Shenk, T. Adenovirus type 5 virions can be assembled in vivo in the absence of detectable polypeptide IX. *J. Virol.* **1981**, *39*, 977–980. [CrossRef]
58. Ostapchuk, P.; Suomalainen, M.; Zheng, Y.; Boucke, K.; Greber, U.F.; Hearing, P. The adenovirus major core protein VII is dispensable for virion assembly but is essential for lytic infection. *PLoS Pathog.* **2017**, *13*, e1006455. [CrossRef]
59. Cusack, S. Adenovirus complex structures. *Curr. Opin. Struct. Biol.* **2005**, *15*, 237–243. [CrossRef]
60. Berk, A.J.; Lee, F.; Harrison, T.; Williams, J.; Sharp, P.A. Pre-early adenovirus 5 gene product regulates synthesis of early viral messenger RNAs. *Cell* **1979**, *17*, 935–944. [CrossRef]
61. Zhang, W.; Imperiale, M.J. Requirement of the adenovirus IVa2 Protein for virus assembly. *J. Virol.* **2003**, *77*, 3586–3594. [CrossRef]
62. Ahi, Y.S.; Mittal, S.K. Components of adenovirus genome packaging. *Front. Microbiol.* **2016**, *7*, 1–15. [CrossRef]
63. Lehmberg, E.; Traina, J.A.; Chakel, J.A.; Chang, R.J.; Parkman, M.; McCaman, M.T.; Murakami, P.K.; Lahidji, V.; Nelson, J.W.; Hancock, W.S.; et al. Reversed-phase high-performance liquid chromatographic assay for the adenovirus type 5 proteome. *J. Chromatogr. B Biomed. Sci. Appl.* **1999**, *732*, 411–423. [CrossRef]
64. Stewart, P.L.; Fuller, S.D.; Burnett, R.M. Difference imaging of adenovirus: Bridging the resolution gap between X-ray crystallography and electron microscopy. *EMBO J.* **1993**, *12*, 2589–2599. [CrossRef] [PubMed]
65. Rancourt, C.; Tihanyi, K.; Bourbonniere, M.; Weber, J.M. Identification of active-site residues of the adenovirus endopeptidase. *Proc. Natl. Acad. Sci. USA* **1994**, *91*, 844–847. [CrossRef] [PubMed]
66. Mangel, W.F.; Martín, C.S. Structure, function and dynamics in adenovirus maturation. *Viruses* **2014**, *6*, 4536–4570. [CrossRef]
67. Perez-Berna, A.J.; Mangel, W.F.; McGrath, W.J.; Graziano, V.; Flint, J.; San Martin, C. Processing of the L1 52/55k protein by the adenovirus protease: A new substrate and new insights into virion maturation. *J. Virol.* **2014**, *88*, 1513–1524. [CrossRef] [PubMed]
68. Cotten, M.; Weber, J.M. The adenovirus protease is required for virus entry into host cells. *Virology* **1995**, *213*, 494–502. [CrossRef]
69. Pérez-Berná, A.J.; Marabini, R.; Scheres, S.H.W.; Menéndez-Conejero, R.; Dmitriev, I.P.; Curiel, D.T.; Mangel, W.F.; Flint, S.J.; Martín, C.S. Structure and uncoating of immature adenovirus. *J. Mol. Biol.* **2009**, *392*, 547–557. [CrossRef] [PubMed]

70. Wu, E.; Pache, L.; von Seggern, D.J.; Mullen, T.-M.; Mikyas, Y.; Stewart, P.L.; Nemerow, G.R. Flexibility of the adenovirus fiber is required for efficient receptor interaction. *J. Virol.* **2003**, *77*, 7225–7235. [CrossRef]
71. Seki, T.; Dmitriev, I.; Kashentseva, E.; Takayama, K.; Rots, M.; Suzuki, K.; Curiel, D.T. Artificial extension of the adenovirus fiber shaft inhibits infectivity in coxsackievirus and adenovirus receptor-positive cell lines. *J. Virol.* **2002**, *76*, 1100–1108. [CrossRef] [PubMed]
72. Shayakhmetov, D.M.; Lieber, A. Dependence of adenovirus infectivity on length of the fiber shaft domain. *J. Virol.* **2000**, *74*, 10274–10286. [CrossRef]
73. Fender, P.; Ruigrok, R.W.H.; Gout, E.; Buffet, S.; Chroboczek, J. Adenovirus dodecahedron, a new vector for human gene transfer. *Nat. Biotechnol.* **1997**, *15*, 52–56. [CrossRef]
74. Fender, P.; Schoehn, G.; Foucaud-Gamen, J.; Gout, E.; Garcel, A.; Drouet, E.; Chroboczek, J. Adenovirus dodecahedron allows large multimeric protein transduction in human cells. *J. Virol.* **2003**, *77*, 4960–4964. [CrossRef]
75. Pied, N.; Wodrich, H. Imaging the adenovirus infection cycle. *FEBS Lett.* **2019**, *593*, 3419–3448. [CrossRef]
76. Nemerow, G.R.; Pache, L.; Reddy, V.; Stewart, P.L. Insights into adenovirus host cell interactions from structural studies. *Virology* **2009**, *384*, 380–388. [CrossRef]
77. Kremer, E.J.; Nemerow, G.R. Adenovirus tales: From the cell surface to the nuclear pore complex. *PLoS Pathog.* **2015**, *11*, e1004821. [CrossRef] [PubMed]
78. Philipson, L.; Lonberg-holm, K.; Pettersson, U.L.F. Virus-receptor interaction in an adenovirus system. *J. Virol.* **1968**, *2*, 1064–1075. [CrossRef]
79. Bergelson, J.M.; Cunningham, J.A.; Droguett, G.; Kurt-Jones, E.A.; Krithivas, A.; Hong, J.S.; Horwitz, M.S.; Crowell, R.L.; Finberg, R.W. Isolation of a common receptor for coxsackie B viruses and adenoviruses 2 and 5. *Science* **1997**, *275*, 1320–1323. [CrossRef] [PubMed]
80. Roelvink, P.W.; Lizonova, A.; Lee, J.G.M.; Li, Y.; Bergelson, J.M.; Finberg, R.W.; Brough, D.E.; Kovesdi, I.; Wickham, T.J. The coxsackievirus-adenovirus receptor protein can function as a cellular attachment protein for adenovirus serotypes from subgroups A, C, D, E, and F. *J. Virol.* **1998**, *72*, 7909–7915. [CrossRef] [PubMed]
81. Mese, K.; Bunz, O.; Schellhorn, S.; Volkwein, W.; Jung, D.; Gao, J.; Zhang, W.; Baiker, A.; Ehrhardt, A. Identification of novel human adenovirus candidates using the coxsackievirus and adenovirus receptor for cell entry. *Virol. J.* **2020**, *17*, 1–7. [CrossRef]
82. Gaggar, A.; Shayakhmetov, D.M.; Lieber, A. CD46 is a cellular receptor for group B adenoviruses. *Nat. Med.* **2003**, *9*, 1408–1412. [CrossRef]
83. Wang, H.; Li, Z.Y.; Liu, Y.; Persson, J.; Beyer, I.; Möller, T.; Koyuncu, D.; Drescher, M.R.; Strauss, R.; Zhang, X.B.; et al. Desmoglein 2 is a receptor for adenovirus serotypes 3, 7, 11 and 14. *Nat. Med.* **2011**, *17*, 96–104. [CrossRef]
84. Arnberg, N.; Pring-Åkerblom, P.; Wadell, G. Adenovirus type 37 uses sialic acid as a cellular receptor on chang C cells. *J. Virol.* **2002**, *76*, 8834–8841. [CrossRef]
85. Arnberg, N.; Kidd, A.H.; Edlund, K.; Nilsson, J.; Pring-Åkerblom, P.; Wadell, G. Adenovirus type 37 binds to cell surface sialic acid through a charge-dependent interaction. *Virology* **2002**, *302*, 33–43. [CrossRef]
86. Arnberg, N.; Kidd, A.H.; Edlund, K.; Olfat, F.; Wadell, G. Initial interactions of subgenus D adenoviruses with A549 cellular receptors: Sialic acid versus αVintegrins. *J. Virol.* **2000**, *74*, 7691–7693. [CrossRef]
87. Wu, H.; Curiel, D.T. Fiber-modified adenoviruses for targeted gene therapy. *Methods Mol. Biol.* **2008**, *434*, 113–132. [CrossRef]
88. Noureddini, S.C.; Curiel, D.T. Genetic targeting strategies for adenovirus. *Mol. Pharm.* **2005**, *2*, 341–347. [CrossRef] [PubMed]
89. Nakayama, M.; Both, G.W.; Banizs, B.; Tsuruta, Y.; Yamamoto, S.; Kawakami, Y.; Douglas, J.T.; Tani, K.; Curiel, D.T.; Glasgow, J.N. An adenovirus serotype 5 vector with fibers derived from ovine atadenovirus demonstrates CAR-independent tropism and unique biodistribution in mice. *Virology* **2006**, *350*, 103–115. [CrossRef] [PubMed]
90. Wickham, T.J.; Mathias, P.; Cheresh, D.A.; Nemerow, G.R. Integrins aJS and U& promote adenovirus internalization but not virus attachment. *Cell* **1993**, *73*, 309–319. [PubMed]
91. Bai, M.; Harfe, B.; Freimuth, P. Mutations that alter an Arg-Gly-Asp (RGD) sequence in the adenovirus type 2 penton base protein abolish its cell-rounding activity and delay virus reproduction in flat cells. *J. Virol.* **1993**, *67*, 5198–5205. [CrossRef] [PubMed]
92. Albinsson, B.; Kidd, A.H. Adenovirus type 41 lacks an RGD α(v)-integrin binding motif on the penton base and undergoes delayed uptake in A549 cells. *Virus Res.* **1999**, *64*, 125–136. [CrossRef]
93. Chiu, C.Y.; Mathias, P.; Nemerow, G.R.; Stewart, P.L. Structure of adenovirus complexed with its internalization receptor, αvβ5 integrin. *J. Virol.* **1999**, *73*, 6759–6768. [CrossRef]
94. Li, E.; Stupack, D.; Bokoch, G.M.; Nemerow, G.R. Adenovirus endocytosis requires actin cytoskeleton reorganization mediated by rho family GTPases. *J. Virol.* **1998**, *72*, 8806–8812. [CrossRef]
95. Li, E.; Stupack, D.G.; Brown, S.L.; Klemke, R.; Schlaepfer, D.D.; Nemerow, G.R. Association of p130(CAS) with phosphatidylinositol-3-OH kinase mediates adenovirus cell entry. *J. Biol. Chem.* **2000**, *275*, 14729–14735. [CrossRef]
96. Wang, K.; Huang, S.; Kapoor-Munshi, A.; Nemerow, G. Adenovirus internalization and infection require dynamin. *J. Virol.* **1998**, *72*, 3455–3458. [CrossRef] [PubMed]
97. Meier, O.; Boucke, K.; Hammer, S.V.; Keller, S.; Stidwill, R.P.; Hemmi, S.; Greber, U.F. Adenovirus triggers macropinocytosis and endosomal leakage together with its clathrin-mediated uptake. *J. Cell Biol.* **2002**, *158*, 1119–1131. [CrossRef] [PubMed]

98. Lindert, S.; Silvestry, M.; Mullen, T.-M.; Nemerow, G.R.; Stewart, P.L. Cryo-electron microscopy structure of an adenovirus-integrin complex indicates conformational changes in both penton base and integrin. *J. Virol.* **2009**, *83*, 11491–11501. [CrossRef] [PubMed]
99. Snijder, J.; Reddy, V.S.; May, E.R.; Roos, W.H.; Nemerow, G.R.; Wuite, G.J.L. Integrin and defensin modulate the mechanical properties of adenovirus. *J. Virol.* **2013**, *87*, 2756–2766. [CrossRef]
100. Greber, U.F.; Willetts, M.; Webster, P.; Helenius, A. Stepwise dismantling of adenovirus 2 during entry into cells. *Cell* **1993**, *75*, 477–486. [CrossRef]
101. Martin-Fernandez, M.; Longshaw, S.V.; Kirby, I.; Santis, G.; Tobin, M.J.; Clarke, D.T.; Jones, G.R. Adenovirus type-5 entry and disassembly followed in living cells by FRET, fluorescence anisotropy, and FLIM. *Biophys. J.* **2004**, *87*, 1316–1327. [CrossRef]
102. Burckhardt, C.J.; Suomalainen, M.; Schoenenberger, P.; Boucke, K.; Hemmi, S.; Greber, U.F. Drifting motions of the adenovirus receptor CAR and immobile integrins initiate virus uncoating and membrane lytic protein exposure. *Cell Host Microbe* **2011**, *10*, 105–117. [CrossRef]
103. Wiethoff, C.M.; Wodrich, H.; Gerace, L.; Nemerow, G.R. Adenovirus protein VI mediates membrane disruption following capsid disassembly. *J. Virol.* **2005**, *79*, 1992–2000. [CrossRef]
104. Wodrich, H.; Henaff, D.; Jammart, B.; Segura-Morales, C.; Seelmeir, S.; Coux, O.; Ruzsics, Z.; Wiethoff, C.M.; Kremer, E.J. A capsid-encoded PPxY-motif facilitates adenovirus entry. *PLoS Pathog.* **2010**, *6*, e1000808. [CrossRef]
105. Maier, O.; Galan, D.L.; Wodrich, H.; Wiethoff, C.M. An N-terminal domain of adenovirus protein VI fragments membranes by inducing positive membrane curvature. *Virology* **2010**, *402*, 11–19. [CrossRef] [PubMed]
106. Maier, O.; Wiethoff, C.M. N-terminal α-helix-independent membrane interactions facilitate adenovirus protein VI induction of membrane tubule formation. *Virology* **2010**, *408*, 31–38. [CrossRef] [PubMed]
107. Montespan, C.; Wiethoff, C.M.; Wodrich, H. A small viral PPxY peptide motif to control antiviral autophagy. *J. Virol.* **2017**, *91*, 1–7. [CrossRef] [PubMed]
108. Montespan, C.; Marvin, S.A.; Austin, S.; Burrage, A.M.; Roger, B.; Rayne, F.; Faure, M.; Campell, E.M.; Schneider, C.; Reimer, R.; et al. Multi-layered control of Galectin-8 mediated autophagy during adenovirus cell entry through a conserved PPxY motif in the viral capsid. *PLoS Pathog.* **2017**, *13*, e1006217. [CrossRef]
109. Daussy, C.F.; Wodrich, H. "Repair me if you can": Membrane damage, response, and control from the viral perspective. *Cells* **2020**, *9*, 2042. [CrossRef]
110. Gastaldelli, M.; Imelli, N.; Boucke, K.; Amstutz, B.; Meier, O.; Greber, U.F. Infectious adenovirus type 2 transport through early but not late endosomes. *Traffic* **2008**, *9*, 2265–2278. [CrossRef] [PubMed]
111. Miyazawa, N.; Leopold, P.L.; Hackett, N.R.; Ferris, B.; Worgall, S.; Falck-Pedersen, E.; Crystal, R.G. Fiber swap between adenovirus subgroups B and C alters intracellular trafficking of adenovirus gene transfer vectors. *J. Virol.* **1999**, *73*, 6056–6065. [CrossRef]
112. Shayakhmetov, D.M.; Li, Z.-Y.; Ternovoi, V.; Gaggar, A.; Gharwan, H.; Lieber, A. The interaction between the fiber knob domain and the cellular attachment receptor determines the intracellular trafficking route of adenoviruses. *J. Virol.* **2003**, *77*, 3712–3723. [CrossRef] [PubMed]
113. Teigler, J.E.; Kagan, J.C.; Barouch, D.H. Late endosomal trafficking of alternative serotype adenovirus vaccine vectors augments antiviral innate immunity. *J. Virol.* **2014**, *88*, 10354–10363. [CrossRef] [PubMed]
114. Barlan, A.U.; Griffin, T.M.; Mcguire, K.A.; Wiethoff, C.M. Adenovirus membrane penetration activates the NLRP3 inflammasome. *J. Virol.* **2011**, *85*, 146–155. [CrossRef] [PubMed]
115. Bremner, K.; Scherer, J.; Yi, J. Adenovirus transport through a direct cytoplasmic dynein-hexon interaction. *Cell Host* **2009**, *6*, 523–535. [CrossRef] [PubMed]
116. Leopold, P.L.; Kreitzer, G.; Miyazawa, N.; Rempel, S.; Pfister, K.K.; Rodriguez-Boulan, E.; Crystal, R.G. Dynein- and microtubule-mediated translocation of adenovirus serotype 5 occurs after endosomal lysis. *Hum. Gene Ther.* **2000**, *11*, 151–165. [CrossRef]
117. Suomalainen, M.; Nakano, M.Y.; Keller, S.; Boucke, K.; Stidwill, R.P.; Greber, U.F. Microtubule-dependent plus- and minus end-directed motilities are competing processes for nuclear targeting of adenovirus. *J. Cell Biol.* **1999**, *144*, 657–672. [CrossRef]
118. Scherer, J.; Vallee, R.B. Conformational changes in the adenovirus hexon subunit responsible for regulating cytoplasmic dynein recruitment. *J. Virol.* **2015**, *89*, 1013–1023. [CrossRef]
119. Kelkar, S.A.; Pfister, K.K.; Crystal, R.G.; Leopold, P.L. Cytoplasmic dynein mediates adenovirus binding to microtubules. *J. Virol.* **2004**, *78*, 10122–10132. [CrossRef]
120. Zhou, J.; Scherer, J.; Yi, J.; Vallee, R.B. Role of kinesins in directed adenovirus transport and cytoplasmic exploration. *PLoS Pathog.* **2018**, *14*, e1007055. [CrossRef]
121. Trotman, L.C.; Mosberger, N.; Fornerod, M.; Stidwill, R.P.; Greber, U.F. Import of adenovirus DNA involves the nuclear pore complex receptor CAN/Nup214 and histone H1. *Nat. Cell Biol.* **2001**, *3*, 1092–1100. [CrossRef]
122. Cassany, A.; Ragues, J.; Guan, T.; Bégu, D.; Wodrich, H.; Kann, M.; Nemerow, G.R.; Gerace, L. Nuclear import of adenovirus DNA involves direct interaction of hexon with an N-terminal domain of the nucleoporin Nup214. *J. Virol.* **2015**, *89*, 1719–1730. [CrossRef]
123. Strunze, S.; Engelke, M.F.; Wang, I.H.; Puntener, D.; Boucke, K.; Schleich, S.; Way, M.; Schoenenberger, P.; Burckhardt, C.J.; Greber, U.F. Kinesin-1-mediated capsid disassembly and disruption of the nuclear pore complex promote virus infection. *Cell Host Microbe* **2011**, *10*, 210–223. [CrossRef]

124. Puntener, D.; Engelke, M.F.; Ruzsics, Z.; Strunze, S.; Wilhelm, C.; Greber, U.F. Stepwise loss of fluorescent core protein V from human adenovirus during entry into cells. *J. Virol.* **2011**, *85*, 481–496. [CrossRef]
125. Crisostomo, L.; Soriano, A.M.; Mendez, M.; Graves, D.; Pelka, P. Temporal dynamics of adenovirus 5 gene expression in normal human cells. *PLoS ONE* **2019**, *14*, e0211192. [CrossRef] [PubMed]
126. Thomas, G.P.; Mathews, M.B. DNA replication and the early to late transition in adenovirus infection. *Cell* **1980**, *22*, 523–533. [CrossRef]
127. Nevins, J.R. Mechanism of activation of early viral transcription by the adenovirus E1A gene product. *Cell* **1981**, *26*, 213–220. [CrossRef]
128. Flint, S.J. Regulation of adenovirus mrna formation. *Adv. Virus Res.* **1986**, *31*, 169–228. [CrossRef]
129. Nevins, J.R. Regulation of early adenovirus gene expression. *Microbiol. Rev.* **1987**, *51*, 419–430. [CrossRef] [PubMed]
130. Farley, D.C.; Brown, J.L.; Leppard, K.N. Activation of the early-late switch in adenovirus type 5 major late transcription unit expression by L4 gene products. *J. Virol.* **2004**, *78*, 1782–1791. [CrossRef] [PubMed]
131. Hidalgo, P.; Gonzalez, R.A. Formation of adenovirus DNA replication compartments. *FEBS Lett.* **2019**, *593*, 3518–3530. [CrossRef] [PubMed]
132. Mondersert, G.; Tribouley, C.; Kedinger, C. Identification of a novel downstream bingind protein implicated in late-phase-specific activation of the adenovirus major late promoter. *Nucleic Acids Res.* **1992**, *20*, 3881–3889. [CrossRef] [PubMed]
133. Shaw, A.R.; Ziff, E.B. Transcripts from the adenovirus-2 major late promoter yield a single early family of 3′ coterminal mRNAs and five late families. *Cell* **1980**, *22*, 905–916. [CrossRef]
134. Weber, J.M.; Déry, C.V.; Amin Mirza, M.; Horvath, J. Adenovirus DNA synthesis is coupled to virus assembly. *Virology* **1985**, *140*, 351–359. [CrossRef]
135. Tollefson, A.E.; Ryerse, J.S.; Scaria, A.; Hermiston, T.W.; Wold, W.S.M. The E3-11.6-kDa adenovirus death protein (ADP) is required for efficient cell death: Characterization of cells infected with adp mutants. *Virology* **1996**, *220*, 152–162. [CrossRef] [PubMed]
136. Tollefson, A.E.; Scaria, A.; Hermiston, T.W.; Ryerse, J.S.; Wold, L.J.; Wold, W.S. The adenovirus death protein (E3-11.6K) is required at very late stages of infection for efficient cell lysis and release of adenovirus from infected cells. *J. Virol.* **1996**, *70*, 2296–2306. [CrossRef]
137. Weber, J. Genetic analysis of adenovirus type 2 III. Temperature sensitivity of processing viral proteins. *J. Virol.* **1976**, *17*, 462–471. [CrossRef]
138. Greber, U.F.; Webster, P.; Weber, J.; Helenius, A. The role of the adenovirus protease in virus entry into cells. *EMBO J.* **1996**, *15*, 1766–1777. [CrossRef] [PubMed]
139. Nguyen, E.K.; Nemerow, G.R.; Smith, J.G. Direct evidence from single-cell analysis that human α-defensins block adenovirus uncoating to neutralize infection. *J. Virol.* **2010**, *84*, 4041–4049. [CrossRef]
140. Worgall, S.; Wolff, G.; Falck-Pedersen, E.; Crystal, R.G. Innate immune mechanisms dominate elimination of adenoviral vectors following in vivo administration. *Hum. Gene Ther.* **1997**, *8*, 37–44. [CrossRef]
141. Yang, Y.; Nunes, F.A.; Berencsi, K.; Furth, E.E.; Gönczöl, E.; Wilson, J.M. Cellular immunity to viral antigens limits E1-deleted adenoviruses for gene therapy. *Proc. Natl. Acad. Sci. USA* **1994**, *91*, 4407–4411. [CrossRef]
142. Tatsis, N.; Ertl, H.C.J. Adenoviruses as vaccine vectors. *Mol. Ther.* **2004**, *10*, 616–629. [CrossRef] [PubMed]
143. Li, J.X.; Hou, L.H.; Meng, F.Y.; Wu, S.P.; Hu, Y.M.; Liang, Q.; Chu, K.; Zhang, Z.; Xu, J.J.; Tang, R.; et al. Immunity duration of a recombinant adenovirus type-5 vector-based Ebola vaccine and a homologous prime-boost immunisation in healthy adults in China: Final report of a randomised, double-blind, placebo-controlled, phase 1 trial. *Lancet Glob. Health* **2017**, *5*, e324–e334. [CrossRef]
144. Raper, S.E.; Chirmule, N.; Lee, F.S.; Wivel, N.A.; Bagg, A.; Gao, G.P.; Wilson, J.M.; Batshaw, M.L. Fatal systemic inflammatory response syndrome in a ornithine transcarbamylase deficient patient following adenoviral gene transfer. *Mol. Genet. Metab.* **2003**, *80*, 148–158. [CrossRef] [PubMed]
145. Schnell, M.A.; Zhang, Y.; Tazelaar, J.; Gao, G.P.; Yu, Q.C.; Qian, R.; Chen, S.J.; Varnavski, A.N.; LeClair, C.; Raper, S.E.; et al. Activation of innate immunity in nonhuman primates following intraportal administration of adenoviral vectors. *Mol. Ther.* **2001**, *3*, 708–722. [CrossRef] [PubMed]
146. Hendrickx, R.; Stichling, N.; Koelen, J.; Kuryk, L.; Lipiec, A.; Greber, U.F. Innate immunity to adenovirus. *Hum. Gene Ther.* **2014**, *25*, 265–284. [CrossRef] [PubMed]
147. Atasheva, S.; Shayakhmetov, D.M. *Innate Immune Response to Adenovirus Vector Administration in Vivo*, 2nd ed.; Elsevier Inc.: Amsterdam, The Netherlands, 2016; ISBN 9780128002766.
148. Atasheva, S.; Yao, J.; Shayakhmetov, D.M. Innate immunity to adenovirus: Lessons from mice. *FEBS Lett.* **2019**, *593*, 3461–3483. [CrossRef] [PubMed]
149. Gregory, S.M.; Nazir, S.A.; Metcalf, J.P. Implications of the innate immune response to adenovirus and adenoviral vectors. *Future Virol.* **2011**, *6*, 357–374. [CrossRef]
150. Borgland, S.L.; Bowen, G.P.; Wong, N.C.W.; Libermann, T.A.; Muruve, D.A. Adenovirus vector-induced expression of the C-X-C chemokine IP-10 is mediated through capsid-dependent activation of NF-κB. *J. Virol.* **2000**, *74*, 3941–3947. [CrossRef]
151. Bowen, G.P.; Borgland, S.L.; Lam, M.; Libermann, T.A.; Wong, N.C.W.; Muruve, D.A. Adenovirus vector-induced inflammation: Capsid-dependent induction of the C-C chemokine RANTES requires NF-κB. *Hum. Gene Ther.* **2002**, *13*, 367–379. [CrossRef]

152. Higginbotham, J.N.; Seth, P.; Blaese, R.M.; Ramsey, W.J. The release of inflammatory cytokines from human peripheral blood mononuclear cells in vitro following exposure to adenovirus variants and capsid. *Hum. Gene Ther.* **2002**, *13*, 129–141. [CrossRef] [PubMed]
153. Mccoy, R.D.; Roessler, B.J.; Huffnagle, G.A.R.Y.B.; Janich, S.L.; Laing, T.J.; Sevion, R.H. Pulmonary inflammation induced by incomplete or inactivated adenoviral particles. *Hum. Gene Ther.* **1995**, *6*, 1553–1560. [CrossRef] [PubMed]
154. Muruve, D.A.; Barnes, M.J.; Stillman, I.E.; Libermann, T.A. Adenoviral gene therapy leads to rapid induction of multiple chemokines and acute neutrophil-dependent hepatic injury in vivo. *Hum. Gene Ther.* **1999**, *10*, 965–976. [CrossRef] [PubMed]
155. Muruve, D.A. The innate immune response to adenovirus vectors. *Hum. Gene Ther.* **2004**, *15*, 1157–1166. [CrossRef] [PubMed]
156. Zhang, Y.; Chirmule, N.; Gao, G.P.; Qian, R.; Croyle, M.; Joshi, B.; Tazelaar, J.; Wilson, J.M. Acute cytokine response to systemic adenoviral vectors in mice is mediated by dendritic cells and macrophages. *Mol. Ther.* **2001**, *3*, 697–707. [CrossRef] [PubMed]
157. Tamanini, A.; Nicolis, E.; Bonizzato, A.; Bezzerri, V.; Melotti, P.; Assael, B.M.; Cabrini, G. Interaction of adenovirus type 5 fiber with the coxsackievirus and adenovirus receptor activates inflammatory response in human respiratory cells. *J. Virol.* **2006**, *80*, 11241–11254. [CrossRef] [PubMed]
158. Farmer, C.; Morton, P.E.; Snippe, M.; Santis, G.; Parsons, M. Coxsackie adenovirus receptor (CAR) regulates integrin function through activation of p44/42 MAPK. *Exp. Cell Res.* **2009**, *315*, 2637–2647. [CrossRef]
159. Iacobelli-Martinez, M.; Nepomuceno, R.R.; Connolly, J.; Nemerow, G.R. CD46-utilizing adenoviruses inhibit C/EBPβ-dependent expression of proinflammatory cytokines. *J. Virol.* **2005**, *79*, 11259–11268. [CrossRef]
160. Di Paolo, N.C.; Miao, E.A.; Iwakura, Y.; Kaja, M.; Flavell, R.A.; Papayannopoulou, T.; Shayakhmetov, D.M. Virus sensing at the plasma membrane triggers interleukin-1α-mediated pro-inflammatory macrophage response in vivo. *Thalia Stud. Lit. Humor* **2010**, *31*, 110–121. [CrossRef]
161. Suomalainen, M.; Nakano, M.Y.; Boucke, K.; Keller, S.; Greber, U.F. Adenovirus-activated PKA and p38/MAPK pathways boost microtubule-mediated nuclear targeting of virus. *EMBO J.* **2001**, *20*, 1310–1319. [CrossRef]
162. Tibbles, L.A.; Spurrell, J.C.L.; Bowen, G.P.; Liu, Q.; Lam, M.; Zaiss, A.K.; Robbins, S.M.; Hollenberg, M.D.; Wickham, T.J.; Muruve, D.A. Activation of p38 and ERK signaling during adenovirus vector cell entry lead to expression of the C-X-C chemokine IP-10. *J. Virol.* **2002**, *76*, 1559–1568. [CrossRef]
163. Li, E.; Stupack, D.; Klemke, R.; Cheresh, D.A.; Nemerow, G.R. Adenovirus endocytosis via αvintegrins requires phosphoinositide-3-OH kinase. *J. Virol.* **1998**, *72*, 2055–2061. [CrossRef]
164. Philpott, N.J.; Nociari, M.; Elkon, K.B.; Falck-Pedersen, E. Adenovirus-induced maturation of dendritic cells through a PI3 kinase-mediated TNF-α induction pathway. *Proc. Natl. Acad. Sci. USA* **2004**, *101*, 6200–6205. [CrossRef]
165. Zsengellér, Z.; Otake, K.; Hossain, S.-A.; Berclaz, P.-Y.; Trapnell, B.C. Internalization of adenovirus by alveolar macrophages initiates early proinflammatory signaling during acute respiratory tract infection. *J. Virol.* **2000**, *74*, 9655–9667. [CrossRef]
166. Liu, Q.; Muruve, D.A. Molecular basis of the inflammatory response to adenovirus vectors. *Gene Ther.* **2003**, *10*, 935–940. [CrossRef]
167. Cerullo, V.; Seiler, M.P.; Mane, V.; Brunetti-Pierri, N.; Clarke, C.; Bertin, T.K.; Rodgers, J.R.; Lee, B. Toll-like receptor 9 triggers an innate immune response to helper-dependent adenoviral vectors. *Mol. Ther.* **2007**, *15*, 378–385. [CrossRef]
168. Fejer, G.; Drechsel, L.; Liese, J.; Schleicher, U.; Ruzsics, Z.; Imelli, N.; Greber, U.F.; Keck, S.; Hildenbrand, B.; Krug, A.; et al. Key role of splenic myeloid DCs in the IFN-αβ response to adenoviruses in vivo. *PLoS Pathog.* **2008**, *4*, e1000208. [CrossRef] [PubMed]
169. Zhu, J.; Huang, X.; Yang, Y. Innate immune response to adenoviral vectors is mediated by both toll-like receptor-dependent and -independent pathways. *J. Virol.* **2007**, *81*, 3170–3180. [CrossRef]
170. Stein, S.C.; Falck-Pedersen, E. Sensing adenovirus infection: Activation of interferon regulatory factor 3 in RAW 264.7 cells. *J. Virol.* **2012**, *86*, 4527–4537. [CrossRef] [PubMed]
171. Nociari, M.; Ocheretina, O.; Murphy, M.; Falck-Pedersen, E. Adenovirus induction of IRF3 occurs through a binary trigger targeting jun N-terminal kinase and TBK1 kinase cascades and type I interferon autocrine signaling. *J. Virol.* **2009**, *83*, 4081–4091. [CrossRef] [PubMed]
172. Muruve, D.A.; Pétrilli, V.; Zaiss, A.K.; White, L.R.; Clark, S.A.; Ross, P.J.; Parks, R.J.; Tschopp, J. The inflammasome recognizes cytosolic microbial and host DNA and triggers an innate immune response. *Nature* **2008**, *452*, 103–107. [CrossRef]
173. Lam, E.; Falck-Pedersen, E. Unabated adenovirus replication following activation of the cGAS/STING-dependent antiviral response in human cells. *J. Virol.* **2014**, *88*, 14426–14439. [CrossRef]
174. Pahl, J.H.W.; Verhoeven, D.H.J.; Kwappenberg, K.M.C.; Vellinga, J.; Lankester, A.C.; van Tol, M.J.D.; Schilham, M.W. Adenovirus type 35, but not type 5, stimulates NK cell activation via plasmacytoid dendritic cells and TLR9 signaling. *Mol. Immunol.* **2012**, *51*, 91–100. [CrossRef]
175. Wohlfart, C.E.; Svensson, U.K.; Everitt, E. Interaction between HeLa cells and adenovirus type 2 virions neutralized by different antisera. *J. Virol.* **1985**, *56*, 896–903. [CrossRef]
176. Watson, G.; Burdon, M.G.; Russell, W.C. An antigenic analysis of the adenovirus type 2 fibre polypeptide. *J. Gen. Virol.* **1988**, *69*, 525–535. [CrossRef] [PubMed]
177. Chen, H.; Xiang, Z.Q.; Li, Y.; Kurupati, R.K.; Jia, B.; Bian, A.; Zhou, D.M.; Hutnick, N.; Yuan, S.; Gray, C.; et al. Adenovirus-based vaccines: Comparison of vectors from three species of adenoviridae. *J. Virol.* **2010**, *84*, 10522–10532. [CrossRef]

178. Roberts, D.M.; Nanda, A.; Havenga, M.J.E.; Abbink, P.; Lynch, D.M.; Ewald, B.A.; Liu, J.; Thorner, A.R.; Swanson, P.E.; Gorgone, D.A.; et al. Hexon-chimaeric adenovirus serotype 5 vectors circumvent pre-existing anti-vector immunity. *Nature* **2006**, *441*, 239–243. [CrossRef]
179. Sumida, S.M.; Truitt, D.M.; Lemckert, A.A.C.; Vogels, R.; Custers, J.H.H.V.; Addo, M.M.; Lockman, S.; Peter, T.; Peyerl, F.W.; Kishko, M.G.; et al. Neutralizing antibodies to adenovirus serotype 5 vaccine vectors are directed primarily against the adenovirus hexon protein. *J. Immunol.* **2005**, *174*, 7179–7185. [CrossRef]
180. Bradley, R.R.; Lynch, D.M.; Iampietro, M.J.; Borducchi, E.N.; Barouch, D.H. Adenovirus serotype 5 neutralizing antibodies target both hexon and fiber following vaccination and natural infection. *J. Virol.* **2012**, *86*, 625–629. [CrossRef]
181. Ahi, S.Y.; Bangari, D.S.; Mittal, S.K. Adenoviral vector immunity: Its implications and circumvention strategies. *Curr. Gene Ther.* **2011**, *11*, 307–320. [CrossRef]
182. Mast, T.C.; Kierstead, L.; Gupta, S.B.; Nikas, A.A.; Kallas, E.G.; Novitsky, V.; Mbewe, B.; Pitisuttithum, P.; Schechter, M.; Vardas, E.; et al. International epidemiology of human pre-existing adenovirus (Ad) type-5, type-6, type-26 and type-36 neutralizing antibodies: Correlates of high Ad5 titers and implications for potential HIV vaccine trials. *Vaccine* **2010**, *28*, 950–957. [CrossRef]
183. Zaiss, A.K.; Machado, H.B.; Herschman, H.R. The influence of innate and pre-existing immunity on adenovirus therapy. *J. Cell. Biochem.* **2009**, *108*, 778–790. [CrossRef]
184. Colloca, S.; Barnes, E.; Folgori, A.; Ammendola, V.; Capone, S.; Cirillo, A.; Siani, L.; Naddeo, M.; Grazioli, F.; Esposito, M.L.; et al. Generation and screening of a large collection of novel simian Adenovirus allows the identification of vaccine vectors inducing potent cellular immunity in humans. *Sci. Transl. Med.* **2012**, *4*, 115ra2. [CrossRef]
185. Nébié, I.; Edwards, N.J.; Tiono, A.B.; Ewer, K.J.; Sanou, G.S.; Soulama, I.; Sanon, S.; Diarra, A.; Yaro, J.B.; Kangoye, D.; et al. Assessment of chimpanzee adenovirus serotype 63 neutralizing antibodies prior to evaluation of a candidate malaria vaccine regimen based on viral vectors. *Clin. Vaccine Immunol.* **2014**, *21*, 901–903. [CrossRef]
186. De Santis, O.; Audran, R.; Pothin, E.; Warpelin-Decrausaz, L.; Vallotton, L.; Wuerzner, G.; Cochet, C.; Estoppey, D.; Steiner-Monard, V.; Lonchampt, S.; et al. Safety and immunogenicity of a chimpanzee adenovirus-vectored Ebola vaccine in healthy adults: A randomised, double-blind, placebo-controlled, dose-finding, phase 1/2a study. *Lancet Infect. Dis.* **2016**, *16*, 311–320. [CrossRef]
187. Zhu, F.C.; Guan, X.H.; Li, Y.H.; Huang, J.Y.; Jiang, T.; Hou, L.H.; Li, J.X.; Yang, B.F.; Wang, L.; Wang, W.J.; et al. Immunogenicity and safety of a recombinant adenovirus type-5-vectored COVID-19 vaccine in healthy adults aged 18 years or older: A randomised, double-blind, placebo-controlled, phase 2 trial. *Lancet* **2020**, *396*, 479–488. [CrossRef]
188. Olive, M.; Eisenlohr, L.C.; Flomenberg, P. T-cell responses from healthy adults. *Viral Immunol.* **2001**, *14*, 403–413. [CrossRef]
189. Flomenberg, P.; Piaskowski, V.; Truitt, R.L.; Casper, J.T. Characterization of human proliferative t cell responses to adenovirus. *J. Infect. Dis.* **1995**, *171*, 1090–1096. [CrossRef]
190. Pédron, B.; Guérin, V.; Cordeiro, D.J.; Masmoudi, S.; Dalle, J.H.; Sterkers, G. Development of cytomegalovirus and adenovirus-specific memory CD4 T-cell functions from birth to adulthood. *Pediatr. Res.* **2011**, *69*, 106–111. [CrossRef]
191. Flomenberg, P.; Piaskowski, V.; Truitt, R.L.; Casper, J.T. Human adenovirus-specific CD8+ T-cell responses are not inhibited by E3-19K in the presence of gamma interferon. *J. Virol.* **1996**, *70*, 6314–6322. [CrossRef] [PubMed]
192. Leen, A.M.; Sili, U.; Savoldo, B.; Jewell, A.M.; Piedra, P.A.; Brenner, M.K.; Rooney, C.M. Fiber-modified adenoviruses generate subgroup cross-reactive, adenovirus-specific cytotoxic T lymphocytes for therapeutic applications. *Blood* **2004**, *103*, 1011–1019. [CrossRef]
193. Smith, C.A.; Woodruff, L.S.; Rooney, C.; Kitchingman, G.R. Extensive cross-reactivity of cytotoxic T cells overview summary introduction. *Hum. Gene Ther.* **1998**, *1427*, 1419–1427. [CrossRef]
194. Veltrop-Duits, L.A.; Heemskerk, B.; Sombroek, C.C.; van Vreeswijk, T.; Gubbels, S.; Toes, R.E.M.; Melief, C.J.M.; Franken, K.L.M.C.; Havenga, M.; van Tol, M.J.D.; et al. Human CD4+ T cells stimulated by conserved adenovirus 5 hexon peptides recognize cells infected with different species of human adenovirus. *Eur. J. Immunol.* **2006**, *36*, 2410–2423. [CrossRef]
195. Zandvliet, M.L.; Falkenburg, J.h.F.; van Liempt, E.V.; Veltrop-Duits, L.A.; Lankester, A.C.; Kalpoe, J.S.; Kester, M.G.D.; van der Steen, D.M.; van Tol, M.J.; Willemze, R.; et al. Combined cd8+ and cd4+ adenovirus hexon-specific T cells associated with viral clearance after stem cell transplantation as treatment for adenovirus infection. *Haematologica* **2010**, *95*, 1943–1951. [CrossRef] [PubMed]
196. Yang, T.C.; Dayball, K.; Wan, Y.H.; Bramson, J. Detailedanalysis of the CD8+ T-cell response followingadenovirusvaccination. *J. Virol.* **2003**, *77*, 13407–13411. [CrossRef] [PubMed]
197. Teng, C.Y.; Millar, J.B.; Grinshtein, N.; Bassett, J.; Finn, J.; Bramson, J.L. T-cell immunity generated by recombinant adenovirus vaccines. *Expert Rev. Vaccines* **2007**, *6*, 347–356.
198. Ouédraogo, A.; Tiono, A.B.; Kargougou, D.; Yaro, J.B.; Ouédraogo, E.; Kaboré, Y.; Kangoye, D.; Bougouma, E.C.; Gansane, A.; Henri, N.; et al. A phase 1b randomized, controlled, double-blinded dosage-escalation trial to evaluate the safety, reactogenicity and immunogenicity of an adenovirus type 35 based circumsporozoite malaria vaccine in burkinabe healthy adults 18 to 45 years of age. *PLoS ONE* **2013**, *8*, e78679. [CrossRef]
199. Xiang, Z.Q.; Yang, Y.; Wilson, J.M.; Ertl, H.C.J. A replication-defective human adenovirus recombinant serves as a highly efficacious vaccine carrier. *Virology* **1996**, *219*, 220–227. [CrossRef] [PubMed]

200. Bassett, J.D.; Swift, S.L.; Bramson, J.L. Optimizing vaccine-induced CD8 + T-cell immunity: Focus on recombinant adenovirus vectors. *Expert Rev. Vaccines* **2011**, *10*, 1307–1319. [CrossRef] [PubMed]
201. Tatsis, N.; Fitzgerald, J.C.; Reyes-Sandoval, A.; Harris-McCoy, K.C.; Hensley, S.E.; Zhou, D.; Lin, S.W.; Bian, A.; Zhi, Q.X.; Iparraguirre, A.; et al. Adenoviral vectors persist in vivo and maintain activated CD8+ T cells: Implications for their use as vaccines. *Blood* **2007**, *110*, 1916–1923. [CrossRef]
202. Heemskerk, B.; Veltrop-Duits, L.A.; van Vreeswijk, T.; ten Dam, M.M.; Heidt, S.; Toes, R.E.M.; van Tol, M.J.D.; Schilham, M.W. extensive cross-reactivity of CD4+ adenovirus-specific T cells: Implications for immunotherapy and gene therapy. *J. Virol.* **2003**, *77*, 6562–6566. [CrossRef]
203. Rock, K.L.; Reits, E.; Neefjes, J. Present yourself! By MHC class I and MHC class II molecules. *Trends Immunol.* **2016**, *37*, 724–737. [CrossRef]
204. Münz, C. Autophagy beyond intracellular MHC class II antigen presentation. *Trends Immunol.* **2016**, *37*, 755–763. [CrossRef] [PubMed]
205. Cadwell, K. Crosstalk between autophagy and inflammatory signalling pathways: Balancing defence and homeostasis. *Nat. Rev. Immunol.* **2016**, *16*, 661–675. [CrossRef] [PubMed]
206. Weidberg, H.; Elazar, Z. TBK1 mediates crosstalk between the innate immune response and autophagy. *Sci. Signal.* **2011**, *4*, 1–4. [CrossRef]
207. Deretic, V.; Levine, B. Autophagy balances inflammation in innate immunity. *Autophagy* **2018**, *14*, 243–251. [CrossRef] [PubMed]
208. Parzych, K.R.; Klionsky, D.J. An overview of autophagy: Morphology, mechanism, and regulation. *Antioxid. Redox Signal.* **2014**, *20*, 460–473. [CrossRef]
209. Deretic, V. Autophagy as an immune defense mechanism. *Curr. Opin. Immunol.* **2006**, *18*, 375–382. [CrossRef] [PubMed]
210. Münz, C. Autophagy proteins influence endocytosis for MHC restricted antigen presentation. *Semin. Cancer Biol.* **2020**, *66*, 110–115. [CrossRef]
211. Nimmerjahn, F.; Milosevic, S.; Behrends, U.; Jaffee, E.M.; Pardoll, D.M.; Bornkamm, G.W.; Mautner, J. Major histocompatibility complex class II-restricted presentation of a cytosolic antigen by autophagy. *Eur. J. Immunol.* **2003**, *33*, 1250–1259. [CrossRef] [PubMed]
212. Paludan, C.; Schmid, D.; Landthaler, M.; Vockerodt, M.; Kube, D.; Tuschl, T.; Münz, C. Endogenous MHC class II processing of a viral nuclear antigen after autophagy. *Science* **2005**, *307*, 593–596. [CrossRef]
213. Leung, C.S.; Haigh, T.A.; Mackay, L.K.; Rickinson, A.B.; Taylor, G.S. Nuclear location of an endogenously expressed antigen, EBNA1, restricts access to macroautophagy and the range of CD4 epitope display. *Proc. Natl. Acad. Sci. USA* **2010**, *107*, 2165–2170. [CrossRef]
214. Lee, H.K.; Mattei, L.M.; Steinberg, B.E.; Alberts, P.; Lee, Y.H.; Chervonsky, A.; Mizushima, N.; Grinstein, S.; Iwasaki, A. In vivo requirement for Atg5 in antigen presentation by dendritic cells. *Immunity* **2010**, *32*, 227–239. [CrossRef] [PubMed]
215. Comber, J.D.; Robinson, T.M.; Siciliano, N.A.; Snook, A.E.; Eisenlohr, L.C. Functional macroautophagy induction by influenza a virus without a contribution to major histocompatibility complex class II-restricted presentation. *J. Virol.* **2011**, *85*, 6453–6463. [CrossRef] [PubMed]
216. Blanchet, F.P.; Moris, A.; Nikolic, D.S.; Lehmann, M.; Cardinaud, S.; Stalder, R.; Garcia, E.; Dinkins, C.; Leuba, F.; Wu, L.; et al. Human immunodeficiency virus-1 inhibition of immunoamphisomes in dendritic cells impairs early innate and adaptive immune responses. *Immunity* **2010**, *32*, 654–669. [CrossRef]
217. Heckmann, B.L.; Green, D.R. LC3-associated phagocytosis at a glance. *J. Cell Sci.* **2019**, *132*, 1–2. [CrossRef] [PubMed]
218. Romao, S.; Gasser, N.; Becker, A.C.; Guhl, B.; Bajagic, M.; Vanoaica, D.; Ziegler, U.; Roesler, J.; Dengjel, J.; Reichenbach, J.; et al. Autophagy proteins stabilize pathogen-containing phagosomes for prolonged MHC II antigen processing. *J. Cell Biol.* **2013**, *203*, 757–766. [CrossRef] [PubMed]
219. Tey, S.K.; Khanna, R. Autophagy mediates transporter associated with antigen processing-independent presentation of viral epitopes through MHC class I pathway. *Blood* **2012**, *120*, 994–1004. [CrossRef]
220. Das, M.; Kaveri, S.V.; Bayry, J. Cross-presentation of antigens by dendritic cells: Role of autophagy. *Oncotarget* **2015**, *6*, 28527–28528. [CrossRef]
221. Uhl, M.; Kepp, O.; Jusforgues-Saklani, H.; Vicencio, J.M.; Kroemer, G.; Albert, M.L. Autophagy within the antigen donor cell facilitates efficient antigen cross-priming of virus-specific CD8+ T cells. *Cell Death Differ.* **2009**, *16*, 991–1005. [CrossRef]
222. Yi, Y.; Zhou, Z.; Shu, S.; Fang, Y.; Twitty, C.; Hilton, T.L.; Aung, S.; Urba, W.J.; Fox, B.A.; Hu, H.M.; et al. Autophagy-assisted antigen cross-presentation: Autophagosome as the argo of shared tumor-specifc antigens and DAMPs. *Oncoimmunology* **2012**, *1*, 976–978. [CrossRef]
223. Crotzer, V.L.; Blum, J.S. Autophagy and its role in MHC-mediated antigen presentation. *J. Immunol.* **2009**, *182*, 3335–3341. [CrossRef]
224. Windheim, M.; Hilgendorf, A.; Burgert, H.G. Immune evasion by adenovirus E3 proteins: Exploitation of intracellular trafficking pathways. *Curr. Top. Microbiol. Immunol.* **2004**, *273*, 29–85.
225. Ginsberg, H.S. The ups and downs of adenovirus vectors. *Bull. N. Y. Acad. Med. J. Urban. Health* **1996**, *73*, 53–58.
226. Bennett, E.M.; Bennink, J.R.; Yewdell, J.W.; Brodsky, F.M. Cutting edge: Adenovirus E19 has two mechanisms for affecting class I MHC expression. *J. Immunol.* **1999**, *162*, 5049–5052.

227. Oliveira, E.R.A.; Bouvier, M. Immune evasion by adenoviruses: A window into host-virus adaptation. *FEBS Lett.* **2019**, *593*, 3496–3503. [CrossRef]
228. Burgert, H.G. Subversion of the MHC class I antigen-presentation pathway by adenoviruses and herpes simplex viruses. *Trends Microbiol.* **1996**, *4*, 107–112. [CrossRef]
229. Klein, S.R.; Jiang, H.; Hossain, M.B.; Fan, X.; Gumin, J.; Dong, A.; Alonso, M.M.; Gomez-Manzano, C.; Fueyo, J. Critical role of autophagy in the processing of adenovirus capsid-incorporated cancer-specific antigens. *PLoS ONE* **2016**, *11*, e0153814. [CrossRef] [PubMed]
230. Neukirch, L.; Fougeroux, C.; Andersson, A.M.C.; Holst, P.J. *The Potential of Adenoviral Vaccine Vectors with Altered Antigen Presentation Capabilities*; Taylor & Francis: Milton Park, UK, 2020; Volume 19, ISBN 0000000275668.
231. Tang, D.; Kang, R.; Livesey, K.M.; Cheh, C.W.; Farkas, A.; Loughran, P.; Hoppe, G.; Bianchi, M.E.; Tracey, K.J.; Zeh, H.J.; et al. Endogenous HMGB1 regulates autophagy. *J. Cell Biol.* **2010**, *190*, 881–892. [CrossRef] [PubMed]
232. Degenhardt, K.; Mathew, R.; Beaudoin, B.; Bray, K.; Anderson, D.; Chen, G.; Mukherjee, C.; Shi, Y.; Gélinas, C.; Fan, Y.; et al. Autophagy promotes tumor cell survival and restricts necrosis, inflammation, and tumorigenesis. *Cancer Cell* **2006**, *10*, 51–64. [CrossRef] [PubMed]
233. Liu, M.; Jiang, L.; Fu, X.; Wang, W.; Ma, J.; Tian, T.; Nan, K.; Liang, X. Cytoplasmic liver kinase B1 promotes the growth of human lung adenocarcinoma by enhancing autophagy. *Cancer Sci.* **2018**, *109*, 3055–3067. [CrossRef] [PubMed]
234. Sharma, P.; Martis, P.C.; Excoffon, K.J.D.A. Adenovirus transduction: More complicated than receptor expression. *Virology* **2017**, *502*, 144–151. [CrossRef]
235. Tritel, M.; Stoddard, A.M.; Flynn, B.J.; Darrah, P.A.; Wu, C.; Wille, U.; Shah, J.A.; Huang, Y.; Xu, L.; Betts, M.R.; et al. Prime-boost vaccination with HIV-1 gag protein and cytosine phosphate guanosine oligodeoxynucleotide, followed by adenovirus, induces sustained and robust humoral and cellular immune responses. *J. Immunol.* **2003**, *171*, 2538–2547. [CrossRef] [PubMed]
236. Ertl, H.C. Viral vectors as vaccine carriers. *Curr. Opin. Virol.* **2016**, *21*, 1–8. [CrossRef]
237. Ginn, S.L.; Amaya, A.K.; Alexander, I.E.; Edelstein, M.; Abedi, M.R. Gene therapy clinical trials worldwide to 2017: An update. *J. Gene Med.* **2018**, *20*, e3015. [CrossRef] [PubMed]
238. Gao, J.; Mese, K.; Bunz, O.; Ehrhardt, A. State-of-the-art human adenovirus vectorology for therapeutic approaches. *FEBS Lett.* **2019**, *593*, 3609–3622. [CrossRef]
239. McGrory, W.J.; Bautista, D.S.; Graham, F.L. A simple technique for the rescue of early region I mutations into infectious human adenovirus type 5. *Virology* **1988**, *163*, 614–617. [CrossRef]
240. Wold, W.S.M.; Doronin, K.; Toth, K.; Kuppuswamy, M.; Lichtenstein, D.L.; Tollefson, A.E. Immune responses to adenoviruses: Viral evasion mechanisms and their implications for the clinic. *Curr. Opin. Immunol.* **1999**, *11*, 380–386. [CrossRef]
241. Crystal, R.G. Adenovirus: The first effective in vivo gene delivery vector. *Hum. Gene Ther.* **2014**, *25*, 3–11. [CrossRef] [PubMed]
242. Bett, A.J.; Prevec, L.; Graham, F.L. Packaging capacity and stability of human adenovirus type 5 vectors. *J. Virol.* **1993**, *67*, 5911–5921. [CrossRef]
243. Bett, A.J.; Haddara, W.; Prevec, L.; Graham, F.L. An efficient and flexible system for construction of adenovirus vectors with insertions or deletions in early regions 1 and 3. *Proc. Natl. Acad. Sci. USA* **1994**, *91*, 8802–8806. [CrossRef]
244. Akusjärvi, G. Proteins with transcription regulatory properties encoded by human adenoviruses. *Trends Microbiol.* **1993**, *1*, 163–170. [CrossRef]
245. Graham, F.L.; Smiley, J.; Russell, W.C.; Nairn, R. Characteristics of a human cell line transformed by DNA from human adenovirus type 5. *J. Gen. Virol.* **1977**, *36*, 59–72. [CrossRef] [PubMed]
246. Shaw, G.; Morse, S.; Ararat, M.; Graham, F.L. Preferential transformation of human neuronal cells by human adenoviruses and the origin of HEK 293 cells. *FASEB J.* **2002**, *16*, 869–871. [CrossRef] [PubMed]
247. Benihoud, K.; Yeh, P.; Perricaudet, M. Adenovirus vectors for gene delivery. *Curr. Opin. Biotechnol.* **1999**, *10*, 440–447. [CrossRef]
248. Hehir, K.M.; Armentano, D.; Cardoza, L.M.; Choquette, T.L.; Berthelette, P.B.; White, G.A.; Couture, L.A.; Everton, M.B.; Keegan, J.; Martin, J.M.; et al. Molecular characterization of replication-competent variants of adenovirus vectors and genome modifications to prevent their occurrence. *J. Virol.* **1996**, *70*, 8459–8467. [CrossRef]
249. Kreppel, F.; Hagedorn, C. Capsid and genome modification strategies to reduce the immunogenicity of adenoviral vectors. *Int. J. Mol. Sci.* **2021**, *22*, 2417. [CrossRef]
250. Yang, Y.; Ertl, H.C.J.; Wilson, J.M. MHC class I-cestricted cytotoxic T lymphocytes to viral antigens destroy hepatocytes in mice infected with E1-deleted recombinant adenoviruses. *Immunity* **1994**, *1*, 433–442. [CrossRef]
251. Engelhardt, J.F.; Litzky, L.; Wilson, J.M. Prolonged transgene expression in cotton rat lung with recombinant adenoviruses defective in E2a. *Hum. Gene Ther.* **1994**, *5*, 1217–1229. [CrossRef]
252. Lusky, M.; Christ, M.; Rittner, K.; Dieterle, A.; Dreyer, D.; Mourot, B.; Schultz, H.; Stoeckel, F.; Pavirani, A.; Mehtali, M. In vitro and in vivo biology of recombinant adenovirus with E1, E1/E2A, or E1/E4 deleted. *J. Virol.* **1998**, *72*, 11. [CrossRef]
253. Schaack, J. Induction and inhibition of innate inflammatory responses by adenovirus early region proteins. *Viral Immunol.* **2005**, *18*, 79–88. [CrossRef]
254. Janssen, J.M.; Liu, J.; Skokan, J.; Gonçalves, M.A.F.V.; de Vries, A.A.F. Development of an AdEasy-based system to produce first- and second-generation adenoviral vectors with tropism for CAR- or CD46-positive cells. *J. Gene Med.* **2013**, *15*, 1–11. [CrossRef]

255. Dedieu, J.F.; Vigne, E.; Torrent, C.; Jullien, C.; Mahfouz, I.; Caillaud, J.M.; Aubailly, N.; Orsini, C.; Guillaume, J.M.; Opolon, P.; et al. Long-term gene delivery into the livers of immunocompetent mice with E1/E4-defective adenoviruses. *J. Virol.* **1997**, *71*, 4626–4637. [CrossRef] [PubMed]
256. Kochanek, S. High-capacity adenoviral vectors for gene transfer and somatic gene therapy. *Hum. Gene Ther.* **1999**, *10*, 2451–2459. [CrossRef] [PubMed]
257. Segura, M.; Alba, R.; Bosch, A.; Chillon, M. Advances in helper-dependent adenoviral vector research. *Curr. Gene Ther.* **2008**, *8*, 222–235. [CrossRef]
258. Brunetti-Pierri, N.; Ng, T.; Iannitti, D.; Cioffi, W.; Stapleton, G.; Law, M.; Breinholt, J.; Palmer, D.; Grove, N.; Rice, K.; et al. Transgene expression up to 7 years in nonhuman primates following hepatic transduction with helper-dependent adenoviral vectors. *Hum. Gene Ther.* **2013**, *24*, 761–765. [CrossRef] [PubMed]
259. Sandig, V.; Youil, R.; Bett, A.J.; Franlin, L.L.; Oshima, M.; Maione, D.; Wang, F.; Metzker, M.L.; Savino, R.; Caskey, C.T. Optimization of the helper-dependent adenovirus system for production and potency in vivo. *Proc. Natl. Acad. Sci. USA* **2000**, *97*, 1002–1007. [CrossRef]
260. Fisher, K.J.; Kelley, W.M.; Burda, J.F.; Wilson, J.M. A novel adenovirus-adeno-associated virus hybrid vector that displays efficient rescue and delivery of the AAV genome. *Hum. Gene Ther.* **1996**, *7*, 2079–2087. [CrossRef]
261. Hardy, S.; Kitamura, M.; Harris-Stansil, T.; Dai, Y.; Phipps, M.L. Construction of adenovirus vectors through Cre-lox recombination. *J. Virol.* **1997**, *71*, 1842–1849. [CrossRef]
262. Hartigan-O'Connor, D.; Amalfitano, A.; Chamberlain, J.S. Improved production of gutted adenovirus in cells expressing adenovirus preterminal protein and DNA polymerase. *J. Virol.* **1999**, *73*, 7835–7841. [CrossRef]
263. Morral, N.; Parks, R.J.; Zhou, H.; Langston, C.; Schiedner, G.; Quinones, J.; Graham, F.L.; Kochanek, S.; Beaudet, A.L. High doses of a helper-dependent adenoviral vector yield supraphysiological levels of α1-antitrypsin with negligible toxicity. *Hum. Gene Ther.* **1998**, *9*, 2709–2716. [CrossRef]
264. Schiedner, G.; Morral, N.; Parks, R.J.; Wu, Y.; Koopmans, S.C.; Langston, C.; Graham, F.L.; Beaudet, A.L.; Kochanek, S. Genomic DNA transfer with a high-capacity adenovirus vector results in improved in vivo gene expression and decreased toxicity. *Nat. Genet.* **1998**, *18*, 180–183. [CrossRef] [PubMed]
265. Brunetti-Pierri, N.; Palmer, D.J.; Mane, V.; Finegold, M.; Beaudet, A.L.; Ng, P. Increased hepatic transduction with reduced systemic dissemination and proinflammatory cytokines following hydrodynamic injection of helper-dependent adenoviral vectors. *Mol. Ther.* **2005**, *12*, 99–106. [CrossRef] [PubMed]
266. Brunetti-Pierri, N.; Liou, A.; Patel, P.; Palmer, D.; Grove, N.; Finegold, M.; Piccolo, P.; Donnachie, E.; Rice, K.; Beaudet, A.; et al. Balloon catheter delivery of helper-dependent adenoviral vector results in sustained, therapeutic hFIX expression in rhesus macaques. *Mol. Ther.* **2012**, *20*, 1863–1870. [CrossRef]
267. Ehrhardt, A.; Xu, H.; Dillow, A.M.; Bellinger, D.A.; Nichols, T.C.; Kay, M.A. A gene-deleted adenoviral vector results in phenotypic correction of canine hemophilia B without liver toxicity or thrombocytopenia. *Blood* **2003**, *102*, 2403–2411. [CrossRef]
268. Top, F.H.; Buescher, E.L.; Bancroft, W.H.; Russell, P.K. Immunization with live types 7 and 4 adenovirus vaccines.Ii. antibody response and protective effect against acute respiratory disease due to adenovirus type 7. *J. Infect. Dis.* **1971**, *124*, 155–160. [CrossRef] [PubMed]
269. Deal, C.; Pekosz, A.; Ketner, G. Prospects for oral replicating adenovirus-vectored vaccines. *Vaccine* **2013**, *31*, 3236–3243. [CrossRef]
270. Radin, J.M.; Hawksworth, A.W.; Blair, P.J.; Faix, D.J.; Raman, R.; Russell, K.L.; Gray, G.C. Dramatic decline of respiratory illness among us military recruits after the renewed use of adenovirus vaccines. *Clin. Infect. Dis.* **2014**, *59*, 962–968. [CrossRef] [PubMed]
271. Broderick, M.; Myers, C.; Balansay, M.; Vo, S.; Osuna, A.; Russell, K. Adenovirus 4/7 vaccine's effect on disease rates is associated with disappearance of adenovirus on building surfaces at a military recruit base. *Mil. Med.* **2017**, *182*, e2069–e2072. [CrossRef] [PubMed]
272. Home—ClinicalTrials.gov. Available online: https://clinicaltrials.gov/ (accessed on 11 May 2021).
273. Sullivan, N.J.; Sanchez, A.; Rollin, P.E.; Yang, Z.Y.; Nabel, G.J. Development of a preventive vaccine for Ebola virus infection in primates. *Nature* **2000**, *408*, 605–609. [CrossRef]
274. Ewer, K.; Sebastian, S.; Spencer, A.J.; Gilbert, S.; Hill, A.V.S.; Lambe, T. Chimpanzee adenoviral vectors as vaccines for outbreak pathogens. *Hum. Vaccines Immunother.* **2017**, *13*, 3020–3032. [CrossRef]
275. Priddy, F.H.; Brown, D.; Kublin, J.; Monahan, K.; Wright, D.P.; Lalezari, J.; Santiago, S.; Marmor, M.; Lally, M.; Novak, R.M.; et al. Safety and immunogenicity of a replication-incompetent adenovirus type 5 HIV-1 clade B gag/pol/nef vaccine in healthy adults. *Clin. Infect. Dis.* **2008**, *46*, 1769–1781. [CrossRef]
276. Shiver, J.W.; Emini, E.A. Recent advances in the development of HIV-1 vaccines using replication-incompetent adenovirus vectors. *Annu. Rev. Med.* **2004**, *55*, 355–372. [CrossRef]
277. Buchbinder, S.P.; Mehrotra, D.V.; Duerr, A.; Fitzgerald, D.W.; Mogg, R.; Li, D.; Gilbert, P.B.; Lama, J.R.; Marmor, M.; del Rio, C.; et al. Efficacy assessment of a cell-mediated immunity HIV-1 vaccine (the Step Study): A double-blind, randomised, placebo-controlled, test-of-concept trial. *Lancet* **2008**, *372*, 1881–1893. [CrossRef]
278. Cohen, J. AIDS research: Promising AIDS vaccine's failure leaves field reeling. *Science* **2007**, *318*, 28–29. [CrossRef] [PubMed]
279. Press Release: Merck Halts Study of "Ineffective" HIV Vaccine | FierceBiotech. Available online: https://www.fiercebiotech.com/biotech/press-release-merck-halts-study-of-ineffective-hiv-vaccine (accessed on 11 May 2021).

280. Harro, C.; Sun, X.; Stek, J.E.; Leavitt, R.Y.; Mehrotra, D.V.; Wang, F.; Bett, A.J.; Casimiro, D.R.; Shiver, J.W.; DiNubile, M.J.; et al. Safety and immunogenicity of the merck adenovirus serotype 5 (MRKAd5) and MRKAd6 human immunodeficiency virus type 1 trigene vaccines alone and in combination in healthy adults. *Clin. Vaccine Immunol.* **2009**, *16*, 1285–1292. [CrossRef] [PubMed]
281. Harro, C.D.; Robertson, M.N.; Lally, M.A.; O'Neill, L.D.; Edupuganti, S.; Goepfert, P.A.; Mulligan, M.J.; Priddy, F.H.; Dubey, S.A.; Kierstead, L.S.; et al. Safety and immunogenicity of adenovirus-vectored near-consensus HIV type 1 clade B gag vaccines in healthy adults. *AIDS Res. Hum. Retroviruses* **2009**, *25*, 103–114. [CrossRef] [PubMed]
282. Patterson, L.J. The "STEP-Wise" Future of adenovirus-based HIV vaccines. *Curr. Med. Chem.* **2011**, *18*, 3981–3986. [CrossRef]
283. Duerr, A.; Huang, Y.; Buchbinder, S.; Coombs, R.W.; Sanchez, J.; del Rio, C.; Casapia, M.; Santiago, S.; Gilbert, P.; Corey, L.; et al. Extended follow-up confirms early vaccine-enhanced risk of HIV acquisition and demonstrates waning effect over time among participants in a randomized trial of recombinant adenovirus HIV vaccine (Step Study). *J. Infect. Dis.* **2012**, *206*, 258–266. [CrossRef]
284. Koblin, B.A.; Mayer, K.H.; Noonan, E.; Wang, C.Y.; Marmor, M.; Sanchez, J.; Brown, S.J.; Robertson, M.N.; Buchbinder, S.P. Sexual risk behaviors, circumcision status, and preexisting immunity to adenovirus type 5 among men who have sex with men participating in a randomized HIV-1 vaccine efficacy trial: Step study. *J. Acquir. Immune Defic. Syndr.* **2012**, *60*, 405–413. [CrossRef]
285. Curlin, M.E.; Cassis-Ghavami, F.; Magaret, A.S.; Spies, G.A.; Duerr, A.; Celum, C.L.; Sanchez, J.L.; Margolick, J.B.; Detels, R.; McElrath, M.J.; et al. Serological immunity to adenovirus serotype 5 is not associated with risk of HIV infection: A case-control study. *AIDS* **2011**, *25*, 153–158. [CrossRef]
286. Stephenson, K.E.; Hural, J.; Buchbinder, S.P.; Sinangil, F.; Barouch, D.H. Preexisting adenovirus seropositivity is not associated with increased HIV-1 acquisition in three HIV-1 vaccine efficacy trials. *J. Infect. Dis.* **2012**, *205*, 1806–1810. [CrossRef]
287. Zak, D.E.; Andersen-Nissen, E.; Peterson, E.R.; Sato, A.; Hamilton, M.K.; Borgerding, J.; Krishnamurty, A.T.; Chang, J.T.; Adams, D.J.; Hensley, T.R.; et al. Merck Ad5/HIV induces broad innate immune activation that predicts CD8+ T-cell responses but is attenuated by preexisting Ad5 immunity. *Proc. Natl. Acad. Sci. USA* **2012**, *109*, E3503–E3512. [CrossRef] [PubMed]
288. Huang, Y.; Duerr, A.; Frahm, N.; Zhang, L.; Moodie, Z.; de Rosa, S.; Mcelrath, M.J.; Gilbert, P.B. Immune-correlates analysis of an HIV-1 vaccine efficacy trial reveals an association of nonspecific interferon-C secretion with increased HIV-1 infection risk: A cohort- Based modeling study. *PLoS ONE* **2014**, *9*, e108631. [CrossRef] [PubMed]
289. Gilbert, S.C. Adenovirus-vectored Ebola vaccines. *Expert Rev. Vaccines* **2015**, *14*, 1347–1357. [CrossRef] [PubMed]
290. Vanderzanden, L.; Bray, M.; Fuller, D.; Roberts, T.; Custer, D.; Spik, K.; Jahrling, P.; Huggins, J.; Schmaljohn, A.; Schmaljohn, C. DNA vaccines expressing either the GP or NP genes of Ebola virus protect mice from lethal challenge. *Virology* **1998**, *246*, 134–144. [CrossRef] [PubMed]
291. Sullivan, N.J.; Geisbert, T.W.; Gelsbert, J.B.; Xu, L.; Yang, Z.Y.; Roederer, M.; Koup, R.A.; Jahrling, P.B.; Nabel, G.J. Accelerated vaccination for Ebola virus haemorrhagic fever in non-human primates. *Nature* **2003**, *424*, 681–684. [CrossRef]
292. Sullivan, N.J.; Geisbert, T.W.; Geisbert, J.B.; Shedlock, D.J.; Xu, L.; Lamoreaux, L.; Custers, J.H.H.V.; Popernack, P.M.; Yang, Z.Y.; Pau, M.G.; et al. Immune protection of nonhuman primates against Ebola virus with single low-dose adenovirus vectors encoding modified GPs. *PLoS Med.* **2006**, *3*, e177. [CrossRef]
293. Ledgerwood, J.E.; Costner, P.; Desai, N.; Holman, L.; Enama, M.E.; Yamshchikov, G.; Mulangu, S.; Hu, Z.; Andrews, C.A.; Sheets, R.A.; et al. A replication defective recombinant Ad5 vaccine expressing Ebola virus GP is safe and immunogenic in healthy adults. *Vaccine* **2010**, *29*, 304–313. [CrossRef]
294. Quinn, K.M.; da Costa, A.; Yamamoto, A.; Berry, D.; Lindsay, R.W.B.; Darrah, P.A.; Wang, L.; Cheng, C.; Kong, W.-P.; Gall, J.G.D.; et al. Comparative analysis of the magnitude, quality, phenotype, and protective capacity of simian immunodeficiency virus gag-specific CD8 + T cells following human-, simian-, and chimpanzee-derived recombinant adenoviral vector immunization. *J. Immunol.* **2013**, *190*, 2720–2735. [CrossRef]
295. Zhu, F.C.; Hou, L.H.; Li, J.X.; Wu, S.P.; Liu, P.; Zhang, G.R.; Hu, Y.M.; Meng, F.Y.; Xu, J.J.; Tang, R.; et al. Safety and immunogenicity of a novel recombinant adenovirus type-5 vector-based Ebola vaccine in healthy adults in China: Preliminary report of a randomised, double-blind, placebo-controlled, phase 1 trial. *Lancet* **2015**, *385*, 2272–2279. [CrossRef]
296. Wu, S.; Kroeker, A.; Wong, G.; He, S.; Hou, L.; Audet, J.; Wei, H.; Zhang, Z.; Fernando, L.; Soule, G.; et al. An adenovirus vaccine expressing Ebola virus variant makona glycoprotein is efficacious in Guinea pigs and nonhuman primates. *J. Infect. Dis.* **2016**, *214*, S326–S332. [CrossRef] [PubMed]
297. Wu, L.; Zhang, Z.; Gao, H.; Li, Y.; Hou, L.; Yao, H.; Wu, S.; Liu, J.; Wang, L.; Zhai, Y.; et al. Open-label phase I clinical trial of Ad5-EBOV in Africans in China. *Hum. Vaccines Immunother.* **2017**, *13*, 2078–2085. [CrossRef]
298. Geisbert, T.W.; Bailey, M.; Hensley, L.; Asiedu, C.; Geisbert, J.; Stanley, D.; Honko, A.; Johnson, J.; Mulangu, S.; Pau, M.G.; et al. Recombinant adenovirus serotype 26 (Ad26) and Ad35 vaccine vectors bypass immunity to Ad5 and protect nonhuman primates against ebolavirus challenge. *J. Virol.* **2011**, *85*, 4222–4233. [CrossRef] [PubMed]
299. Milligan, I.D.; Gibani, M.M.; Sewell, R.; Clutterbuck, E.A.; Campbell, D.; Plested, E.; Nuthall, E.; Voysey, M.; Silva-Reyes, L.; McElrath, M.J.; et al. Safety and immunogenicity of novel adenovirus type 26-and modified vaccinia Ankara-vectored Ebola vaccines: A randomized clinical trial. *JAMA J. Am. Med. Assoc.* **2016**, *315*, 1610–1623. [CrossRef] [PubMed]
300. Singh, S.; Kumar, R.; Agrawal, B. Adenoviral vector-based vaccines and gene therapies: Current status and future prospects. In *Adenoviruses*; IntechOpen: London, UK, 2019.
301. Matz, K.M.; Marzi, A.; Feldmann, H. Ebola vaccine trials: Progress in vaccine safety and immunogenicity. *Expert Rev. Vaccines* **2019**, *18*, 1229–1242. [CrossRef] [PubMed]

302. Kobinger, G.P.; Feldmann, H.; Zhi, Y.; Schumer, G.; Gao, G.; Feldmann, F.; Jones, S.; Wilson, J.M. Chimpanzee adenovirus vaccine protects against Zaire Ebola virus. *Virology* **2006**, *346*, 394–401. [CrossRef]
303. Stanley, D.A.; Honko, A.N.; Asiedu, C.; Trefry, J.C.; Lau-Kilby, A.W.; Johnson, J.C.; Hensley, L.; Ammendola, V.; Abbate, A.; Grazioli, F.; et al. Chimpanzee adenovirus vaccine generates acute and durable protective immunity against ebolavirus challenge. *Nat. Med.* **2014**, *20*, 1126–1129. [CrossRef]
304. Ledgerwood, J.E.; DeZure, A.D.; Stanley, D.A.; Coates, E.E.; Novik, L.; Enama, M.E.; Berkowitz, N.M.; Hu, Z.; Joshi, G.; Ploquin, A.; et al. Chimpanzee adenovirus vector ebola vaccine. *N. Engl. J. Med.* **2017**, *376*, 928–938. [CrossRef] [PubMed]
305. Ewer, K.; Rampling, T.; Venkatraman, N.; Bowyer, G.; Wright, D.; Lambe, T.; Imoukhuede, E.B.; Payne, R.; Fehling, S.K.; Strecker, T.; et al. A monovalent chimpanzee adenovirus Ebola vaccine boosted with MVA. *N. Engl. J. Med.* **2016**, *374*, 1635–1646. [CrossRef]
306. Kennedy, S.B.; Bolay, F.; Kieh, M.; Grandits, G.; Badio, M.; Ballou, R.; Eckes, R.; Feinberg, M.; Follmann, D.; Grund, B.; et al. Phase 2 placebo-controlled trial of two vaccines to prevent Ebola in Liberia. *N. Engl. J. Med.* **2017**, *377*, 1438–1447. [CrossRef] [PubMed]
307. Mu, J.; Jeyanathan, M.; Shaler, C.R.; Horvath, C.; Damjanovic, D.; Zganiacz, A.; Kugathasan, K.; McCormick, S.; Xing, Z. Respiratory mucosal immunization with adenovirus gene transfer vector induces helper CD4 T cell-independent protective immunity. *J. Gene Med.* **2010**, *12*, 693–704. [CrossRef] [PubMed]
308. Forbes, E.K.; Sander, C.; Ronan, E.O.; McShane, H.; Hill, A.V.S.; Beverley, P.C.L.; Tchilian, E.Z. Multifunctional, high-level cytokine-producing Th1 cells in the lung, but not spleen, correlate with protection against mycobacterium tuberculosis aerosol challenge in mice. *J. Immunol.* **2008**, *181*, 4955–4964. [CrossRef]
309. Wang, J.; Thorson, L.; Stokes, R.W.; Santosuosso, M.; Huygen, K.; Zganiacz, A.; Hitt, M.; Xing, Z. Single mucosal, but not parenteral, immunization with recombinant adenoviral-based vaccine provides potent protection from pulmonary tuberculosis. *J. Immunol.* **2004**, *173*, 6357–6365. [CrossRef]
310. Smaill, F.; Jeyanathan, M.; Smieja, M.; Medina, M.F.; Thanthrige-Don, N.; Zganiacz, A.; Yin, C.; Heriazon, A.; Damjanovic, D.; Puri, L.; et al. A human type 5 adenovirus-based tuberculosis vaccine induces robust T cell responses in humans despite preexisting anti-adenovirus immunity. *Sci. Transl. Med.* **2013**, *5*. [CrossRef] [PubMed]
311. Raviprakash, K.; Wang, D.; Ewing, D.; Holman, D.H.; Block, K.; Woraratanadharm, J.; Chen, L.; Hayes, C.; Dong, J.Y.; Porter, K. A tetravalent dengue vaccine based on a complex adenovirus vector provides significant protection in rhesus monkeys against all four serotypes of dengue virus. *J. Virol.* **2008**, *82*, 6927–6934. [CrossRef] [PubMed]
312. Jaiswal, S.; Khanna, N.; Swaminathan, S. Replication-defective adenoviral vaccine vector for the induction of immune responses to dengue virus type 2. *J. Virol.* **2003**, *77*, 12907–12913. [CrossRef]
313. Khanam, S.; Pilankatta, R.; Khanna, N.; Swaminathan, S. An adenovirus type 5 (AdV5) vector encoding an envelope domain III-based tetravalent antigen elicits immune responses against all four dengue viruses in the presence of prior AdV5 immunity. *Vaccine* **2009**, *27*, 6011–6021. [CrossRef]
314. Hung, P.P.; Chanda, P.K.; Natuk, R.J.; Mason, B.B.; Chengalvala, M.; Bhat, B.M.; Molnar-Kimber, K.L.; Dheer, S.K.; Morin, J.E.; Mizutani, S.; et al. Adenovirus vaccine strains genetically engineered to express HIV-1 or HBV antigens for use as live recombinant vaccines. *Nat. Immun. Cell Growth Regul.* **1990**, *9*, 160–164.
315. Lubeck, M.D.; Davis, A.R.; Chengalvala, M.; Natuk, R.J.; Morin, J.E.; Molnar-Kimber, K.; Mason, B.B.; Bhat, B.M.; Mizutani, S.; Hung, P.P.; et al. Immunogenicity and efficacy testing in chimpanzees of an oral hepatitis B vaccine based on live recombinant adenovirus. *Proc. Natl. Acad. Sci. USA* **1989**, *86*, 6763–6767. [CrossRef]
316. Makimura, M.; Miyake, S.; Akino, N.; Takamori, K.; Matsuura, Y.; Miyamura, T.; Saito, I. Induction of antibodies against structural proteins of hepatitis C virus in mice using recombinant adenovirus. *Vaccine* **1996**, *14*, 28–34. [CrossRef]
317. Park, S.H.; Yang, S.H.; Lee, C.G.; Youn, J.W.; Chang, J.; Sung, Y.C. Efficient induction of T helper 1 CD4+ T-cell responses to hepatitis C virus core and E2 by a DNA prime-adenovirus boost. *Vaccine* **2003**, *21*, 4555–4564. [CrossRef]
318. Matsui, M.; Moriya, O.; Akatsuka, T. Enhanced induction of hepatitis C virus-specific cytotoxic T lymphocytes and protective efficacy in mice by DNA vaccination followed by adenovirus boosting in combination with the interleukin-12 expression plasmid. *Vaccine* **2003**, *21*, 1629–1639. [CrossRef]
319. Arribillaga, L.; de Cerio, A.L.D.; Sarobe, P.; Casares, N.; Gorraiz, M.; Vales, A.; Bruna-Romero, O.; Borrás-Cuesta, F.; Paranhos-Baccala, G.; Prieto, J.; et al. Vaccination with an adenoviral vector encoding hepatitis C virus (HCV) NS3 protein protects against infection with HCV-recombinant vaccinia virus. *Vaccine* **2002**, *21*, 202–210. [CrossRef]
320. Seong, Y.R.; Choi, S.; Lim, J.S.; Lee, C.H.; Lee, C.K.; Im, D.S. Immunogenicity of the E1E2 proteins of hepatitis C virus expressed by recombinant adenoviruses. *Vaccine* **2001**, *19*, 2955–2964. [CrossRef]
321. Prevec, L.; Campbell, J.B.; Christie, B.S.; Belbeck, L.; Graham, F.L. A recombinant human adenovirus vaccine against rabies. *J. Infect. Dis.* **1990**, *161*, 27–30. [CrossRef] [PubMed]
322. Vos, A.; Neubert, A.; Pommerening, E.; Müller, T.; Döhner, L.; Neubert, L.; Hughes, K. Immunogenicity of an E1-deleted recombinant human adenovirus against rabies by different routes of administration. *J. Gen. Virol.* **2001**, *82*, 2191–2197. [CrossRef] [PubMed]
323. Bouet-Cararo, C.; Contreras, V.; Fournier, A.; Jallet, C.; Guibert, J.M.; Dubois, E.; Thiery, R.; Bréard, E.; Tordo, N.; Richardson, J.; et al. Canine adenoviruses elicit both humoral and cell-mediated immune responses against rabies following immunisation of sheep. *Vaccine* **2011**, *29*, 1304–1310. [CrossRef]

324. Xiang, Z.Q.; Greenberg, L.; Ertl, H.C.; Rupprecht, C.E. Protection of non-human primates against rabies with an adenovirus recombinant vaccine. *Virology* **2014**, *450–451*, 243–249. [CrossRef]
325. Van Kampen, K.R.; Shi, Z.; Gao, P.; Zhang, J.; Foster, K.W.; Chen, D.T.; Marks, D.; Elmets, C.A.; Tang, D.C.C. Safety and immunogenicity of adenovirus-vectored nasal and epicutaneous influenza vaccines in humans. *Vaccine* **2005**, *23*, 1029–1036. [CrossRef]
326. Peters, W.; Brandl, J.R.; Lindbloom, J.D.; Martinez, C.J.; Scallan, C.D.; Trager, G.R.; Tingley, D.W.; Kabongo, M.L.; Tucker, S.N. Oral administration of an adenovirus vector encoding both an avian influenza A hemagglutinin and a TLR3 ligand induces antigen specific granzyme B and IFN-γ T cell responses in humans. *Vaccine* **2013**, *31*, 1752–1758. [CrossRef]
327. Gurwith, M.; Lock, M.; Taylor, E.M.; Ishioka, G.; Alexander, J.; Mayall, T.; Ervin, J.E.; Greenberg, R.N.; Strout, C.; Treanor, J.J.; et al. Safety and immunogenicity of an oral, replicating adenovirus serotype 4 vector vaccine for H5N1 influenza: A randomised, double-blind, placebo-controlled, phase 1 study. *Lancet Infect. Dis.* **2013**, *13*, 238–250. [CrossRef]
328. Gao, W.; Soloff, A.C.; Lu, X.; Montecalvo, A.; Nguyen, D.C.; Matsuoka, Y.; Robbins, P.D.; Swayne, D.E.; Donis, R.O.; Katz, J.M.; et al. Protection of mice and poultry from lethal H5N1 avian influenza virus through adenovirus-based immunization. *J. Virol.* **2006**, *80*, 1959–1964. [CrossRef] [PubMed]
329. Zhang, C.; Zhou, D. Adenoviral vector-based strategies against infectious disease and cancer. *Hum. Vaccines Immunother.* **2016**, *12*, 2064–2074. [CrossRef] [PubMed]
330. Ebola Virus Disease Democratic Republic of Congo: External Situation Report 94/2019. Available online: https://www.who.int/publications/i/item/10665-332654 (accessed on 12 May 2021).
331. Der Li, Y.; Chi, W.Y.; Su, J.H.; Ferrall, L.; Hung, C.F.; Wu, T.C. Coronavirus vaccine development: From SARS and MERS to COVID-19. *J. Biomed. Sci.* **2020**, *27*, 104.
332. Dai, L.; Gao, G.F. Viral targets for vaccines against COVID-19. *Nat. Rev. Immunol.* **2021**, *21*, 73–82. [CrossRef]
333. Ou, X.; Liu, Y.; Lei, X.; Li, P.; Mi, D.; Ren, L.; Guo, L.; Guo, R.; Chen, T.; Hu, J.; et al. Characterization of spike glycoprotein of SARS-CoV-2 on virus entry and its immune cross-reactivity with SARS-CoV. *Nat. Commun.* **2020**, *11*, 1–12. [CrossRef]
334. Zhu, F.C.; Li, Y.H.; Guan, X.H.; Hou, L.H.; Wang, W.J.; Li, J.X.; Wu, S.P.; Wang, B.S.; Wang, Z.; Wang, L.; et al. Safety, tolerability, and immunogenicity of a recombinant adenovirus type-5 vectored COVID-19 vaccine: A dose-escalation, open-label, non-randomised, first-in-human trial. *Lancet* **2020**, *395*, 1845–1854. [CrossRef]
335. Mercado, N.B.; Zahn, R.; Wegmann, F.; Loos, C.; Chandrashekar, A.; Yu, J.; Liu, J.; Peter, L.; McMahan, K.; Tostanoski, L.H.; et al. Single-shot Ad26 vaccine protects against SARS-CoV-2 in rhesus macaques. *Nature* **2020**, *586*, 583–588. [CrossRef]
336. Van Doremalen, N.; Lambe, T.; Spencer, A.; Belij-Rammerstorfer, S.; Purushotham, J.N.; Port, J.R.; Avanzato, V.A.; Bushmaker, T.; Flaxman, A.; Ulaszewska, M.; et al. ChAdOx1 nCoV-19 vaccine prevents SARS-CoV-2 pneumonia in rhesus macaques. *Nature* **2020**, *586*, 578–582. [CrossRef] [PubMed]
337. Graham, S.P.; McLean, R.K.; Spencer, A.J.; Belij-Rammerstorfer, S.; Wright, D.; Ulaszewska, M.; Edwards, J.C.; Hayes, J.W.P.; Martini, V.; Thakur, N.; et al. Evaluation of the immunogenicity of prime-boost vaccination with the replication-deficient viral vectored COVID-19 vaccine candidate ChAdOx1 nCoV-19. *Vaccines* **2020**, *5*. [CrossRef]
338. Logunov, D.Y.; Dolzhikova, I.V.; Zubkova, O.V.; Tukhvatullin, A.I.; Shcheblyakov, D.V.; Dzharullaeva, A.S.; Grousova, D.M.; Erokhova, A.S.; Kovyrshina, A.V.; Botikov, A.G.; et al. Safety and immunogenicity of an rAd26 and rAd5 vector-based heterologous prime-boost COVID-19 vaccine in two formulations: Two open, non-randomised phase 1/2 studies from Russia. *Lancet* **2020**, *396*, 887–897. [CrossRef]
339. Logunov, D.Y.; Dolzhikova, I.V.; Shcheblyakov, D.V.; Tukhvatulin, A.I.; Zubkova, O.V.; Dzharullaeva, A.S.; Kovyrshina, A.V.; Lubenets, N.L.; Grousova, D.M.; Erokhova, A.S.; et al. Safety and efficacy of an rAd26 and rAd5 vector-based heterologous prime-boost COVID-19 vaccine: An interim analysis of a randomised controlled phase 3 trial in Russia. *Lancet* **2021**, *397*, 671–681. [CrossRef]
340. Lanzi, A.; Youssef, G.B.; Perricaudet, M.; Benihoud, K. Anti-adenovirus humoral responses influence on the efficacy of vaccines based on epitope display on adenovirus capsid. *Vaccine* **2011**, *29*, 1463–1471. [CrossRef] [PubMed]
341. Vujadinovic, M.; Vellinga, J. Progress in adenoviral capsid-display vaccines. *Biomedicines* **2018**, *6*, 81. [CrossRef]
342. Barouch, D.H.; Kik, S.V.; Weverling, G.J.; Dilan, R.; King, S.L.; Maxfield, L.F.; Clark, S.; Ng'ang'a, D.; Brandariz, K.L.; Abbink, P.; et al. International seroepidemiology of adenovirus serotypes 5, 26, 35, and 48 in pediatric and adult populations. *Vaccine* **2011**, *29*, 5203–5209. [CrossRef]
343. Thorner, A.R.; Lemckert, A.A.C.; Goudsmit, J.; Lynch, D.M.; Ewald, B.A.; Denholtz, M.; Havenga, M.J.E.; Barouch, D.H. Immunogenicity of heterologous recombinant adenovirus prime-boost vaccine regimens is enhanced by circumventing vector cross-reactivity. *J. Virol.* **2006**, *80*, 12009–12016. [CrossRef]
344. Wang, D.; Chen, L.; Ding, Y.; Zhang, J.; Hua, J.; Geng, Q.; Ya, X.; Zeng, S.; Wu, J.; Jiang, Y.; et al. Viral etiology of medically attended influenza-like illnesses in children less than five years old in Suzhou, China, 2011–2014. *J. Med. Virol.* **2016**, *88*, 1334–1340. [CrossRef]
345. Ma, J.; Duffy, M.R.; Deng, L.; Dakin, R.S.; Uil, T.; Custers, J.; Kelly, S.M.; McVey, J.H.; Nicklin, S.A.; Baker, A.H. Manipulating adenovirus hexon hypervariable loops dictates immune neutralisation and coagulation factor X-dependent cell interaction in vitro and in vivo. *PLoS Pathog.* **2015**, *11*, e1004673. [CrossRef]
346. Zhang, S.; Huang, W.; Zhou, X.; Zhao, Q.; Wang, Q.; Jia, B. Seroprevalence of neutralizing antibodies to human adenoviruses type-5 and type-26 and chimpanzee adenovirus type-68 in healthy Chinese adults. *J. Med. Virol.* **2013**, *85*, 1077–1084. [CrossRef]

347. McCoy, K.; Tatsis, N.; Korioth-Schmitz, B.; Lasaro, M.O.; Hensley, S.E.; Lin, S.-W.; Li, Y.; Giles-Davis, W.; Cun, A.; Zhou, D.; et al. Effect of preexisting immunity to adenovirus human serotype 5 antigens on the immune responses of nonhuman primates to vaccine regimens based on human- or chimpanzee-derived adenovirus vectors. *J. Virol.* **2007**, *81*, 6594–6604. [CrossRef]
348. Moffatt, S.; Hays, J.; Hogenesch, H.; Mittal, S.K. Circumvention of vector-specific neutralizing antibody response by alternating use of human and non-human adenoviruses: Implications in gene therapy. *Virology* **2000**, *272*, 159–167. [CrossRef]
349. Hammond, J.M.; McCoy, R.J.; Jansen, E.S.; Morrissy, C.J.; Hodgson, A.L.M.; Johnson, M.A. Vaccination with a single dose of a recombinant porcine adenovirus expressing the classical swine fever virus gp55 (E2) gene protects pigs against classical swine fever. *Vaccine* **2000**, *18*, 1040–1050. [CrossRef]
350. Wüest, T.; Both, G.W.; Prince, A.M.; Hofmann, C.; Löser, P. Recombinant ovine atadenovirus induces a strong and sustained T cell response against the hepatitis C virus NS3 antigen in mice. *Vaccine* **2004**, *22*, 2717–2721. [CrossRef]
351. Barlan, A.U.; Danthi, P.; Wiethoff, C.M. Lysosomal localization and mechanism of membrane penetration influence nonenveloped virus activation of the NLRP3 inflammasome. *Virology* **2011**, *412*, 306–314. [CrossRef] [PubMed]
352. Defer, C.; Belin, M.T.; Caillet-Boudin, M.L.; Boulanger, P. Human adenovirus-host cell interactions: Comparative study with members of subgroups B and C. *J. Virol.* **1990**, *64*, 3661–3673. [CrossRef] [PubMed]
353. Miyazawa, N.; Crystal, R.G.; Leopold, P.L. Adenovirus serotype 7 retention in a late endosomal compartment prior to cytosol escape is modulated by fiber protein. *J. Virol.* **2001**, *75*, 1387–1400. [CrossRef]
354. Iacobelli-Martinez, M.; Nemerow, G.R. Preferential activation of toll-like receptor nine by CD46-utilizing adenoviruses. *J. Virol.* **2007**, *81*, 1305–1312. [CrossRef]
355. Campden, R.I.; Zhang, Y. The role of lysosomal cysteine cathepsins in NLRP3 inflammasome activation. *Arch. Biochem. Biophys.* **2019**, *670*, 32–42. [CrossRef]
356. Kelley, N.; Jeltema, D.; Duan, Y.; He, Y. The NLRP3 inflammasome: An overview of mechanisms of activation and regulation. *Int. J. Mol. Sci.* **2019**, *20*, 3328. [CrossRef] [PubMed]
357. Bossy-Wetzel, E.; Newmeyer, D.D.; Green, D.R. Mitochondrial cytochrome c release in apoptosis occurs upstream of DEVD-specific caspase activation and independently of mitochondrial transmembrane depolarization. *EMBO J.* **1998**, *17*, 37–49. [CrossRef] [PubMed]
358. Kupgan, G.; Hentges, D.C.; Muschinske, N.J.; Picking, W.D.; Picking, W.L.; Ramsey, J.D. The effect of fiber truncations on the stability of adenovirus type 5. *Mol. Biotechnol.* **2014**, *56*, 979–991. [CrossRef] [PubMed]
359. Wiethoff, C.M.; Nemerow, G.R. Adenovirus membrane penetration: Tickling the tail of a sleeping dragon. *Virology* **2015**, *479–480*, 591–599. [CrossRef] [PubMed]
360. Nociari, M.; Ocheretina, O.; Schoggins, J.W.; Falck-Pedersen, E. Sensing infection by adenovirus: Toll-like receptor-independent viral DNA recognition signals activation of the interferon regulatory factor 3 master regulator. *J. Virol.* **2007**, *81*, 4145–4157. [CrossRef] [PubMed]
361. Favier, A.L.; Burmeister, W.P.; Chroboczek, J. Unique physicochemical properties of human enteric Ad41 responsible for its survival and replication in the gastrointestinal tract. *Virology* **2004**, *322*, 93–104. [CrossRef] [PubMed]
362. V'kovski, P.; Gultom, M.; Kelly, J.N.; Steiner, S.; Russeil, J.; Mangeat, B.; Cora, E.; Pezoldt, J.; Holwerda, M.; Kratzel, A.; et al. Disparate temperature-dependent virus-host dynamics for SARS-CoV-2 and SARS-CoV in the human respiratory epithelium. *PLoS Biol.* **2021**, *19*, e3001158. [CrossRef]
363. Brown, M.; Wilson-Friesen, H.L.; Doane, F. A block in release of progeny virus and a high particle-to-infectious unit ratio contribute to poor growth of enteric adenovirus types 40 and 41 in cell culture. *J. Virol.* **1992**, *66*, 3198–3205. [CrossRef]
364. Schoggins, J.W.; Nociari, M.; Philpott, N.; Falck-Pedersen, E. Influence of fiber detargeting on adenovirus-mediated innate and adaptive immune activation. *J. Virol.* **2005**, *79*, 11627–11637. [CrossRef]
365. Yu, X.; Veesler, D.; Campbell, M.G.; Barry, M.E.; Asturias, F.J.; Barry, M.A.; Reddy, V.S. Cryo-EM structure of human adenovirus D26 reveals the conservation of structural organization among human adenoviruses. *Sci. Adv.* **2017**, *3*, e1602670. [CrossRef]
366. Abbink, P.; Lemckert, A.A.C.; Ewald, B.A.; Lynch, D.M.; Denholtz, M.; Smits, S.; Holterman, L.; Damen, I.; Vogels, R.; Thorner, A.R.; et al. Comparative seroprevalence and immunogenicity of six rare serotype recombinant adenovirus vaccine vectors from subgroups B and D. *J. Virol.* **2007**, *81*, 4654–4663. [CrossRef]
367. Peruzzi, D.; Dharmapuri, S.; Cirillo, A.; Bruni, B.E.; Nicosia, A.; Cortese, R.; Colloca, S.; Ciliberto, G.; la Monica, N.; Aurisicchio, L. A novel Chimpanzee serotype-based adenoviral vector as delivery tool for cancer vaccines. *Vaccine* **2009**, *27*, 1293–1300. [CrossRef]
368. Roy, S.; Gao, G.; Clawson, D.S.; Vandenberghe, L.H.; Farina, S.F.; Wilson, J.M. Complete nucleotide sequences and genome organization of four chimpanzee adenoviruses. *Virology* **2004**, *324*, 361–372. [CrossRef]
369. Dicks, M.D.J.; Spencer, A.J.; Coughlan, L.; Bauza, K.; Gilbert, S.C.; Hill, A.V.S.; Cottingham, M.G. Differential immunogenicity between HAdV-5 and chimpanzee adenovirus vector ChAdOx1 is independent of fiber and penton RGD loop sequences in mice. *Sci. Rep.* **2015**, *5*. [CrossRef]
370. Kremer, E.J. CAR chasing: Canine adenovirus vectors—All bite and no bark? *J. Gene Med.* **2004**, *6*, S139–S151. [CrossRef] [PubMed]
371. Soudais, C.; Laplace-Builhe, C.; Kissa, K.; Kremer, E.J. Preferential transduction of neurons by canine adenovirus vectors and their efficient retrograde transport in vivo. *FASEB J.* **2001**, *15*, 2283–2285. [CrossRef] [PubMed]

372. Perreau, M.; Mennechet, F.; Serratrice, N.; Glasgow, J.N.; Curiel, D.T.; Wodrich, H.; Kremer, E.J. Contrasting effects of human, canine, and hybrid adenovirus vectors on the phenotypical and functional maturation of human dendritic cells: Implications for clinical efficacy. *J. Virol.* **2007**, *81*, 3272–3284. [CrossRef]
373. Soudais, C.; Boutin, S.; Kremer, E.J. Characterization of cis-acting sequences involved in canine adenovirus packaging. *Mol. Ther.* **2001**, *3*, 631–640. [CrossRef] [PubMed]
374. Matthews, D.A.; Russell, W.C. Adenovirus protein-protein interactions: Molecular parameters governing the binding of protein VI to hexon and the activation of the adenovirus 23K protease. *J. Gen. Virol.* **1995**, *76*, 1959–1969. [CrossRef] [PubMed]
375. Hernando-Pérez, M.; Martín-González, N.; Pérez-Illana, M.; Suomalainen, M.; Condezo, G.N.; Ostapchuk, P.; Gallardo, J.; Menéndez, M.; Greber, U.F.; Hearing, P.; et al. Dynamic competition for hexon binding between core protein VII and lytic protein VI promotes adenovirus maturation and entry. *Proc. Natl. Acad. Sci. USA* **2020**, *117*, 13699–13707. [CrossRef]
376. Moyer, C.L.; Besser, E.S.; Nemerow, G.R. A single maturation cleavage site in adenovirus impacts cell entry and capsid assembly. *J. Virol.* **2016**, *90*, 521–532. [CrossRef] [PubMed]
377. Van Rosmalen, M.G.M.; Nemerow, G.R.; Wuite, G.J.L.; Roos, W.H. A single point mutation in precursor protein VI doubles the mechanical strength of human adenovirus. *J. Biol. Phys.* **2018**, *44*, 119–132. [CrossRef] [PubMed]
378. Moyer, C.L.; Nemerow, G.R. Disulfide-bond formation by a single cysteine mutation in adenovirus protein VI impairs capsid release and membrane lysis. *Virology* **2012**, *428*, 41–47. [CrossRef] [PubMed]
379. Martinez, R.; Schellenberger, P.; Vasishtan, D.; Aknin, C.; Austin, S.; Dacheux, D.; Rayne, F.; Siebert, A.; Ruzsics, Z.; Gruenewald, K.; et al. The amphipathic helix of adenovirus capsid protein VI contributes to penton release and postentry sorting. *J. Virol.* **2015**, *89*, 2121–2135. [CrossRef] [PubMed]
380. Moyer, C.L.; Wiethoff, C.M.; Maier, O.; Smith, J.G.; Nemerow, G.R. Functional genetic and biophysical analyses of membrane disruption by human adenovirus. *J. Virol.* **2011**, *85*, 2631–2641. [CrossRef]
381. Takahashi, H. Antigen presentation in vaccine development. *Comp. Immunol. Microbiol. Infect. Dis.* **2003**, *26*, 309–328. [CrossRef]
382. Maier, O.; Marvin, S.A.; Wodrich, H.; Campbell, E.M.; Wiethoff, C.M. Spatiotemporal dynamics of adenovirus membrane rupture and endosomal escape. *J. Virol.* **2012**, *86*, 10821–10828. [CrossRef] [PubMed]
383. Wang, F.; Gómez-Sintes, R.; Boya, P. Lysosomal membrane permeabilization and cell death. *Traffic* **2018**, *19*, 918–931. [CrossRef]

viruses

MDPI

Case Report

DIC-Like Syndrome Following Administration of ChAdOx1 nCov-19 Vaccination

Gerardo Casucci * and Domenico Acanfora

Unit of Internal Medicine, San Francesco Hospital, Viale Europa 21, 82037 Telese Terme, Italy; domenico.acanfora29@gmail.com
* Correspondence: segr.dott.gerardocasucci@virgilio.it

Abstract: In recent weeks, adverse reactions have been reported after administration of Oxford–AstraZeneca chimpanzee adenovirus vectored vaccine ChAdOx1 nCoV-19 (AZD1222), in particular thrombus formation, which has led several European Countries to discontinue administration of this vaccine. On March 8, 2021, the European Medicines Agency Safety Committee did not confirm this probable association. We report the case of a patient who developed disseminated intravascular coagulation after the first dose of Oxford-Astra Zeneca vaccine, which resolved in a few days with the administration of dexamethasone and enoxaparin. This work demonstrates the safety of the Oxford-Astra Zeneca vaccine and that any development of side effects can be easily managed with a prompt diagnosis and in a short time with a few commonly used drugs.

Keywords: SARS-CoV-2; chimpanzee adenovirus vectored vaccine; disseminated intravascular coagulation

1. Introduction

The Severe Acute Respiratory Syndrome Coronavirus 2 (SARS-CoV2) has spread quickly around the world, causing clusters of prevalent respiratory Coronavirus Disease 2019 (COVID-19), including Acute Respiratory Distress Syndrome (ARDS), and becoming a serious public health concern. The development and subsequent use of anti-COVID-19 vaccines is the winning weapon to slow down and stop the Sars-CoV-2 pandemic.

Currently, 4 anti-COVID-19 vaccines are available in Europe that use different platforms to deliver mRNA SARS-CoV-2 Spike (S) protein (Table 1).

Table 1. COVID-19 Vaccines available in Europe.

Manufacturer	Name of Vaccine	Platform
BioNTech/Fosun Pharma/Pfizer	BNT162b2/COMIRNATY (INN tozinameran)	Nucleoside modified mNRA
Moderna/National Institute of Allergy and Infectious Diseases	mRNA-1273	mNRA-based vaccine encapsulated in lipid nanoparticle (LNP)
University of Oxford/AstraZeneca	AZD1222	Recombinant replication defective chimpanzee adenovirus expressing the SARS-CoV-2 S surface glycoprotein
Janssen Pharmaceutical Companies	Ad26.COV2.S	Recombinant, replication incompetent adenovirus type 26 (Ad26) vectored vaccine encoding the (SARS-CoV-2) Spike (S) protein

From 7 to 18 March 2021, some European Countries suspended the inoculation of the AstraZeneca anti-SARS-CoV-2 vaccine. This measure came after some serious cases of blood clots were reported in people who were vaccinated with AstraZeneca's SARS-CoV-2 vaccine. At the moment, a link between the vaccine and blood clots is not ascertained; however, it has been decided to perform further investigations. The first country to do so was Austria which was then joined by several other countries (Bulgaria, Cyprus, Denmark, Estonia, France, Greece, Iceland, Ireland, Latvia, Luxembourg, Malta, Netherlands, Poland, Spain, Sweden, Estonia, Lithuania, Luxembourg and Italy). On March 18, 2021, the European Medicines Agency Safety Committee reported that the benefits of AstraZeneca's SARS-CoV-2 vaccine far outweigh the risks of side effects and above all the thrombotic risk. The vaccine may be associated with rare cases of thrombocytopenia, with or without bleeding, including rare cases of cerebral venous thrombosis. Various platforms for vaccine development are available namely: virus vectored vaccines, protein subunit vaccines, genetic vaccines, and monoclonal antibodies for passive immunization which are under evaluation for SARS-CoV-2, each having demonstrated discrete benefits and hindrances. Oxford AstraZeneca consists of a replication-deficient chimpanzee adenoviral vector ChAdOx1, containing the SARS-CoV-2 structural surface glycoprotein antigen (spike protein; nCoV-19) gene (AZD1222). A replication-deficient chimpanzee adenoviral vector ChAdOx1 has already been used for the development of several vaccines [1–8]. Feng-Cai Zhu et al. report that the occurrence of side effects in subjects inoculated with Ad5-vectored SARS-CoV-2 vaccine was associated with decreasing age and low pre-existing immunity to the vaccine vector Ad5virus [9]. Phase 1/2 and 3 studies report very rare cases of arterial or venous thrombosis after administration of Oxford AstraZeneca chimpanzee adenovirus vectored vaccine ChAdOx1 nCoV-19 (AZD1222) [10]. We report the case of a 52-year-old woman with disseminated intravascular coagulation (DIC) [10,11] after administration of the first dose of AstraZeneca anti-SARS-CoV-2 vaccine (Lot ABV4678) in the absence of thrombus in the circulatory system.

2. Case History

A 52-year-old woman with a medical history of hepatitis B, headache, left breast cancer treated with bilateral mastectomy, left ovarian cyst treated with oophorectomy and salpingectomy, no prescription of chemotherapy or estrogen–progestogen treatment. The patient remained asymptomatic for the past five years and without any noteworthy diseases; among other things, she was a blood donor. On February 22, 2021, she received the first dose of AstraZeneca anti-SARS-CoV-2 vaccine (Lot ABV4678). Within hours of the administration of the vaccine, the patient presented with throbbing headache, photophobia, nausea, chills, fever (39 °C), muscle and joint pain, and inability to walk mainly due to severe asthenia. The patient was treated at home with paracetamol, ibuprofen and ketoralac. After 48 h, the fever regressed with persistence of all other symptoms (especially an intractable headache). Due to the worsening of symptoms and the appearance of a large ecchymosis of the left buttock on March 9, 2021, the patient was hospitalized.

On admission, vital signs were as follows: blood pressure, 130/80 mm Hg; heart rate, 64 beats/min; respiratory rate, 18/min, and temperature 36.8 °C. The patient tested negative for SARS-Cov-2 on molecular swab.

No abnormalities were detected on echocardiography, arterial and venous ultrasound Doppler in the lower limbs, abdominal ultrasound, ultrasound Doppler of the supra-aortic vessels, and CT scan of the chest and brain. At MRI angiography of the intra and extracranial vessels, no relevant abnormalities were found, while in the venous phase only a hypoplasia of the left transverse sinus was found, in the absence of thrombus (Figure 1).

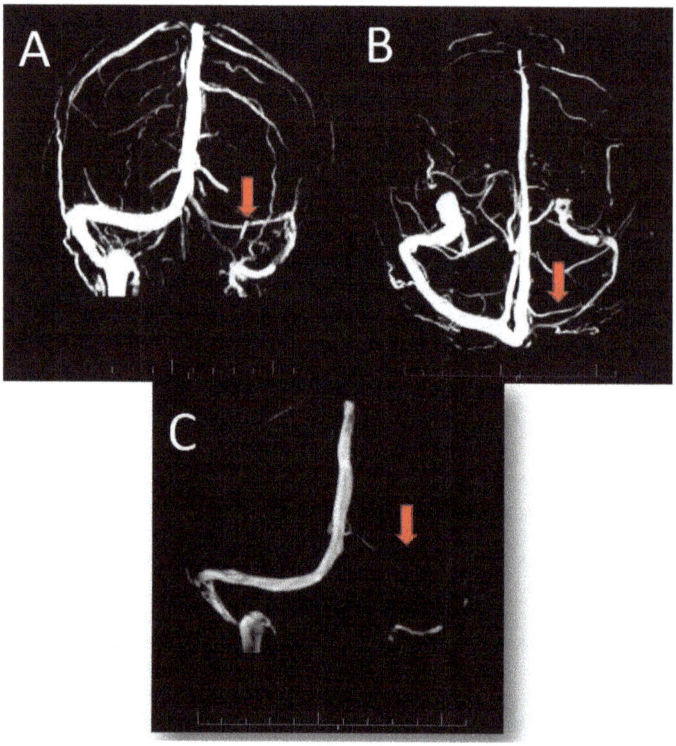

Figure 1. 3D reconstruction of venous angio-RM showing hypoplasia of left transverse sinus, indicated in (**A**) (Coronal), (**B**) (Caudocranial) and (**C**) (Sagittal) figures by the red arrows, in the absence of thrombus.

Based on laboratory data, the disseminated intravascular coagulation score (DIC Score = 5) was calculated [12], suggestive of overt DIC.

On March 9, 2021, the patient was started on subcutaneous enoxaparin (55 mg/day) and intravenous dexamethasone (8 mg/day).

Table 2 summarizes laboratory data before and during the disease. After six days of treatment, all laboratory parameters associated with DIC returned to normal and the patient was asymptomatic. On the seventh day, she was discharged without any home therapy. At discharge, the total IgG assay directed against the Spike protein RBD receptor was 15.1 U/mL (reference range < 0.80 /mL).

Table 2. Summary of Patient Laboratory Data Related to Disseminated Intravascular Coagulation.

Laboratory Values (Reference Range)	13 December 2020		9 March 2021	10 March 2021	11 March 2021	12 March 2021	13 March 2021	15 March 2021	16 March 2021
White Blood Cells count (3.7–10.3), $\times 10^9$/L	6.88	**22 February 2021 Administration of First dose of AstraZeneca anti-SARS-CoV2 vaccine (Lot ABV4678)**	12.52	13.15	10.78	8.11	9.63	10.71	
Red Blood Cells count (4.0–10.0), $\times 10^6$/L	4.22		4.3	3.80	3.53	3.79	3.69	3.86	
Haemoglobin (13.7–17.5), g/dL	13.3		12.6	11.2	10.4	11.1	10.7	11.4	
Platelet count (155–369), $\times 10^9$/L	347		77	91	79	108	137	265	330
Prothrombin time (9.6–12.5), s			15.0	15.9				14.1	
International normalized ratio (0.9–1.2)			1.12	1.19				1.05	
Activated partial Thromboplastin time (19–30), s			28.0	28.1				25.5	
Fibrinogen (150–450), mg/dL			100	85	78	90	111	123	248
Lactate dehydrogenase (140–280), U/L			579	587	485	514	461	316	
Creatinine (0.8–1.30), mg/dL	0.8		0.54	0.54	0.53	0.76	0.67		
Aspartate Aminotrasferase (0–31), U/L	26		37	32	27	20	20		
Alanine Aminotrasferase (0–34), U/L	18		43	38	36	28	28		
High Sensitivity C Reactive Protein (0–45), mg/L			2.42	2.05	1.21	0.85	0.62	0.29	
Sodium (135–155), mEq/L			140						
Potassium (3.5–5.5), mEq/L			4.1						
D-dimer (250–500), ng/mL			8298	6481	5280	6187	5128	1624	460
DIC Score			5	6				0	

Venous thrombosis, including deep vein thrombosis (DVT), rare cases of cerebral venous thrombosis (CVT) and pulmonary embolism (PE), occur at an annual incidence of about 1 per 1000 adults [13]. After administration of Oxford AstraZeneca chimpanzee adenovirus vectored vaccine ChAdOx1 nCoV-19, only 660 thrombotic events were reported in around 11 million people, with an incidence of 6 per 100,000 vaccinated subjects [14,15]. DIC is an acquired syndrome characterized by disordered blood coagulation. Insults or injuries that can lead to DIC can be infectious or non-infectious disorders. To the best of our knowledge, this is the first report of DIC after administration of the first dose of Oxford AstraZeneca chimpanzee adenovirus vectored vaccine ChAdOx1 nCoV-19.

In the patient observed at our Institute we observed a rapid remission of the symptoms and normalization of the blood chemical parameters, in particular those relating to DIC.

Oxford AstraZeneca chimpanzee adenovirus vectored vaccine ChAdOx1 nCoV-19 (AZD1222) can potentially induce a DIC-like syndrome due to the vector used to transfer the genetic substrate for SARS-CoV-2 immunization [1–8]. The chimpanzee adenovirus may induce a pro-coagulative state and a subsequent DIC due to the lack of previous exposure to the specific adenovirus. Binding of adenovirus to platelets can cause platelet activation and thrombosis [16].

Previous studies [1–8] have shown that the surface antigenic structure of chimpanzee adenovirus could influence the immune response of the vector and induce an abnormal immune reaction to immunization. Chimpanzee adenovirus vaccine platforms offer a good safety profile in phase I in humans, efficient bio-manufacture of millions of doses of non-replicating adenoviral vectors is possible using Good Manufacturing Practice (GMP)-approved cell lines. They avoid pre-existing immunity to human adenoviruses, and a single non-adjuvated dose of ChAdOx1 nCoV-19 vaccine is enough to achieve remarkable breadth, durability and potency of both humoral and cellular immune response.

The DIC in our patient is likely to be associated with an age under 60 and low pre-existing immunity to the vaccine vector Ad5 virus [16].

The Oxford AstraZeneca chimpanzee adenovirus vectored vaccine ChAdOx1 nCoV-19 appears to be effective and safe for SARS-CoV-2 immunization [10] and the development of a DIC can be easily controlled by the administration of dexamethasone and enoxaparin for a short time. The timeliness of diagnosis is therefore the best way to reassure patients. For this purpose, blood levels of fibrinogen and d-dimer and platelet counts may constitute more than reliable biological markers of onset or ongoing DIC.

Author Contributions: All the authors contributed equally to this work. All authors have read and agreed to the published version of the manuscript.

Funding: This research received no external funding.

Institutional Review Board Statement: This case was approved (March 22, 2021) by the Institutional Review Board of the San Francesco Hospital of Telese Terme (BN).

Informed Consent Statement: The patient signed the consent for this publication.

Data Availability Statement: Data sharing is not applicable to this case report.

Conflicts of Interest: The authors declare no conflict of interest.

References

1. Antrobus, R.D.; Coughlan, L.; Berthoud, T.K.; Dicks, M.; Hill, A.V.; Lambe, T.; Gilbert, S.C. Clinical Assessment of a Novel Recombinant Simian Adenovirus ChAdOx1 as a Vectored Vaccine Expressing Conserved Influenza A Antigens. *Mol. Ther.* **2014**, *22*, 668–674. [CrossRef] [PubMed]
2. López-Camacho, C.; Kim, Y.C.; Blight, J.; Moreli, M.L.; Montoya-Diaz, E.; Huiskonen, J.T.; Kümmerer, B.M.; Reyes-Sandoval, A. Assessment of Immunogenicity and Neutralisation Efficacy of Viral-Vectored Vaccines Against Chikungunya Virus. *Viruses* **2019**, *11*, 322. [CrossRef] [PubMed]

3. López-Camacho, C.; Abbink, P.; LaRocca, R.A.; Dejnirattisai, W.; Boyd, M.; Badamchi-Zadeh, A.; Wallace, Z.R.; Doig, J.; Velazquez, R.S.; Neto, R.D.L.; et al. Rational Zika vaccine design via the modulation of antigen membrane anchors in chimpanzee adenoviral vectors. *Nat. Commun.* **2018**, *9*, 2441. [CrossRef] [PubMed]
4. Warimwe, G.M.; Lorenzo, G.; Lopez-Gil, E.; Reyes-Sandoval, A.; Cottingham, M.G.; Spencer, A.J.; A Collins, K.; Dicks, M.D.J.; Milicic, A.; Lall, A.; et al. Immunogenicity and efficacy of a chimpanzee adenovirus-vectored Rift Valley Fever vaccine in mice. *Virol. J.* **2013**, *10*, 349. [CrossRef] [PubMed]
5. López-Camacho, C.; De Lorenzo, G.; Slon-Campos, J.L.; Dowall, S.; Abbink, P.; LaRocca, R.A.; Kim, Y.C.; Poggianella, M.; Graham, V.; Findlay-Wilson, S.; et al. Immunogenicity and Efficacy of Zika Virus Envelope Domain III in DNA, Protein, and ChAdOx1 Adenoviral-Vectored Vaccines. *Vaccines* **2020**, *8*, 307. [CrossRef] [PubMed]
6. McMahon, M.; Arunkumar, G.A.; Liu, W.-C.; Stadlbauer, D.; Albrecht, R.; Pavot, V.; Aramouni, M.; Lambe, T.; Gilbert, S.C.; Krammer, F. Vaccination With Viral Vectors Expressing Chimeric Hemagglutinin, NP and M1 Antigens Protects Ferrets Against Influenza Virus Challenge. *Front. Immunol.* **2019**, *10*, 2005. [CrossRef] [PubMed]
7. Stylianou, E.; Griffiths, K.; Poyntz, H.; Harrington-Kandt, R.; Dicks, M.; Stockdale, L.; Betts, G.; McShane, H. Improvement of BCG protective efficacy with a novel chimpanzee adenovirus and a modified vaccinia Ankara virus both expressing Ag85A. *Vaccine* **2015**, *33*, 6800–6808. [CrossRef] [PubMed]
8. Stylianou, E.; Harrington-Kandt, R.; Beglov, J.; Bull, N.; Pinpathomrat, N.; Swarbrick, G.M.; Lewinsohn, D.A.; Lewinsohn, D.M.; McShane, H. Identification and Evaluation of Novel Protective Antigens for the Development of a Candidate Tuberculosis Subunit Vaccine. *Infect. Immun.* **2018**, *86*, e00014-18. [CrossRef] [PubMed]
9. Zhu, F.C.; Guan, X.H.; Li, Y.H.; Huang, J.Y.; Jiang, T.; Hou, L.H.; Li, J.X.; Yang, B.F.; Wang, L.; Wang, W.J.; et al. Immunogenicity and safety of a recombinant adenovirus type-5-vectored COVID-19 vaccine in healthy adults aged 18 years or older: A randomised, double-blind, placebo-controlled, phase 2 trial. *Lancet* **2020**, *396*, 479–488. [CrossRef]
10. Ramasamy, M.N.; Minassian, A.M.; Ewer, K.J.; Flaxman, A.L.; Folegatti, P.M.; Owens, D.R.; Voysey, M.; Aley, P.K.; Angus, B.; Babbage, G.; et al. Safety and immunogenicity of ChAdOx1 nCoV-19 vaccine administered in a prime-boost regimen in young and old adults (COV002): A single-blind, randomised, controlled, phase 2/3 trial. *Lancet* **2020**, *396*, 1979–1993. [CrossRef]
11. Gando, S.; Levi, M.; Toh, C.H. Disseminated intravascular coagulation. *Nat. Rev. Dis. Primers* **2016**, *2*, 16037. [CrossRef] [PubMed]
12. Taylor, F.B., Jr.; Toh, C.H.; Hoots, W.K.; Wada, H.; Levi, M.; Scientific Subcommittee on Disseminated Intravascular Coagulation (DIC) of the International Society on Thrombosis and Haemostasis (ISTH). Towards definition, clinical and laboratory criteria, and a scoring system for disseminated intravascular coagulation. *Thromb Haemost.* **2001**, *86*, 1327–1330. [CrossRef] [PubMed]
13. Cushman, M. Epidemiology and risk factors for venous thrombosis. *Semin. Hematol.* **2007**, *44*, 62–69. [CrossRef] [PubMed]
14. Mahase, E. Covid-19: AstraZeneca vaccine is not linked to increased risk of blood clots, finds European Medicine Agency. *BMJ* **2021**, *372*, n774. [CrossRef] [PubMed]
15. European Medicines Agency. Covid-19 vaccine AstraZeneca: PRAC Investigating Cases of Thromboembolic Events—Vaccine's Benefits Currently Still Outweigh Risks: Update. Mar 2021. Available online: https://www.ema.europa.eu/en/news/covid-19-vaccine-astrazeneca-prac-investigating-casesthromboembolic-events-vaccines-benefits (accessed on 31 March 2021).
16. Signal Assessment Report on Embolic and Thrombotic Events (SMQ) with COVID-19 Vaccine (ChAdOx1-S [recombinant])—COVID-19 Vaccine AstraZeneca (Other Viral Vaccines) EMA/PRAC/157045/2021. Available online: https://www.ema.europa.eu/en/documents/prac-recommendation/signal-assessment-report-embolic-thrombotic-events-smq-covid-19-vaccine-chadox1-s-recombinant-covid_en.pdf (accessed on 31 March 2021).

MDPI AG
Grosspeteranlage 5
4052 Basel
Switzerland
Tel.: +41 61 683 77 34

Viruses Editorial Office
E-mail: viruses@mdpi.com
www.mdpi.com/journal/viruses

www.ingramcontent.com/pod-product-compliance
Lightning Source LLC
Chambersburg PA
CBHW061600210926
51850CB00009B/2164